BOSTON STUDIES IN THE PHILOSOPHY OF SCIENCE

Editor

ROBERT S. COHEN, *Boston University*

Editorial Advisory Board

ADOLF GRÜNBAUM, *University of Pittsburgh*

SYLVAN S. SCHWEBER, *Brandeis University*

JOHN J. STACHEL, *Boston University*

MARX W. WARTOFSKY, *Baruch College of the City University of New York*

VOLUME 103

THE COMPARATIVE RECEPTION OF RELATIVITY

Edited by

THOMAS F. GLICK
Department of History, Boston University

D. REIDEL PUBLISHING COMPANY

A MEMBER OF THE KLUWER  ACADEMIC PUBLISHERS GROUP

DORDRECHT / BOSTON / LANCASTER / TOKYO

Library of Congress Cataloging-in-Publication Data

The Comparative reception of relativity.

(Boston studies in the philosophy of science ; 103)
"The present volume grew out of a double session of the Boston Colloquium for
the Philosophy of Science held on March 25, 1983"—Pref.
Includes bibliographies and index.
1. Relativity (Physics)—Congresses. 2. Science—Philosophy—Congresses.
3. Einstein, Albert, 1879–1955—Influence—Congresses. 4. Europe—Intellectual
life—20th century—Congresses. 5. United States—Intellectual life—20th century—
Congresses. I. Glick, Thomas F. II. Boston Colloquium for the Philosophy of Science
(1983 : Boston, Mass.) III. Series: Boston studies in the philosophy of science ; v. 103.
Q174.B67 vol. 103 [QC173.5]   001′01 s   87–23388
ISBN 90–277–2498–9   [530.1′1]

Published by D. Reidel Publishing Company,
P.O. Box 17, 3300 AA Dordrecht, Holland.

Sold and Distributed in the U.S.A. and Canada
by Kluwer Academic Publishers,
101 Philip Drive, Norwell, MA 02061, U.S.A.

In all other countries, sold and distributed
by Kluwer Academic Publishers Group,
P.O. Box 322, 3300 AH Dordrecht, Holland.

All Rights Reserved

© 1987 by D. Reidel Publishing Company, Dordrecht, Holland
No part of the material protected by this copyright notice
may be reproduced or utilized in any form or by any means, electronic
or mechanical including photocopying, recording or by any information
storage and retrieval system, without written permission from
the copyright owner

Printed in the Netherlands

TABLE OF CONTENTS

Einstein in Toledo, Spain, March 4, 1923. Left to right: Manuel B. Cossio (art historian), Einstein, Lina Kocher-thaler, Julio Kocherthaler, Maria Luisa Cazurla (art historian), Kuno Kocherthaler, José Ortega y Gasset (Fundación José Ortega y Gasset).

PREFACE

The present volume grew out of a double session of the Boston Colloquium for the Philosophy of Science held in Boston on March 25, 1983. The papers presented there (by Biezunski, Glick, Goldberg, and Judith Goodstein[1]) offered both sufficient comparability to establish regularities in the reception of relativity and Einstein's impact in France, Spain, the United States and Italy, and sufficient contrast to suggest the salience of national inflections in the process. The interaction among the participants and the added perspectives offered by members of the audience suggested the interest of commissioning articles for a more inclusive volume which would cover as many national cases as we could muster.

Only general guidelines were given to the authors: to treat the special or general theories, or both, hopefully in a multidisciplinary setting, to examine the popular reception of relativity, or Einstein's personal impact, or to survey all these topics. In a previous volume, on the comparative reception of Darwinism,[2] one of us devised a detailed set of guidelines which in general were not followed. In our opinion, the studies in this collection offer greater comparability, no doubt because relativity by its nature and its complexity offers a sharper, more easily bounded target. As in the Darwinism volume, this book concludes with an essay intended to draw together in comparative perspective some of many themes addressed by the participants.

The cost of translating chapters from French and Russian was kindly provided by the Dean of the Boston University College of Liberal Arts and its Humanities Foundation and by the Dean of the Graduate School.

THOMAS F. GLICK
Boston, October 1986
ROBERT S. COHEN

NOTES

[1] Judith Goodstein, 'The Italian Mathematicians of Relativity,' *Centaurus* **26** (1983), 241–261.

[2] Thomas F. Glick, ed., *The Comparative Reception of Darwinism* (Austin, University of Texas Press, 1974; new ed., University of Chicago Press, 1988).

vii

STANLEY GOLDBERG

PUTTING NEW WINE IN OLD BOTTLES*
The Assimilation of Relativity in America

NATIONAL STYLES OF SCIENCE

At the beginning of the twentieth century, J.T. Merz reviewed the
development of intellectual thought in Europe in the nineteenth century
and argued that one of the changes that took place in the course of that
century was the gradual universalization of scientific knowledge.[1] According to Merz, the decided national differences that one might have
expected to find in Europe earlier, had been eradicated by improvements in communication and transportation.

Merz notwithstanding, for a number of years I have worked with the
notion of national styles of science as a way of understanding, in part,
how scientific innovations are received in different scientific cultures.
The case I concentrated on was a comparison of the reception of
Einstein's special theory of relativity in France, Germany, England and
the United States in the years immediately following publication of the
theory by Einstein in 1905.[2] In this paper I use the concept of national
styles of science to examine how Einstein's special theory of relativity
was introduced into the American scientific community, how the community initially responded to the theory and then how the theory has
been assimilated since that time.

One of the premises which underlies my analysis of the content of the
theory of relativity is that the relationship between evidence and belief
in science is precisely like the relationship between evidence and belief
in other areas of human endeavor. If that is the case then national
difference in science should be no more or no less pronounced than
national differences in other spheres. Furthermore, if the social aspects
of science, its organization and its relationship to other social institutions are important to how a particular scientific theory is constructed
and propagated, then differences in understanding in different countries
should be related to differences in the social organization of science.

I am, therefore, skeptical of Merz's claims *vis-à-vis* the disappearance
of national styles. With regard to the theory of relativity, I am not
suggesting that physicists in a given country acted together, lock-step,

1

Thomas F. Glick (ed.), The Comparative Reception of Relativity, 1–26.
© *1987 by D. Reidel Publishing Company.*

according to a preordained party line. Nor do group traits have mystical origins. Their basis is social in nature. The drawing of national boundaries cannot in and of itself be the important variable that results in different responses to scientific innovation. The situation is very complex. Physicists *do* move from one country to another. They *are* in communication with each other with an ease not earlier thought possible. Such experiences will affect individual scientists in different ways. But national boundaries do bespeak differences in social institutions and traditions.

Since knowledge in science is no more and no less subject to social processes than knowledge in other spheres of human endeavor, there are many factors that help to shape a scientist's point of view. In the first instance there will be differences which result from genetic inheritance, upbringing, and experiences unique to the individual. There will also be local characteristics identified with local customs or local social structures and beliefs. And there will be national characteristics defined by national customs and social institutions – a kind of national intelligence. The reader will have to judge for herself or himself how significant each factor is in any particular case. And though the process is *post hoc*, that in itself does not mean that the identification of style is not a useful tool. Most historical analysis, like most innovative analysis in science, is *post hoc*.

I am never impressed by the argument that knowledge in science is universal. This is usually based on at least one of two premises:

1. Knowledge about the physical world is not subject to any interpretation. It is ultimately confirmed by sense experience.
2. Understanding of concepts in science is universal to the extent that science employs the language of mathematics.

It is my contention that empirical evidence never entails or excludes any kind of explanation. For example, it is possible to construct a geocentric theory to account for *all* the evidence that is now accounted for by assuming that the earth orbits the sun and spins on its axis. Such a theory might look clumsy, appear to be adorned with a myriad of seemingly arbitrary and ad-hoc hypotheses and seem contrary to common sense, but judgments such as these are themselves normative and based on criteria which are subject to change over time. And while it is true that there might be universal agreement on how mathematical symbols and operators are to be manipulated, in order for the symbols and operations to have meaning outside the universe of mathematics itself, the symbols have to be translated into ordinary language. As you might expect, there is about as much agreement on such translations as there is when musicians attempt to translate a musical score into ordinary language.

In fact, one of the interesting features of studying the reception of the theory of relativity is that we are presented with two competing theories – the Lorentz theory of the electron and Einstein's special theory of relativity – which have identical formalisms but whose meanings are as different as night and day. The Lorentz theory was a dynamical theory intended to account for the behavior of all matter, radiation and their interactions, using the overarching premises of what has been termed the electromagnetic world view.[3]

Einstein's theory, on the other hand, was a kinematical theory, not wedded to the electromagnetic world view at all, but reflecting Einstein's already well formed convictions that a good theory was a theory which did what it had to do with a minimum number of postulates. Einstein's theory contained only two such postulates: The principle of relativity and the principle of the invariance of the speed of light. Behind those two postulates was the realization that all temporal and spatial measurements involved judgments of simultaneity and that in order to make such judgments, an observer had to stipulate the isotropy of space for at least one finite signal speed.[4]

Whereas the Lorentz theory was intended to modify and replace Newton's dynamics, Einstein's theory was a critique not of Newtonian dynamics, but of Newton's theory of measurement (contained in the *Principia* in the Scholium just prior to the Axioms) which implicitly assumed the possibility of an infinite signal speed.[5]

RESPONSE TO EINSTEIN'S THEORY IN EUROPE

As I suggested earlier, national styles of science, if they exist, are an artifact of differing social institutions. With regard to an innovation such as the theory of relativity, the most obvious social institutions to examine *outside* the scientific community would be the educational systems in the countries being compared. The evidence suggests that there is a close relationship between the nature of the response to the theory of relativity in different countries and the *structure* of the educational system in those countries. There also exist close relationships between the views of individual scientists who can be identified with the culture of a particular country and the traditional concerns of that culture with regard to the status of scientific knowledge.

In Germany (i.e. among German-university-system-trained European physicists occupying positions in German-speaking universities), Einstein's theory was closely scrutinized and the question of its merits was a matter of hot debate. Though there were few supporters, by 1911, Einstein's theory was no longer considered to be at the cutting edge of physics. The formalism, often confused with the Lorentz theory, was widely accepted. Can that process possibly be related to national differences? Indeed it can. It was *only* in Germany that such a debate took place, and it was only in Germany that scientific and educational institutions could have tolerated such debate. In fact, the debate entailed elaboration of the theory and it was during the process of elaboration, that one by one, Germany's leading physicists perceived that Einstein had produced not a theory of matter, but rather a theory of measurement.[6]

In France, during the same period of time, there was no mention of Einstein's theory. There was utter silence. That silence must certainly be related to the fact that Henri Poincaré, France's leading theoretician, and a man who simultaneously occupied several of the most important chairs in mathematical physics, theoretical physics, and experimental physics, as well as in theoretical and experimental astronomy, chose to ignore Einstein's innovation. Given the conical, rigid structure of both the Academy of Sciences and the Ministry of Education and how they were related, it is no surprise not only that no one else mentioned Einstein's theory, but also that the French physics curriculum was virtually silent as well through the sixth decade of this century.[7]

In England, while there was not total silence, three years passed before mention of Einstein's theory made its way into the literature. The British did not begin the process of assimilating the theory of relativity until a way had been found to make it compatible with British concepts of the ether. The theory was then introduced as a principle supplementary to the electron theory. British views on this subject were closely related to the manner in which British theoretical physicists were being trained. While somewhat more diverse than the French institutions, there were filters through which those being trained in physics had to pass which virtually ensured that a theory such as Einstein's would not get a very sympathetic reception. In fact, there were open and frank admissions on the part of leading theoretical scientists that they could not understand what it was that Einstein was talking about.[8]

THE CHARACTER OF AMERICAN SCIENCE

It might be thought that national response to relativity would not be likely in America. After all it is a relatively young country without the longstanding social and cultural traditions of its European counterparts. And even though there was from the beginning a close identification with European culture – especially with British culture – there can be no question that America was "A Nation of Nations."

The most pervasive theme that one finds in the literature on American attitudes toward science and scientific research during the nineteenth and early twentieth centuries is the so-called indifference theme. Beginning in the second third of the nineteenth century it was argued by American scientists such as Joseph Henry that while there was a great deal of activity devoted to inventions and the creation of practical implements, Americans were indifferent to basic science and basic research. This point of view was reinforced by foreign observers such as de Tocqueville near the middle of the century who saw no science being done in America,[9] or by Tyndall in the last quarter of the century who feared that American fascination with application would not only lead to intellectual stagnation but to the end of industrial progress.[10] In his view one needed to create a pool of ideas gleaned from pure research in order to keep technology advancing.

There has been considerable controversy among historians concerning the extent to which there was indifference to science in American culture.[11] It is not a subject that will be pursued here. But there is no question that Americans did respond to the prodding of Henry, de Tocqueville and Tyndall. In fact those peculiarly American phenomena of the last decades of the nineteenth and first decades of the twentieth centuries, the independent research laboratories, often explained their mission to a curious public with copious references to both Tyndall and de Tocqueville. But something more was added to these explanations. True enough, technological growth, they said, required the fertilization of knowledge acquired from pure research. But since technological growth meant economic profit, one did pure research for economic profit.[12]

Near the turn of the century, A.D. Little defined the mission of the basic research projects of his independent research laboratories as "the making and saving of money."[13] American attitudes are typified by an editorial which appeared in *The Wall Street Journal* in the spring of 1907. It began by proclaiming science as a source of strength in promoting private wealth and public welfare, went on to underscore this by pointing out that every commercial transaction in the civilized world is based on the chemist's certificate as to the value of gold "which forms the ultimate measure of values," reinforced this by trumpeting that "faith may move mountains but modern science relies on dynamite" and ended by reaffirming the opinion of those Americans who had recently written that the death of Pierre Curie in an automobile accident in the streets of Paris was a greater loss to the world than the loss of property and lives in the earthquake in San Francisco.[14]

The obvious contradiction between the concept of pure or fundamental research as a search for knowledge for its own sake, and the motivation of financial wealth permeates all such writings that one encounters during this period in the growth of the American culture. For example, in their enthusiasm over the awarding of the Nobel Prize for 1907 to Albert A. Michelson for his invention of his interferometer, the editors of *Popular Science Monthly* suggested that if the terms of Nobel's will were strictly heeded, the award should go first to Alexander Graham Bell and next to Thomas Alva Edison.[15]

Such examples are illustrative of a certain kind of tradition with regard to the role of research which had developed in the American culture by the turn of the century and the relationship which was perceived to operate between science, technological progress, and economic power. I would describe it as a mixture of German *Wissenschaft*, British empiricism, and a twist of profit motive. If it was, and is, a somewhat confused philosophy, it was effective in capturing the public imagination and firing the enthusiasm of industry for whatever it was they termed "basic," "pure," or "fundamental" research. One might call the developments I have described above as the *pragmatic tradition* for scientific research in America.

The business and industrial community were not the only source for this tradition. During the same period, the last quarter of the nineteenth century and the first decades of this century, higher education in America was undergoing a rapid and thorough revolution. The education of proper gentlemen and gentlewomen was being replaced with the education of professionals.

There is much evidence that the German model for higher education and research affected the outlook of American scholars, intellectuals, and university officials during the last quarter of the nineteenth century. They became increasingly convinced that the study of science, philosophy and history were aspects of *Wissenschaft* and should be pursued as such. In particular, philosophy had to become a distinct field, no longer identified solely with moral philosophy and ethics. Science also had to be a separate, specialized field, objectified to exclude all that was metaphysical and nonscientific.[16] By 1860 some sort of science was required in the curriculum of most institutions of higher learning in this country. As one might expect, the science taught in the colleges was almost exclusively experimental. Contrary to practice on the Continent and in the United Kingdom, virtually no theoretical science was taught in American colleges and universities. When Charles Elliot became president of Harvard in 1869, science became a more important part of the Harvard curriculum, reflecting the fact that Elliot was trained in and had taught science. But again, the emphasis was on laboratory science. No theoretical study was offered.[17]

From the middle of the nineteenth century to the beginning of the twentieth, the growth of the American physics community paralleled the growth of the American community of scientists as a whole. In 1870, there were no more than seventy-five people in the country who called themselves physicists. Of these, only one can be identified as a theoretical physicist. He was Josiah Willard Gibbs, who taught at Yale and lived in almost total obscurity. The importance of his work in theoretical thermodynamics, statistical mechanics and vector analysis was discovered in Europe, chiefly by Clerk Maxwell, who was responsible for bringing it to the attention of the world.[18]

As interest in science in the American universities developed, and as the popular identification of the importance of science in technological progress emerged, there was a concomitant increase in physics in the universities. By 1890, there were fifty-four Ph.D.'s in physics in the United States. This growth continued unabated through most of this

century: In the year 1909 twenty-five Ph.D.'s in physics were granted to
Americans and by the early 1960's 600 Ph.D.'s in physics were being
granted annually.

Near the turn of the century, outside the scientific community, the
view was widely held that support for science was needed for industrial
growth and that ideas should have practical import. Within the scientific
community, including the growing community of physicists, the criterion for good science was not so much practical outcomes as empirical
foundations. Theory was tolerated, but only those theories whose
premises could be shown to be true.[19] This demand amounted to the
conviction that not only must the predictions of the theory conform to
experience, but so too must the initiating premises of the theory. This
kind of expectation is nowhere better illustrated than in the initial
response and long-term assimilation of the theory of relativity in
America.

THE EARLY AMERICAN RESPONSE, 1905–1911

In America the early response to the theory of relativity followed a
pattern not dissimilar to that in Europe in the sense that the theory was
virtually ignored until 1908. During that period, there was a preoccupation with ether drift experiments, resistance to any suggestions that the
ether be dismissed, and skeptical renunciations of those Europeans who
were perceived as undermining the foundations of the mechanistic
philosophy. Americans were devising new and ingenious ways of per-
forming the Michelson-Morley experiment and other second order ether
drift experiments. The interferometer was placed in basements and on
mountains, it was enclosed in glass or placed out in the open, the base
was built of sandstone, of wood, or of steel. And always the result was
the same: no detectable shift in the fringes. In spite of the fact that the
instruments being used could detect effects several orders of magnitude
less than that being sought after, D.B. Brace, Professor of Physics at the
University of Nebraska, concluded that more accuracy was needed
before a definitive judgment could be made.[20]

The feeling that one gets in reading the papers of American experimenters such as Brace, Morley, Miller, or Trowbridge was of the
sensible reality of the ether, and a feeling of loss in the face of such null
results. One is impressed with the unwillingness on the part of Americans, in comparison to their European colleagues, to allow the ether to
remain an abstract entity.

The first American paper dealing with the subject of relativity was written in 1908 by G.N. Lewis.[21] Lewis, of course, was not a physicist; he was a physical chemist then at MIT. But throughout his career Lewis exhibited wide-ranging interests and he was always willing to express his views in public. Besides publishing papers in physics, he published in mathematics and even economic theory. Lewis seems to have had a knack for coming up with reasonable answers, even when his approach seemed unreasonable, if not nonsensical.

His first paper in relativity theory is an example of that. He was motivated, he said, to develop a system which he called "non-Newtonian mechanics" in order that momentum, energy, and mass be conserved at each moment in time during chemical reactions. His mechanics was premised on the belief that "something in light has mass when it moves at the speed of light." Direct application of classical notions of momentum and energy to that "something" in association with the already familiar concept of light pressure led directly to the result that $E = mc^2$ where m is the mass of the "something." In a footnote, Lewis noted that A. Einstein had come to a similar conclusion but the advantage to Lewis's derivation was, he said, that his was exact whereas Einstein's was an approximation.[22] In truth, Lewis's paper was devoid of physical meaning. Several years later, in a letter to Sommerfeld, Lewis commented that he had written the paper without knowledge of Einstein's work. Someone had pointed it out to him after the fact.[23]

Within a year of his first publication on the subject, Lewis in association with R.C. Tolman, a student in physical chemistry at MIT, had published a highly original paper explicating Einstein's theory.[24] It was the first attempt anywhere to discuss the kinematical character of the special theory of relativity without reference to electrmagnetic theory. The thrust of the paper was to show that the theory of relativity was sensible, practical and thoroughly rooted in experiment. In several subsequent papers, Tolman endeavored to show that both the first and the second postulates of relativity were the direct result of experimental evidence.[25]

While Lewis and Tolman were endeavoring to convince their colleagues that the theory was empirical and practical, those Americans who did comment on the theory ridiculed it as errant nonsense. L.T. More, Professor of Physics at the University of Cincinnati, teased relativists on the grounds that if they were right then the sun was melting away at an incredible rate and that when a man caught a baseball, the mass of both the ball and the man's hand should change.[26] But he became deadly serious when he thundered that the "electronicist theories" were a throwback representing an attempt to undermine the three-century struggle which science had waged to purge itself of metaphysics, that is, nonsense.[27]

The climax of these kinds of remarks came at the 1911 meeting of the American Association for the Advancement of Science when W.F. Magie, in his presidential address, called for a return to Newton. In a ringing declaration, Magie pleaded for a theory that was more than descriptive, that provided a mechanical model. As far as Magie was concerned, any theory that was true had to be comprehensible to everyone.[28]

The concepts of usefulness, practicality or comprehensibility to anyone were undertones of much of this early response to the theory of relativity in America. It is quite consistent with other early twentieth century American attitudes that the worth of any undertaking in science should be judged with an eye toward its usefulness and practicality. In Europe one did not find this kind of appeal to practicality. The British literature on the relationship between research in science and practical affairs did emphasize the importance of the research providing ideas for technological exploitation. But nowhere did one find the appeal put in terms which stressed the notion that theoretical understanding of nature, to be worth anything, should be comprehensible to everyone. In France and Germany such justifications were even rarer.

The response of Magie and More cannot be written off as crackpot. It *was* the response. Almost every one of the individuals to whom I have thus far referred had been starred in the first edition of *American Men of Science*.[29] These *were* the physicists. The appeal to common sense and comprehensibility certainly underpinned More's derisive attack on the theory. Less obviously, it was also a factor in Lewis and Tolman's interpretation of the theory. It would be very misleading to lump together More and Tolman for example in some simple-minded fashion. More did not evidence any kind of sophistication about formal aspects of the special theory of relativity. His view of the nature of the scientific enterprise was probably vastly different from Tolman's. But Tolman's lifelong expressed view that the postulates of the special theory of relativity were experimentally testable, mirrors, I think, an underlying sharing of certain assumptions by More, Tolman, and indeed other

Americans about the importance of practicality and down-to-earth empirical grounding as a criterion in judging scientific theory.[30]

THREE CASE STUDIES: TOLMAN, CARMICHAEL, AND BRIDGMAN

R.C. Tolman

Tolman's attitude toward the theory is particularly revealing on that score. As mentioned earlier, in several papers in the years immediately following his joint effort with Lewis, he undertook to show that not only was the first postulate (the principle of relativity) proven inductively from experiment, but so was the second postulate. In other words, in Tolman's view, not only were the conclusions of the theory tested by experience, but so too were the postulates. Tolman published a monograph on the theory of relativity in 1917.[31] While recognizing that the second postulate could be viewed as the combination of the principle of relativity and the principle of the independence of the velocity of light from the velocity of its source, Tolman argued that this view contained a nasty contradiction. The first postulate effectively denied the existence of the luminiferous ether while the principle of the independence of the velocity of light from the velocity of its source grew out of nineteenth century ether models. This supposed contradiction led O.M. Stewart at Cornell to reject the theory.[32] Tolman, on the other hand, cited the experimental determinations of the speed of light from both terrestrial and celestial sources as falsifying any emission theory and providing positive confirmation of the second postulate. Indeed not only did Tolman repeat this view in his very influential *Relativity, Thermodynamics and Cosmology* published in 1934, but a year or so earlier, in an address given in honor of Michelson, Tolman said,

... I do not feel that Professor Michelson approves of all the strange and bizarre conclusions which have resulted directly and indirectly from his experiment. . . . I do not quite approve of it all myself.[33]

R.D. Carmichael

The first monograph on relativity in the English language, published in 1912, was written by a mathematician, R.D. Carmichael, then at the University of Indiana.[34] It grew out of a course Carmichael gave at a time when, according to him, most people ridiculed the theory. The treatment by Carmichael contains almost all the features found emphasized in the early American response to the theory of relativity. The text begins with a review of first and second order ether drift experiments. The postulates of relativity were introduced as experimentally tested statements. The theory itself, according to Carmichael, had been verified experimentally by the determinations of the specific charge of the electron as a function of electron velocity. This claim was being made at a time when all such experiments were under a cloud because of technical lacks.[35]

Carmichael not only published mathematical studies and studies in relativistic physics, he also published in the philosophy of science. In a series of studies written in the 1920's, he carefully examined the logic of scientific arguments and the relationship between evidence and belief.[36] He recognized, for example, so far as deduction is concerned, postulates are unproven; but he insisted that in the natural sciences there is sometimes direct or indirect empirical evidence for their truth. Carmichael thus made a sharp distinction between the role of postulates in deductive systems and in natural science. There are, according to him, two parts to natural science:

1. An empirical part in which fundamental laws are secured by intuition and subjected to the verification of a preliminary test as to their validity.
2. A second stage in which an attempt is made to set apart a few of the empirical laws as basic and fundamental and to derive all the remaining empirical laws from those by strict deduction. Furthermore the logical consequences of the fundamental laws often point to the existence of new empirical laws not yet recognized.

In fact, Carmichael cited the theory of relativity to support the notion that postulates can be verified experimentally.

What makes Carmichael's work so interesting is that he had a fine grasp of the structure of logical arguments. He understood as well as anyone the importance of what Einstein had referred to as a "finely tuned intuition" in selecting postulates, but unlike Einstein and many other Europeans, Carmichael was not able to take the step which recognized that the basic postulates of a theory are beyond experimental test and are, in fact, the free creation of the human mind.

P.W. Bridgman

In 1914, P.W. Bridgman was turning his attention to Einstein's special theory of relativity.[37] Even though his first widely available public references to Einstein's work were not until the publication of *The Logic of Modern Physics* in 1927, his personal papers show that he devoted considerable time to contemplating not only the details of the formalism, but the philosophical implications of the theory.

Perhaps no other figure in the history of American physics is so identified with empiricism as Bridgman. He is, after all, the father of Operationism, a philosophy which holds that the very meaning of physical terms is conferred on them by specifying how they are measured. Bridgman often denied that this was a philosophy and also denied that he was a philosopher. He once wrote to Rudolph Carnap declining an invitation to join the Vienna Circle on the grounds that he did not believe that science was merely logic. Operationism was not a logical system, it was a technique for doing physics.[38] But like many philosophers, Bridgman confused the demand that science be done logically (i.e. sensibly) with the demand that it conform to the rules of formal logic.

In *The Logic of Modern Physics* Bridgman began by citing Einstein's analysis of the concept of simultaneity as the epitome of what he meant by Operationism. Later in the same book, Bridgman took up an issue which was to trouble him for the rest of his life: what he referred to as the problem of spreading time over space. At this point he was convinced that the invariance of the speed of light was a matter to be determined empirically. This having been established, one could synchronize clocks, not at the same place, using the techniques Einstein had described. Bridgman's published and unpublished writings on relativity in the period between 1920 and 1961 document his conviction that the isotropy of space for the one-way speed of light was a testable claim.

In 1962, Bridgman's *Sophisticate's Primer of Relativity* was posthumously published.[39] It is a book which is dominated by the problem of spreading time over space. At least four times in the book, Bridgman develops the argument that the problem of one-way measurement of the speed of light cannot be separated from the problem of spreading time over space, i.e. the problem of synchronizing clocks in a frame of reference. He argued that, therefore, one-way measurements of the

speed of light were irrelevant. Bridgman left a note in his papers just
two months before he died explaining this change in his attitude:

[A] major change [in my thinking] was my realization, early in 1961 that one-way velocity
can have no physical significance by itself, but is essentially a two-clock concept and has
meaning only when the method has been specified by which time is spread over space.
This realization negatived my effort to find some physical method of measuring one way
velocity . . . The second revision of the primer . . . begun in March 1961 is written in
accordance with this new insight.[40]

This insight came to Bridgman just as he was beginning the final draft
revisions of *Sophisticate's Primer*. Though he tried to repair the damage,
his realization of the impossibility of making a one-way measurement of
the speed of light undermines the foundations on which the entire text is
built. Yet it did not prevent Bridgman from giving a proof (flawed to be
sure) that space is isotropic for the speed of light.[41]

In a paper I published some time ago I pointed out that when Henri
Poincaré *talked* about physics he took a Conventionalist position. But
when Poincaré *worked* as a physicist, his position was more akin to that
of a Realist.[42] Bridgman's confusion between "being operational" and
"Operationism" is related. It is the confusion between being logical, i.e
thinking well as a physicist, and treating physics as if it were a branch of
formal logic.

That confusion is not restricted to physicists who concern themselves
with the philosophical status of the field. Philosophers of science often
fall prey to the same confusion. In fact, in *Sophisticate's Primer* Bridgman's discussion of the issue of one-way measurements of the speed of
light is defined, in part, by his concern over the treatment of the
problem by two philosophers of science, Hans Reichenbach and Adolph
Grünbaum. Reichenbach and Grünbaum were both interested not only
in the logical structure of the theory of relativity but in logical reconstructions of the theory.[43]

There is ample justification for such examinations in their own right.
There are an infinite number of possible logical structures which might
account for any body of empirical evidence. Alternatives to Einstein's
formulation should not be considered better or worse, they are different. But all too often, philosophers use such reconstructions as a
substitute for evidence in *historical* arguments. The possibility of such
reconstructions should not blind us to the fact that science is not logic.

Reichenbach was well aware of the dangers. In 1949 he wrote:

The philosopher of science is not much interested in the thought processes which lead to scientific discoveries; he looks for a logical analysis of the completed theory, including the relationships establishing its validity. That is, he is not interested in the context of discovery, but in the context of justification. . . .

The philosophy of physics . . . is not a product of creed but of analysis. . . . it endeavors to clarify the meaning of physical theories, independently of the interpretations by their authors, and is concerned with logical relations alone.[44]

It is in this spirit that Reichenbach undertook an analysis showing that it was not necessary to assume isotropy of space for the speed of light. Space could be anisotropic. In that case, decisions about distant simultaneity would require calculation. From the point of view of logic, Reichenbach is quite correct; it makes no difference. However, from the point of view of our *intuitions* about simultaneity, stipulating anisotropy makes no sense whatsoever.

With regard to relativity, Reichenbach claimed more:

. . . it appears amazing to what extent the logical analysis of relativity coincides with the original interpretation by its author.[45]

Note that Reichenbach has already limited the possibilities to *the* logical analysis. It is a small step to argue that logical reconstruction serve as a heuristic for historical investigation. Grünbaum, Lakatos and Zahar have all argued in this way.[46] Not only does such a step ignore the multiplicity of possible logical reconstructions, it biases historical accounts to those which follow the rules of formal logic. Just as Bridgman confused logic with physics, philosophers of science such as Grünbaum, Lakatos, Zahar, and others committed to rational reconstruction have confused logic with history.

The cases of Tolman, Carmichael, and Bridgman illustrate a kind of appeal to practicality that a European, even a European empiricist, would blush at. In fact, the view that not only have the conclusions of the theory of relativity been tested, but that the theory is built on postulates which themselves have been proven by experiment is not restricted to Tolman, Carmichael and Bridgman. It is a conviction shared by almost all American physicists during this century. For example, H.P. Robertson explicitly reformulated the premises of special relativity to reflect what he considered to be the empirical basis of the theory. His postulates were the principle of relativity and the Lorentz contraction. That is closer to Poincaré's theory than to

Einstein's.[47] It is this empirical view of the theory which permeates not just the professional literature, but the text literature as well.

THE ROLE OF TEXTBOOKS IN THE ASSIMILATION OF RELATIVITY IN AMERICA

In *Structure of Scientific Revolutions*,[48] Kuhn argues that at the time new ideas are introduced, not only are there few converts, but it takes significant periods of time for the new point of view to take hold. According to Kuhn, after the transition period from the old theory to the new, it is the place of scientific textbooks to transmit the now standard way of theorizing about the class of physical problems in which theory has been revolutionized. My own view is that the role that textbooks play will differ from culture to culture, and that given the importance of textbooks within the structure of the American educational system it would not be too surprising to find that they had played a significant role in shaping American attitudes about new innovations.

Kuhn's characterization of textbooks as the static receptacles of received knowledge is counterintuitive. In fact, the textbook literature is, I think, a significant body of data by which we can understand the interpretation that physicists gave to the theory of relativity. It is my contention that the text literature in America is more valuable for the purpose of studying the evolution of attitudes of the physics community toward the theory of relativity than is the primary literature.

In the primary literature, any number of interpretations can be enveloped within a common formalism. That is nowhere better illustrated than in the work of the distinguished American experimentalist H.E. Ives. His name is associated with a classic experiment, the Ives-Stillwell,[49] which is viewed as important confirmatory evidence for the theory of relativity. What is not widely known is that not only was he adamantly opposed to Einstein's theory, after World War II he was actively working behind the scenes to undermine not just Einstein's physics, but Einstein's reputation as well.[50] The relationship between an author's personal taste and his or her formal treatment of a subject is much more likely to be explicit in a textbook treatment.

Textbook treatments have a special significance in American culture. In the nineteenth century, the distinguishing mark of American education was the textbook. In fact, it was called "The American System" by Europeans.[51] As we all know, texts are still a central feature of American education.

On the other hand, regardless of whether one subscribes to Kuhn's view of revolution in science or to some other view, a commonsense model for the introduction of a new idea such as the special theory of relativity into the text literature would predict that the initial treatments of the theory would occur at the most advanced, graduate levels and gradually filter down first to treatments in intermediate courses and then finally to beginning physics courses. (This ignores the treatment of the theory in popular literature which is a separate problem.) This model does seem to be supported by the evidence with the proviso that regardless of the level that one examines, the introduction of the special theory of relativity bore little resemblance to the theory Einstein introduced, i.e. a theory based on two postulates, the principle of relativity and the invariance of the speed of light, behind which stood the analysis of simultaneity. In fact, *Einstein's* theory did not get any wide hearing until after 1970.[52]

Through 1950, with two exceptions, all graduate texts which explored the theory of relativity (largely texts in electrodynamics) took the position that the postulates of relativity were experimentally verified and in fact, that the Michelson-Morley experiment was direct experimental evidence that the speed of light is invariant. One of the exceptions was Leigh Page's 1922 book, *Introduction to Electrodynamics*. The book did not seem to find an audience. In his later, more widely adopted graduate texts, Page returned to the position he usually argued between 1912 and 1922 that the invariance of the speed of light should be treated as a law of physics and as such, was a special case of the first postulate.[53]

Another exception was Peter Bergmann's 1942 *Introduction to Einstein's Theory of Relativity*.[54] Though written for the advanced student, in introducing the subject Bergmann not only paid homage to Einstein's lay popularization, he replicated that treatment in great detail using no mathematics beyond elementary algebra. Significantly, as far as I can tell that treatment had little effect on other authors until the late 1960's. Even then its influence did not extend to changing in any essential way, the epistemological outlook of Americans. It should be noted that Bergmann, himself, was *not* trained in the American system.

The books I have thus far cited were intended for graduate students. What about intermediate texts?

Of course there are a great number of such treatments. Most of them appear in advanced undergraduate texts in electricity or surveys of modern or atomic physics. Overwhelmingly the postulates of relativity are viewed as being facts confirmed or proved by experiment. Rather than addressing each one of these treatments, it seems more useful to look at one of the more important of these texts for which a case of representativeness can be made. The book commonly referred to as "Richtmyer and Kennard" is an obvious choice. This book was first published in 1928 under the authorship of F.K. Richtmyer with the title *Introduction to Modern Physics*. The only reference to the theory of relativity is as follows:[55]

There has been much controversy and much weighing of both theoretical and experimental evidence to decide whether the relativity law of variation of mass is the correct one. For our purpose, the important point is that *qualitatively* a variation of mass with velocity is to be expected, *irrespective of the principle of relativity*. The expression "relative change of mass" should therefore, refer to the particular formulae used to compute masses at various velocities rather than to the origin of the phenomena.

The second edition was published in 1934 and contained a new appendix, entitled "Concerning Relativity."[56] In that appendix, the Lorentz transformation equations are derived by invoking, on the authority of D.C. Miller, the Lorentz-Fitzgerald contraction hypothesis as explaining the results of the Michelson-Morley experiment and then by citing the Michelson-Morley experiment as justification for assuming the invariance of the speed of light. The only other reference is to Tolman's 1917 treatise for a "correct" treatment of the concept of relativistic mass. There is virtually no discussion of the meaning of the theory or how it might apply to physical systems and there is not one mention of Albert Einstein.

The third edition of this book appeared in 1942, some three years after Richtmyer's death. E.H. Kennard was the new author. Special relativity was the subject of chapter 4 of this edition. After a discussion of Galilean relativity and a review of several of the more popular first and second order ether drift experiments, Kennard introduced special relativity as resting on two postulates – the principle of relativity and the invariance of the speed of light. Kennard characterized the epistemological status of these postulates as follows:

Of these two "postulates" the second is believed to represent a rather simple experimental fact, whereas the first is a generalization from a wide range of physical experience. There is no implication that the first postulate, which contains the new principle of relativity, is in any way self-evident; like the assumptions made in all physical theories, it is intended as a hypothesis to be tested by comparing deductions from it with experimental observations.[57]

These remarks were repeated in the fourth (1947) edition and the fifth (1955) edition. The fifth edition had a new "junior" author, Tom Lauritsen, Professor of Physics at the California Institute of Technology. Correspondence between Lauritsen, Kennard, and the McGraw-Hill Co., the publisher, indicates that Lauritsen's contributions were concentrated in the chapters on cosmic rays. When their collaboration began, Kennard had already revised chapters 1–10. Yet, there was intense discussion concerning the treatment of the theory of relativity (now chapter 2) as an example of the style of the book and the audience to which it was directed.[58]

In a letter to Kennard, the draft of which is dated Feb 2 [1954], Lauritsen said that he had shown some of the manuscript to his colleague R.B. Leighton:

He thinks it is ok, but he wants no part of it for his students! Too much history, too much conversation, not enough facts. I'm afraid that it is getting to be a common attitude among students. They are impatient with the historical approach and want a clean statement of current knowledge to write down in their notebooks. I'm sticking to my guns of course because I think this attitude simply produces handbook engineers, but the going is tough and I need comfort.

Later in the same letter, Lauritsen suggested that any discussion of the second postulate was, "of course redundant as it is merely a special case of the first" and then went on to suggest that one did not simply *assume* that the velocity of light was the same in both directions. He asked if that was not a requirement of relativity and might even have been proven by experiments measuring the speed of light.

Kennard's reply to Lauritsen, dated February 8, 1954, began as follows:

The pedagogical question raised by Prof. Leighton is one on which I don't feel very competent. I firmly believe in a certain acquaintance with the history of physics, i.e. with the pathway trod by prior physicists in arriving at our present fund of knowledge. It may be necessary to feed it to them in rather small doses, tho [sic] my idea was always to select *significant* items, not to mention false starts unless they seemed instructive, and not to describe details that any student would naturally fill in correctly.

Kennard rejected Lauritsen's suggestion that the second postulate was a special case of the first. He did not agree that the equality of the velocity of light in all directions in one frame of reference followed from the theory of relativity. Kennard explained that the isotropy of space for the speed of light required an arbitrary convention to make it so and then said:

Operationally, one way velocity is meaningless until clocks at different places can be synchronized; and how to do this without a new *pure* assumption?

In his correspondence, Kennard showed that he understood the relationship between one-way velocity measurements, synchronization of clocks and the fact that nothing can keep up with a beam of light. In spite of this, in the text, Kennard continued to insist that the invariance of the speed of light was a simple experimental fact.

The attitudes of Leighton, Lauritsen and Kennard concerning the use of historical materials in physics textbooks is also illuminating. Though they disagreed on using historical examples, Leighton and Kennard agreed that there were correct and incorrect explanations. Current belief is the "truth:" it represents "the facts." It is ironic that Lauritsen found precisely the same assumption among students so discouraging.

The sixth edition of Richtmyer and Kennard appeared in 1968, now authored by J.N. Cooper. The second postulate now represented not a "simple experimental fact," but simply an "experimental fact." Be that as it may, like his predecessors, Kennard and Lauritsen, Cooper then undertook a detailed examination of the problem of synchronizing clocks and finally concluded that "the simplest procedure is the usual one of *assuming* the velocity of light to be the same in any one direction as in the opposite direction."

When one turns to introductory texts, the situation is as one might expect: Those books which, prior to World War II, mentioned the theory of relativity, took the position that it was a theory which was suggested by the Michelson-Morley experiment and it was that experiment which proved that the speed of light was an invariant. When one considers the new generations of introductory textbooks which began to appear after the war, the situation is no different. All editions of Sears and Zemansky, Weidner and Sells, and Halliday and Resnick take the same tack. To the extent that special relativity is treated in these and other books, it is an empirical theory, suggested by experiment which serves as the bedrock confirmation of the postulates of the theory.

In many ways the watershed for consideration of the theory of relativity in the American textbook literature seems to have been the 1955 publication of Panofsky and Phillips, *Classical Electricity and*

Magnetism.[59] It was the treatment in this book which seems to have caught the imagination of later writers. Even though Bergmann earlier had treated the basis of the theory of relativity clearly in extraordinarily simple terms, that treatment had not caught the American fancy any more than Einstein's own popularization had. Recall that the treatments by Einstein and Bergmann do not appeal to experiment as the source of the second postulate. Rather the second postulate emerges from consideration of the need to develop criteria for judgments of simultaneity.

Panofsky and Phillips is a graduate text and as such, assumes an understanding of calculus, linear algebra and differential geometry. Yet when the authors came to discuss the theory of relativity, the language was high school algebra. More than that, Panofsky and Phillips examined what they considered to be all the experimental evidence that might pertain to the domain of the special theory of relativity and other competing theories, including the Lorentz theory. In a table that was later to be widely cited and reproduced, they concluded (wrongly) that only the theory of relativity satisfactorily accounts for the results of all first and second order experiments. Emphasizing that no single experiment proves the theory, they did say that the experiments provide evidence for the claim that the existence of the ether is undemonstrable, that various emission theories are untenable, and that a theory based on Einstein's postulates (experimentally tested of course) is "plausible."

Within five years of the publication of Panofsky and Phillips a flurry of monographs on special relativity written for undergraduates began to appear.[60] It had suddenly become fashionable to teach the special theory of relativity to undergraduates. Many of these authors cited Panofsky and Phillips. Some of them reproduced the table Panofsky and Phillips had constructed to show that only the special theory of relativity accounted for all first and second order ether drift experiments. (The table is incorrect in this regard since the Lorentz theory is perfectly adequate to the task.) With one exception (A.P. French's *Special Relativity*), all of these books proclaim the postulates to be generalizations from experience and experimental facts. Many of them justify writing "yet another monograph on relativity" on the grounds that the theory now has practical import for engineering applications to high energy physics.

By the year 1970 in specialty monographs or in elementary texts such as those by Sears and Zemansky or Halliday and Resnick, not only were the kinematical relationships we associate with the special theory of relativity readily available to the novice in undergraduate physics, but the means of deriving those relationships from first principles were also available. Those principles did not happen to be the ones that Einstein proposed. On the other hand they were in an epistemological tradition which recognizes experiment and measurement as the bedrock on which theoretical statements rest. According to Einstein, the source of the premises on which theories are constructed is a finely tuned sensitive intuition in conjunction with the free play of ideas. In the view of most American physicists whose writings I have examined, the source and the proof of the premises is experiment.

Rather than reifying a revolution which had already occurred, as Kuhn has suggested, what the American textbooks seem to have ensured is that in spite of the revolutionary views of physicists such as Einstein, the new formalism could be digested and adapted to fit within an epistemological framework that was familiar and comfortable. As has been the case with so many transitions and swings of the pendulum in the history of ideas: the more things change, the more they remain the same.

NOTES

* An earlier version of this paper was read at the Boston Colloquium for the Philosophy of Science, March 25, 1983. This research was supported, in part, by Grant No. 8206019 from the National Science Foundation.
[1] J.T. Merz, *A History of European Thought in the Nineteenth Century* (New York, rpr., 1965).
[2] S. Goldberg, *Early Response to Einstein's Theory of Relativity: A Case Study in National Differences, 1905–1911* (Harvard University, unpublished Ph.D. Dissertation, 1968).
[3] S. Goldberg, 'The Lorentz Theory of the Electron and Einstein's Theory of Relativity,' *American Journal of Physics*, 1969, **37**: 982–994. Cf. T. Hirosige, 'Origins of the Lorentz Theory of Electrons and the Concept of the Electromagnetic Field,' *Historical Studies in Physical Science*, 1969, **1**: 151–209; R. McCormmach, 'H.A. Lorentz and the Electromagnetic View of Nature,' *Isis*, 1970, **61**:459–497; R. McCormmach, 'Einstein, Lorentz and the Electron Theory,' *Historical Studies in Physical Science*, 1970, **2**: 41–87. Cf. R. McCormmach, 'Hendrik Antoon Lorentz,' in *Dictionary of Scientific Biography* (16 Vols.; New York, 1970–1980) Vol. 8, pp. 487–500.
[4] S. Goldberg, 'Albert Einstein and the Creative Act: The Case of Special Relativity,' in A. Aris *et al, The Springs of Scientific Creativity* (Minneapolis, 1983) pp. 232–253.
[5] S. Goldberg, *Understanding Relativity: Origins and Impact of a Scientific Revolution* (Boston: Birkhaeuser, 1984) especially Chapters 3 and 4.
[6] S. Goldberg, *Early Response* Chap. 4; S. Goldberg, *Understanding Relativity*, Chap. 6. Cf. S. Goldberg, 'Max Planck's Philosophy of Nature and His Early Elaboration of the Special Theory of Relativity,' *Historical Studies in Physical Science*, 1976, **7**: 125–160.

[7] S. Goldberg, *Early Response*, Chap. 3. S. Goldberg, 'Henri Poincaré and Einstein's Relativity,' *American Journal of Physics*, 1967, **35**: 934–944. S. Goldberg, 'Poincaré's Silence and Einstein's Relativity,' *Brit. Jour. Hist. Sci.*, 1970, **5**: 74–84.

[8] S. Goldberg, *Early Response* Chap. 4. S. Goldberg, 'In Defense of Ether: The British Response to Einstein's Relativity,' *Historical Studies in Physical Science*, 1970, **2**: 89–125.

[9] A. de Tocqueville, *Democracy in America*, tr. H. Reeves (2 Vols.; Philadelphia, 1841), Chapters 9 and 10.

[10] John Tyndall, *Lectures on Light: Delivered in the United States, 1872–1873* (New York, 1883) 'Conclusion' and 'Appendix.'

[11] N. Reingold, 'American Indifference to Basic Research: A Reappraisal,' in G.H. Daniels (ed.), *Nineteenth Century American Science: A Reappraisal* (Evanston, 1972), pp. 38–62.

[12] There were many articles and books written by Americans during the first three decades of this century which stress these values. The literature cited here is only suggestive: Raphael Meldora, 'The Relation between Scientific Research and the Chemical Industry,' *Scientific American Supplement*, 1903, **56**: 23301–23303, 23314–23315; David S. Jordan, 'Utilitarian Science,' *Popular Science Monthly*, 1904, **66**: 76–91; N.M. Hopkins, *The Outlook for Research and Invention* (New York, 1919); Arthur D. Little, *The Fifth Estate*.

[13] A.D. Little, *The Relation of Research to Industrial Development* (Boston, 1917).

[14] Quoted by William McMurtie in 'Address at the Dedication of the Walker Laboratory of the Rensselaer Polytechnic Institute,' *Science*, 1907, **26**: 329–332.

[15] 'The Progress of Science,' *Popular Science Monthly*, 1908, **72**: 283–288.

[16] J. Herbst, *The German Historical School in American Scholarship* (Ithaca, 1965). Cf. D. Noble, *America by Design* (New York, 1977); B. Bledstein, *The Culture of Professionalism* (New York, 1976).

[17] J.S. Brubacher and W. Rudy, *Higher Education in Transition. An American History* (New York, 1958) pp. 178–179.

[18] The data on the American physics community and its growth here and below has been gleaned from H.A. Barton *et al.*, 'Survey of Graduate Students in Physics,' *Physics Today*, (June) 1962, **15**: 42; L.R. Harmon, 'Physics Ph.D.s . . . Whence . . . Whither . . . When? *Physics Today*, (Oct) 1962, **15**: 21; and from D.J. Kevles, *The Physicists: The History of a Scientific Community in America* (New York, 1978), pp. 25–101. On Gibbs see Martin J. Klein, 'Josiah Willard Gibbs,' in *Dictionary of Scientific Biography* Vol. 5, pp. 386–393.

[19] Examples include, E.R. Nichols, 'Science and the Practical Problems,' *Science*, 1907, **29**: 1–10; A.G. Webster, 'America's Intellectual Product,' *Popular Science Monthly*, 1908, **78**: 193–210; R.C. MacLaurin, 'The Outlook for Research,' *Publication of the Clark University Library*, (#7) 1911. **2**. C.A. Krause, 'The Future of Science in America,' *Publication of the Clark University Library*, (#3) 1917. **5**.

[20] D.B. Brace, 'The Negative Results of Second and Third Order Tests of the 'Aether Drift' and Possible First Order Methods,' *Philosophical Magazine*, 1905, **10**: 71–80; Cf. Brace, 'A Repetition of Fizeau's Experiment on the Charge Produced by the Earth's Motion in the Rotation of a Refracted Ray,' *Ibid*, 591–599; E.W. Morley and D.C. Miller, 'Report on an Experiment to Detect the FitzGerald–Lorentz Contraction,' *Philosophical Magazine*, 1905, **9**: 680–685; A. Trowbridge, 'The Ether Drift,' *Proceedings of the American Philosophical Society*, 1910, **49**: 52–56.

[21] G.N. Lewis, 'A Revision of the Fundamental Laws of Matter and Energy,' *Philosophical Magazine*, 1908, **16**: 705–717.

[22] Lewis apparently had in mind the fact that Einstein's statement of the relationship was an expression of the first two terms of the series $m_0c^2 (1 - 1/2 (v/c)^2 + \ldots)$.

[23] Lewis to Arnold Sommerfeld, Dec. 12, 1910. The letter is in the Sommerfeld papers, at the Deutsches Museum, Munich.

[24] G.N. Lewis and R.C. Tolman, 'The Principle of Relativity and Non-Newtonian Mechanics,' *Proceedings of the American Academy of Science*, 1909, **44**: 711–730.

[25] R.C. Tolman, 'The Second Postulate of Relativity,' *Physical Review*, 1910, **31**: 26–40.

[26] L.T. More, 'On Theories of Matter and Mass,' *Philosophical Magazine*, 1909, **18**: 17–26.

[27] L.T. More, 'The Metaphysical Tendencies of Modern Physics,' *Hibbert Journal*, 1910; **8**: 800–817; L.T. More, 'Recent Theories of Electricity,' *Philosophical Magazine* 1911, **21**: 196–218.

[28] W.F. Magie, 'Primary Concepts of Physics,' *Science*, 1912, **35**: 281–293.

[29] In the year 1903, with the support of the Carnegie Institution, James M. Cattell began gathering data for the publication of a register of American scientists. *American Men of Science* was published in 1906 and has gone through fourteen editions. It is now titled *American Men and Women of Science*. At the time that he collected the data, Cattell used polling devices to determine who among those in the registry were the most productive and respected. Those names were denoted in the registry by a star; they became known as 'scientists starred.' Almost all the individuals in America who responded to the theory of relativity up to 1911 were starred scientists. See S.S. Visher, *Scientists Starred – 1903–1943 – in American Men of Science* (New York, 1975).

[30] R.C. Tolman, 'The Second Postulate of Relativity'; cf. R.C. Tolman, *Relativity, Thermodynamics and Cosmology*. (Cambridge, 1934).

[31] R.C. Tolman, *The Theory of Relativity of Motion* (Berkeley, 1917).

[32] O.M. Stewart, 'The Second Postulate of Relativity and the Electromagnetism of Light,' *Physical Review*, 1911, **32**: 418–428.

[33] Tolman Papers, California Institute of Technology. Quoted in Jeffrey M. Crelinsten, *The Reception of Einstein's General Theory of Relativity Among American Astronomers, 1910–1930* (University of Montreal: Unpublished Ph.D. Dissertation, 1981) p. 123.

[34] R.D. Carmichael, *The Theory of Relativity* (New York, 1912).

[35] S. Goldberg, 'Max Planck's Philosophy of Nature.'

[36] These were collected together in R.D. Carmichael, *The Logic of Discovery* (Chicago, 1930).

[37] Much of the information concerning the evolution of Bridgman's thought about the theory of relativity is in unpublished material contained in his papers at the Harvard University Archives. I take this opportunity to acknowledge the help of Maila Walter, Harvard University, who was generous in sharing her knowledge of the organizational structure and contents of the Bridgman Papers. I am also indebted to Ms. Walter for sharing with me her draft manuscript, 'Laboratory Practice and the Realities of Physics: The Operational Interpretation of P.W. Bridgman' and for taking the time to discuss with me many of the issues in this section.

[38] Bridgman to Carnap, Sept. 19, 1934. Bridgman Papers, Harvard University, HUG 4234.10. *Cf.* Bridgman to E.B. McGivray, April 8, 1928. Bridgman Papers, HUG 4234.12.

[39] Percy W. Bridgman, *A Sophisticate's Primer of Relativity* (Middletown: Wesleyan University Press, 1st ed., 1963; 2nd ed., 1983). The first edition of the book contained a forward and afterword by Adolph Grünbaum. The fact that Grünbaum's afterword was critical called forth irate responses from some of Bridgman's colleagues in the Harvard Physics Department who objected to such remarks within the posthumously published volume. They attempted to communicate their irritation to potential reviewers of the book. This prompted Grünbaum to reply with a defense of his work. The upshot was, apparently, to discourage any serious reviews of the book in professional or popular journals. The publisher was of the opinion that the dispute accounted for a poor distribution and sales record. See the correspondence between Professor Grünbaum, representatives of the Wesleyan University Press, and Professor Wendel Furry and Professor E.C. Kemble contained in the Bridgman Papers, Harvard University Archives. The second edition of the book contains a new introduction by A.I. Miller, hereafter cited as "Miller, 'Introduction.'" The second edition of *Sophisticate's Primer of Relativity* was published after the first draft of this paper was completed. I have taken note of Miller's contributions and of differences in our analyses.

[40] This note was dated April 7, 1961. The draft to which Bridgman refers has the date 4/4/61 on the first page, but contains pages with earlier dates going back to August 1960. *Cf.* Miller, 'Introduction,' p. xxxiv. Miller did not include the last sentence of this note. Miller argues persuasively that Bridgman explicitly left this note for posterity.

[41] While noting that Bridgman's derivation is in error, Miller says that it is not "merely an error." He describes it as "an act of desperation." See Miller, 'Introduction,' pp. xxxii–xxxv. Since Bridgman said that he had written this draft of *Sophisticate's Primer* in the light of understanding that it was impossible to operationalize one-way measurements of the speed of light, I cannot accept Miller's interpretation.

[42] Goldberg, 'Poincaré's Silence.'

[43] See the discussion by Grünbaum in the 'Afterword,' in *Sophisticate's Primer* and the literature cited therein.

[44] H. Reichenbach, 'The Philosophical Significance of the Theory of Relativity,' in P.A. Schillp (ed.), *Albert Einstein: Philosopher–Scientist* (New York, 1949) pp. 292–293. *Cf.* G. Holton, 'Einstein, Michelson, and the Crucial Experiment,' in G. Holton, *Thematic Origins of Scientific Thought* (Cambridge, 1973), esp. n. 152. Holton's discussion has important bearing on the questions raised here.

[45] Reichenbach, *op. cit.* p. 293.

[46] A. Grünbaum, 'The Special Theory of Relativity as a Case Study in the Importance of the Philosophy of Science for the History of Science,' in B. Baumrin (ed.) *Philosophy of Science* (New York, 1963); A. Grünbaum, 'The Bearing of the Philosophy of Science on the History of Science,' *Science*, 1964, **143**: 1406–1412; I. Lakatos, 'Falsification and the Methodology of Scientific Research Programmes,' in I. Lakatos and A. Musgrave (eds.), *Criticism and the Growth of Knowledge* (Cambridge, England, 1970); E. Zahar, 'Why Did Einstein's Programme Supersede Lorentz'?' *British Journal for the Philosophy of Science*, 1973, **24**: 95–123, 223–262.

[47] This view is reflected in Robertson's notebooks and personal correspondence available in the California Institute of Technology Archives. *Cf.* H.P. Robertson and T.N. Noonan, *Relativity and Cosmology* (Philadelphia, 1960); S. Goldberg, *Understanding Relativity*, p. 301.

[48] T.S. Kuhn, *The Structure of Scientific Revolutions* (2nd ed. Chicago, 1970), p. 144.

[49] For a brief account of Ives' contributions see B. Finn, 'Herbert Eugene Ives,' *Dictionary of Scientific Biography*, Vol. 7, p. 36.

[50] Ives participated in a correspondence with an industrial chemist in which they schemed to defame the reputation of 'The Jew Einstein' and his work. Ives Papers, Manuscript Division, Library of Congress.

[51] F. Fitzgerald, *America Revised: History Schoolbooks in the Twentieth Century* (Boston, 1979).

[52] For a detailed discussion of the text literature in America on relativity, see S. Goldberg, 'Up from Pencil Pushing: The Role of Textbooks in the Assimilation of Relativity in the American Physics Community,' in press.

[53] L. Page, *Introduction to Electrodynamics: From the Standpoint of the Electron Theory* (Boston, 1922). *Cf.* L. Page, 'A Derivation of the Fundamental Relations of Electrodynamics from Those of Electrostatics,' *American Journal of Science*, 1912, **34**: 57–68; L. Page, *Introduction to Theoretical Physics* (New York, 1928; 2nd ed. 1935; 3rd ed. 1952). Page's contributions have been discussed by E. Purcell, in H. Woolf (ed.) *Some Strangeness in the Proportion: A Centennial Symposium to Celebrate the Achievements of Albert Einstein* (Princeton, 1981) p. 109.

[54] P.G. Bergmann, *Introduction to the Theory of Relativity* (Englewood Cliffs, 1942; 2nd ed. 1964).

[55] F.K. Richtmyer, *Introduction to Modern Physics* (New York, 1928), esp. p. 373. I am indebted to Professor Gerald Holton for first suggesting that this might be an interesting book to examine in this context.

[56] F.K. Richtmyer, *Introduction to Modern Physics* (2nd ed., New York, 1934).

[57] Richtmyer and Kennard, *Introduction to Modern Physics* (3rd ed., New York, 1942) pp. 131–132.

[58] This correspondence is in the Kennard Papers at the Center for the History of Physics, American Institute of Physics, New York, and the Lauritsen Papers at the Archives of the California Institute of Physics. I am indebted to Spencer Weart and Joan Warnow at the Center for the History of Physics and to Judith Goodstein at the Archives of the California Institute of Technology for their assistance.

[59] W.F.H. Panofsky and M. Phillips, *Classical Electricity and Magnetism* (Reading, Mass., 1955).

[60] Examples include David Bohm, *The Special Theory of Relativity* (New York, 1953); A.P. French, *Special Relativity* (Boston, 1966); T.M. Helliwell, *Introduction to Special Relativity* (Boston, 1966); Claude Kaeser, *Introduction to the Special Theory of Relativity* (Englewood Cliffs, 1967); R. Katz, *An Introduction to the Special Theory of Relativity* (Princeton, 1964); N.D. Mermin, *Space and Time in Special Relativity* (New York, 1968); R. Resnick, *Introduction to Special Relativity* (New York, 1968); W. Rindler, *Essential Relativity* (New York, 1969); W. Rindler, *Special Relativity* (New York, 1960); H.M. Schwartz, *Introduction to Special Relativity* (New York, 1968); F.W. Sears and R.W. Brehme, *Introducion to Special Relativity* (Reading, 1968); A. Shadowitz, *Special Relativity* (Philadelphia, 1968); J.H. Smith, *Introduction to Special Relativity* (New York, 1965); E.F. Taylor and J.A. Wheeler, *Spacetime Physics* (San Francisco, 1966).

Smithsonian Institution

JOSÉ M. SÁNCHEZ-RON.

THE RECEPTION OF SPECIAL RELATIVITY IN GREAT BRITAIN.

1. INTRODUCTION

One of the main purposes of any historical study dealing with the introduction of a new theory in a given scientific community is the identification of its principal characteristics; that is, of the common traits of such introduction. In the case of the introduction of special relativity in Great Britain, Stanley Goldberg[1] carried out some studies assigning a prominent role to the ether concept. According to him there was widespread acceptance of the ether concept among British physicists during the nineteenth century and the first decades of the twentieth. One of the consequences of that situation was that "the acceptance [of special relativity] hinged upon making it compatible with the concept of the ether. As paradoxical as that might be, there was almost unanimous agreement within the British physics community about such a program".[2]

In this paper I shall argue that although most of what Goldberg says is true, there also existed other trends within British physics that, when taken into account, present a more diversified picture than that one which is dominated by the ether concept.

2. THE ETHER: AN AMBIGUOUS, PROBLEMATIC, AND NOT ALWAYS ACCEPTED CONCEPT

It is true that the ether[3] was one of the most cited and popular concepts in British physics during the times of Maxwell, Kelvin, J.J. Thomson and Rutherford. However, much too often such ether-adherence came from second-rate physicists; that is, from scientists who did not determine the main trends of development in their discipline. If we take this fact into account, then we are left with a much lower number of examples of "real" significance to support contentions like Goldberg's, and therefore the task of identifying some of the first-rate physicists who showed, in an explicit or implicit manner, some misgivings towards the ether concept, or who pointed out some of its limitations, is of special significance. In this section I will consider a few of these cases.

Thomas F. Glick (ed.), The Comparative Reception of Relativity, 27–58.
© 1987 by D. Reidel Publishing Company.

2.a. J.H. Poynting

My first example concerns John Henry Poynting, the famous physicist who was attached to Birmingham for the largest part of his career (first as professor of physics at Mason College, and later on, when in 1900 Mason College became the University of Birmingham, as dean of the Faculty of Science till 1912).

Among Poynting's writings there is one particularly germane to my purposes: his Presidential Address to the Mathematical and Physical Section of the British Association for the Advancement of Science meeting held in Dover in 1899. Poynting's Address[4] reveals an almost Machian standpoint.[5] He said, for instance, that:[6]

No long time ago [physical laws] were quite commonly described as the Fixed Laws of Nature, and were supposed sufficient in themselves to govern the universe. Now we can only assign to them the humble rank of mere descriptions, often tentative, often erroneous, of similarities which we believe we have observed.

Naturally, by 1899 a president of the Mathematical and Physical Section of the BAAS who wished to survey his field had to pay attention to concepts like the ether. Poynting did so by reminding his audience that (p. 605):

While light was regarded as corpuscular – in fact molecular – and while direct action at a distance presented no difficulty, the molecular hypothesis served as the one foundation for the mechanical representation of phenomena. But when it was shown that infinitely the best account of the phenomena of light could be given on the supposition that it consisted of waves, something was needed, as Lord Salisbury has said, to wave, both in the interstellar and in the intermolecular spaces. So the hypothesis of an ether was developed, a necessary complement of that form of the molecular hypothesis in which matter consists of discrete particles with matter-free intervening spaces.

At this point, the questions which Poynting faced were (p. 606):

How are we to regard these hypotheses as to the constitution of matter and the connecting ether? How are we to look upon the explanations they afford? Are we to put atoms and ether on an equal footing with the phenomena observed by our senses, as truths to be investigated for their own sake? Or are they mere tools in the search for truth, liable to be worn out or superseded?

To answer such questions, Poynting began by directing his attention to rather elemental facts (p. 606):

That matter is grained in structure is hardly more than the expression of the fact that in very thin layers it ceases to behave as in thicker layers. But when we pass on from this general statement and give definite form to the granules or assume definite qualities to the intergranular cement we are dealing with pure hypothesis.

It is hardly possible to think that we shall ever see an atom or handle the ether. We make no attempt whatever to render them evident to the senses. We connect observed conditions and changes in gross visible matter by invisible molecular and ethereal machinery. The changes at each end of the machinery of which we seek to give an account are in gross matter, and this gross matter is our only instrument of detection, and we never receive direct sense impressions of the imagined atoms or the intervening ether. To a strictly descriptive physicist their only use and interest would lie in their service in prediction of the changes which are to take place in gross matter.

Poynting's conclusions were rather definite and, in some senses, probably shocking to many among his audience:

It appears quite possible that various types of machinery might be devised to produce the known effects. The type we have adopted is undergoing constant minor changes, as new discoveries suggest new arrangements of the parts. Is it utterly beyond possibility that the type itself should change?

The special molecular and ethereal machinery which we have designed, and which we now generally use, has been designed because our most highly developed sense is our sense of sight. Were we otherwise, had we a sense more delicate than sight, one affording us material for more definite mental presentation, we might quite possibly have constructed very different hypotheses [here Poynting offers a possible example dealing with the sense of smell].

In spite of all these critical statements, Poynting knew too well that (p. 607): "It is merely a true description of ourselves to say that we must believe in the continuity of physical processes, and that we must attempt to form mental pictures of those processes the details of which elude our observation. For such pictures we must frame hypotheses, and we have to use the best material at command in framing then. *At present there is only one fundamental hypothesis – the molecular and ethereal hypothesis* – in some such form as is generally accepted" (emphasis mine). Nevertheless, even accepting this, Poynting wished to point out what was, according to him, the real position of the ether concept, a position which was far from being the same as the one accepted by men like Lord Kelvin, Oliver Lodge, or George Green, to name but a few.

In this country [he continues (pp. 607–8)] there is no need for any defence of the use of the molecular hypothesis. But abroad the movement from the position in which hypothesis is confounded with observed truth has carried many through the position of equilibrium equally far on the other side, and a party has been formed which totally abstains from molecules as a protest against immoderate indulgence in their use . . .

But the protest will have value if it will put us on our guard against using molecules and the ether everywhere and everywhen. There is, I think, some danger that we may get so accustomed to picturing everything in terms of these hypotheses that we may come to suppose that we have no firm basis for the facts of observation until we have given a molecular account of them, that a molecular basis is a firmer foundation than direct experience . . .

There is more danger of confusion of hypothesis with fact in the use of the ether: more risk of failure to see what is accomplished by its aid. In giving an account of light, for instance, the right course, it appears to me, is to describe the phenomena and lay down the laws under which they are grouped, leaving it an open question what it is that waves, until the phenomena oblige us to introduce something more than matter, until we see what properties we must assign to the ether; properties not possessed by matter, in order that it may be competent to afford the explanations we seek. We should then realise more clearly that it is the constitution of matter which we have imagined, the hypothesis of discrete particles, which obliges us to assume an intervening medium to carry on the disturbance from particle to particle [emphasis mine].

There is another argument used by Poynting, which shows that he was "modern enough" as to be able to accept that there is no necessity of an ether for fixing the positions of material bodies, a question that, as we know, is closely related to the foundations of special relativity. The argument goes as follows in Poynting's own words (p. 610):

Another illustration of the illegitimate use of our hypothesis, as it appears to me, is in the attempt to find in the ether a fixed datum for the measurement of material velocities and accelerations, a something in which we can draw our coordinate axes so that they will never turn a bend. But this is as if, discontented with the movement of the earth's pole, we should seek to find our zero lines of latitude and longitude in the Atlantic Ocean. Leaving out of sight the possibility of ethereal currents which we cannot detect, and the motions due to every ray of light which traverses space, we could only fix positions and directions in the ether by buoying them with matter. We know nothing of the ether, except by its effects on matter, and, after all it would be the material buoys which would fix the positions and not the ether in which they float.

2.b. J. Larmor

To include Joseph Larmor among the scientists who did not accept the ether concept would be a gross mistake. However, there is one aspect of Larmor's relationship with the ether which deserves to be considered as it raises, in my opinion, some interesting questions concerning the real status of the ether concept. Let me explain what I mean.

Quite frequently Joseph Larmor is presented as a physicist who only felt at home with nineteenth century physics,[7] an ultra-conservative individual, in this regard. Nevertheless, Larmor's well-known ideas as to the fundamental character of the Least Action Principle,[8] can be interpreted in the sense that had he been faced with the pressing necessity of choosing between the ether and the Least Action Principle, he would have opted for the latter; that is, he would have behaved in what can be considered a "modern" manner.

Take, for instance, the following passage from his Presidential Address to the Mathematical and Physical Science Section of the BAAS meeting held in Bradford during September 1900:[9]

But whatever views may be held as to the ultimate significance of this principle of Action, its importance not only for mathematical analysis, but as a guide to physical exploration, remains fundamental. When the principles of the dynamics of material systems are refined down to their ultimate common basis, this principle of minimum is what remains . . . In so far as we are given the algebraic formula for the time-integral which constitutes the Action, expressed in terms of any suitable coordinates, we know implicitly the whole dynamical constitution and history of the system to which it applies. Two systems in which the Action is expressed by the same formula are mathematically identical, are physically precisely correlated, so that they have all dynamical properties in common. When the structure of a dynamical system is largely concealed from view, the safest and most direct way towards answering the prior question as to whether it is a purely dynamical system at all, is thorough this order of ideas. *The ultimate test that a system is a dynamical one is not that we shall be able to trace mechanical stresses throughout it, but that its relations can be in some way or other consolidated into accordance with this principle of minimum Action* [emphasis mine].

It is true that ether and the Least Action Principle seemed to get along quite well. In Larmor's own words:[10]

Returning to the molecules, it is now verified that the Action Principle forms a valid foundation throughout electrodynamics and optics; the introduction of the aether into the system has not affected its application. It is therefore a reasonable hypothesis that the principle forms an allowable foundation for the dynamical analysis of the radiant vibrations in the system formed by a single molecule and surrounding ether.

But still, the interest of trying to "weight" Larmor's defence of the Action Principle *versus* the ether concept remains. Let me now, in this sense, present the case of a Professor A. McAulay, of the colonial university of Tasmania. As I see it, this example tends to support one of my contentions (namely, that Larmor is not a good example of an ether-believer), but even if we were to persist in including Larmor on the side of those who wished to make of the ether the authentic cornerstone of physics, McAulay's case will show that even within the "ether camp" there existed strong tensions which could touch on rather fundamental issues. The image of a problematic and ambiguous ether would be thus reinforced, opening the way to consider, in a sympathetic manner, the possibility that the ether may not be the only, or most unique, concept underlying the different reactions of British physicists when faced with the special theory of relativity. My thesis, to be developed later on, is that it is not possible to understand the introduction of special relativity in Great Britain if one does not pay special attention to the different research programmes (Larmor's for instance) which were underway at that epoch, research programmes which, of course, were affected by, and also affected, the ether concept. But let us return to McAulay.

In 1910 this physicist published an interesting paper entitled "Spontaneous Generation of Electrons in an Elastic Solid Aether".[11] McAulay was clear in his purpose: "Some steps are taken" – he wrote on p. 134 – "in the direction of rendering intelligible the fundamental assumptions of Sir Joseph Larmor's system of the Universe as explained in his 'Aether and Matter'."

But even more interesting than this quotation is the following one, which summarizes the conflict I am referring to:

Occasionally [McAulay says on pp. 134–135] Sir J. Larmor seems, in a half apologetic manner, to invite us to regard the 'rotationally elastic' property as something of an ultimate nature, behind which there should be no urgent demand to go. Whether or not this invitation is really intended, most of us cannot tolerate the mental attitude here alluded to with confort. We have the same instinctive objection to an elastic resistance to absolute rotation of a medium, as we have to action between bodies at a distance; both alike we seek to 'explain' by something beneath, something much nearer to the direct suggestions of our muscular sense. If we cannot hope and need not seek, as Sir J. Larmor seems sometimes to imply, to probe to a lower depth than the action principle ($\delta \int L \, dt = 0$), by all means let us readmit the discredited action at a distance, as fine an example of the principle as we could desire.

This last comparison, which sets the action principle on an equal footing to actions at a distance, is important. It shows that to some British physicists Larmor's least action principle was considered as "repulsive" as the old, "discredited", actions at a distance; in other words, as a sort of attack on the idea that the basic cornerstone of physics was a "push-pull" dynamic of ether.[12]

2.c. N.R. Campbell

In spite of the fact that Campbell's case has already been studied by Goldberg, I consider it worthwhile to dedicate a few comments to his ideas on the ether concept.

Norman Robert Campbell's most well-known work was his book

Modern Electrical Theory, which went through two editions,[13] the second one enlarged with several appendixes, almost monographs if taken by themselves.[14]

In the first edition (1907), Campbell avoided almost completely any reference to the ether concept until the very last chapter. There he began by referring to the many "confusions and misunderstandings" which affect it. "The amazing pronouncements" – he states on p. 288 – "about the 'ether' which have been made by many philosophers are rivalled by the statements which are to be found in the writings of men of science of the highest repute". As a matter of fact, Campbell was not "acquainted with any authoritative formal definition of the term 'aether'", and consequently he took the course, "not diverg[ing] from current usage", of taking it "to mean a medium in which electromagnetic actions take place in the absence of substances generally recognized as matter." "Of these actions" – he added – "the vibrations which constitute light are the most important: studies of the properties of the aether usually resolve themselves into optical investigations." Actually, Campbell was not ready to accept for the ether any further roles. This is clear in section 11 ("Aether and Energy") of chapter XIV ("The Laws of Electromagnetism") of the first edition of *Modern Electrical Theory*, as well as in later works; for instance, in the paper entitled "The Aether," which he published in 1910.[15] There we find paragraphs like the following one (pp. 181–182):

No doubt much of the dissatisfaction with 'the aether' is based on the recent theories of the atomic nature of radiation and on the proof that the principle of relativity is an adequate foundation for electromagnetic theory, but it is clear that such theories do not provide either a sufficient or a necessary reason for abandoning the conception. Sir J.J. Thomson, the author of the earliest and most far-reaching atomic theory of radiation, devoted much of his presidential address before the British Association to a description of the properties of the aether, while on the other hand, I hope to show that a consideration of no ideas more novel than the elements of electrostatics may lead to grave doubts concerning the utility of that conception.

This last sentence shows that Campbell's arguments against the ether concept did not follow only from his positivistic outlook, or from his adherence, posterior to 1907,[16] to Einstein's special relativity theory.

3. FROM ELECTRODYNAMICS TO RELATIVITY

In the preceding section I have tried to show that, at the turn of the century, the ether concept was not as unproblematic as it is often implied (this view will be reinforced in the present section). One of the consequences of this fact is that the status of the ether concept as *the* key element for understanding the introduction of special relativity in Great Britain must be somewhat lowered. In this section I will consider the role played by electrodynamics in such introduction, arguing that to focus on "electrodynamics research" for explaining the way in which special relativity was introduced in Britain, has the enormous advantage that we are thus dealing with a field of research which permeated practically all the research done in physics, in Britain as elsewhere, and, secondly, that, contrary to what is usually believed, one could do research in electrodynamics and, nevertheless, not accept the ether (Campbell), or, at least, not make use of it in principle (Schott). And this is not to mention the obvious fact that special relativity was the "last" step in a long chain of developments which had in electrodynamics its main protagonist.

Because he has seldom been considered, and also because his case is, to my purposes, very illuminating, I am going to begin my discussion with G.A. Schott.

3.a. G.A. Schott

The works of George Augustus Schott (1868–1937), professor of Applied Mathematics in Aberystwyth from 1910 till his retirement in 1929,[17] show many of the principal characteristics of the introduction and development of special relativity in Great Britain. Specially interesting in this sense is Schott's preface to his great monograph, *Electromagnetic Radiation*,[18] an Adams Prize essay at the University of Cambridge in 1909.[19]

Almost at the end of the preface Schott states his great debt to J. Larmor and H.A. Lorentz, "whose writings have furnished the theoretical foundation of this essay" (p. xv), as well as to – and this is important – his former teacher, J.J. Thomson, "to whose paper on Cathode Rays this investigation owes its inception."

I have said that this is important because one of my points is that it is not possible to understand completely the process of introduction and development of special relativity in Britain without taking into account: (i) the discoveries and researches which were being carried out in the realm of "atomic physics", and (ii) the efforts which were undertaken for explaining those new results of atomic physics within the framework of electrodynamics.

Schott was very clear in this regard. Thus, he wrote (pp. v–vi):

In consequence of the discoveries of the last few years in the fields of radioactivity, vacuum-tube phenomena, magnetism and radiation, a great need has arisen for a comprehensive Electron Theory of Matter,[20] which shall systematize the results already achieved, as well as serve as a guide in future researches. . . . a beginning as has been made by J.J. Thomson in his well known paper on the Structure of the Atom and in his books on Electricity and Matter and the Corpuscular Theory of Matter.

The problem was, however, that Thomson's investigations had been carried out under the restriction that the electrons in the atomic model move with velocities so small, compared with the velocity of light, that they can be treated like the particles of ordinary mechanics. Schott, as well as many others, knew, however, that this was not true. By that time it was well-known that moving electric charges do not behave like the particles of ordinary mechanics; for instance, their mass varies with their speed, and they generate a magnetic field which reacts on their motion in various ways (*radiation-reaction*[21]). When the speeds of the charges are small compared with that of light, "these efforts" – Schott pointed out – "are small but not negligible." For example, Ritz's then well-known theory of the production of Spectrum Series rested on the effect of a magnetic field in the atom on the motion of electrons. There was, moreover, the problem of the assumption of the smallness of the velocity of light. According to Schott (p. vi):

it is by no means certain that all the electrons inside the atom are moving with speeds small compared with that of light; we know that β-particles are expelled from comparatively stable atoms, like that of Radium, with speeds differing from that of light by only 2%. The kinetic energy of such a β-particle amounts to three millionths of an erg, which is five times the mutual electrostatic energy of two negative electrons in contact. It is not easy to imagine an arrangement of negative and positive charges in equilibrium, or in slow stationary motion, which shall be sufficiently permanent and stable to serve as a model of the Radium atom, and at the same time capable of setting free sufficient potential energy to supply the kinetic energy of a β-particle and also overcome the attraction of the positive charges. If on the other hand we suppose the β-particle to be already moving inside the Radium atom with a speed comparable with that of light this difficulty does not arise.

Actually, Schott's endeavour since at least 1906[22] was precisely that one: to explain "atomic phenomena" by means of "classical physics," electrodynamics in particular. It was in that way that he came across problems related to special relativity; but it was, as it were, a sort of offshoot from his specific topic of research: an electrodynamical theory of atomic phenomena (electron theory of matter). My point is that, in this sense, Schott is just one example – an important example, that is true – of a way to approach the "relativity worldview", which was common to many of the physicists – British or not – who by that time were trying to understand the large number of new phenomena which were being discovered. As we shall see, the ether concept was not necessarily one of the premises of such approach.

Still, an important question remains: When did Schott become acquainted with the theory of special relativity? A partial answer to this question can be obtained by means of Schott's paper "On the Radiation from Moving Systems of Electrons, and on the Spectrum of Canal Rays".[23] As note 2 of this paper shows, when Schott wrote the article in 1907, he already knew Einstein's paper of 1905. Thus, he stated (p. 687):

It is incorrect to call (u', v', w') [defined on p. 659 of the paper] the velocity of the system . . . ; the expression for this velocity has been given by Einstein (*Ann. Phys. 322*, p. 916). The results of this paper are, however, not appreciably affected thereby.

Therefore, by 1907 careful readers of the *Philosophical Magazine* could find Einstein's name referred to.

Let us go back to Schott's *Electromagnetic Radiation*. The preface is, once more, illuminative (p. vii):

Let us now consider the fundamental assumptions on which the present investigation is based. . . . In choosing the fundamental assumptions I have throughout aimed at securing the greatest generality consistent with throughly well establish experimental results. Additional assumptions have only been introduced when further progress seemed impossible without them, or when a comparison of results already obtained with experiment clearly indicated that further restrictions were desirable. For these reasons *I have refrained from making any use, either of the Postulate of Relativity, or the Aether Hypothesis, which by some are regarded as inconsistent with each other*. Some of the results obtained in this essay are consistent with the Postulate of Relativity; others cannot be reconciled with it, at least when it is used in the strictest possible sense, and find their natural explanation in terms of the aether. All however are deduced quite independently of either hypothesis. *It does not appear to me that either of the new theories is so well established and so generally accepted yet as to be properly made the basis of an investigation in which the utmost generality is aimed at* [emphasis mine].

One of the problems to which Schott was referring dealt with an objection raised by Max Abraham in the second edition (1908) of the second volume ("Der Elektromagnetische Theorie der Strahlung") of his famous textbook *Theorie der Elektrizität*. There, Abraham pointed out that, according to the theory of relativity, the mass, m_0, and the electric energy, W_0, of a slowly moving electron satisfy the relation $m_0 = W_0/c^2$, while the mass of the Lorentz electron (not a *point* electron) is 4/3 of the amount determined by this equation. Schott was able to show[24] that even for a symmetric electron m_0 always exceeds W_0/c^2, whatever the configuration of the electron may be when it is at rest. He also found that an extended electron, whatever the distribution of its charge may be, cannot exist unless it is subjected to a suitable pressure on its outer surface. According to him, this meant that (p. xiv): "We must either postulate the existence of some external medium which shall produce the required surface pressure, or admit that the elements of charge of the electron exert on each other actions at a distance, which are not electromagnetic and follow quite different laws."

To Schott (p. xiv), the "first hypothesis, that of an external medium, appears to be the more reasonable of the two and amounts to admitting the existence of the electromagnetic aether." His conclusion is clear: "Thus it appears that the acceptance of the Postulate of Relativity in its strictest form almost necessitates the adoption of the hypothesis of the extended charge, while the hypothesis of the extended charge leads naturally to the adoption of the aether hypothesis."

This was, however, only one aspect of the "controversy" *Postulate of Relativity-Aether Hypothesis*; the preface makes it clear that Schott did not think the controversy was settled.[25] How little ether-dependent was Schott's approach can be seen, for instance, when, referring to Maxwell-Hertz's electromagnetic equations, he writes (p. vii): "These equations already imply the existence of a system of axes to which they are referred; *it is immaterial for our purpose whether these axes be regarded as fixed in space, relative to a fixed aether, or as only fixed relative to the observer*" [emphasis mine].

With Schott we have, therefore, one example of a British physicist who carried on high-level research related to relativity, and who, nevertheless, was not *a priori* committed to the ether concept.

3.b. N.R. Campbell

I am going now to return to the case of Norman Robert Campbell to show, as in Schott's case, but in a probably clearer manner, that the ether concept was not necessarily behind all the presentations and/or

researches which were being carried out in the field of electrodynamics.
We have already seen (Section 2.c.) that Campbell was a fierce critic of
the ether concept, but we have not yet discussed the role played by
Einstein's special theory of relativity in the two editions of his book,
Modern Electrical Theory.[26]

As far as special relativity is concerned we have that, in the first
edition (1907) of *Modern Electrical Theory*, Einstein's name or ideas do
not appear. Campbell only refers to Bucherer, Lorentz, Abraham or
Larmor, among others. This fact suggests that, in contrast with the case
of the already mentioned Schott or of Ebenezer Cunningham, by 1907
Campbell was not acquainted with Einstein's 1905 paper. In fact, it
seems that by that time Campbell was rather close to *some* points of the
"Electromagnetic View of Nature".[27] Thus he wrote (p. 302): "Now we
have seen that it is probable that all the properties of material bodies –
chemical, mechanical, optical, thermal and the rest – can be reduced to
electrical properties".[28]

In this sense, it is important to point out that Campbell thought that
Michelson and Morley's experiment supports the electromagnetic
worldview. So, we read (p. 303):

From this point of view, this negative result [Michelson and Morley's] proves that the
optical properties of matter (on which the experiment is based) are electrical in origin – a
conclusion that nobody doubts nowadays. Accordingly the Lorentz-Fitzgerald hypothesis
has met with general acceptance, and it has been concluded that the difficulties which were
raised by the Michelson-Morley experiment have been solved satisfactorily.

We see, therefore, that Campbell's point of departure was electrodynamics (an electrodynamics free from the ether, a concept that he did
not accept). The question is: did he "arrive" at special relativity, and if
so, when?. Indeed, he did. By 1913, when the second edition of *Modern
Electrical Theory* appeared,[29] Campbell already knew about Einstein's
special relativity theory. The new preface is quite clear on this point:
"This volume, nominally, a second edition, is really a new book . . .
The Principle of Relativity and Stark's work on atomic structure have
altered Part III completely."

But Campbell not only knew *about* special relativity, he also understood it quite well. "This last chapter" – we read on p. 351 – "will be
devoted to a consideration of certain problems which in the past have
been the subject of much discussion among physicists. These problems
have little direct connection with those which have concerned us hitherto
and our study of modern electrical theory would be logically complete

with only the barest reference to them." That is, Campbell understood what is one of the fundamental ideas of the special theory of relativity; namely, its basic independence from electrodynamics, or, in other words, that it is a *kinematics* which ought to be applied to any interaction (to the electromagnetic one too, of course). Most probably, his methodological opinions, which led Campbell to reject the ether, would help him in understanding and accepting relativity. In this sense, it is worth quoting from appendix I ("The Aether") to the 2nd edition of *Modern Electrical Theory* (p. 388):

... very shortly Einstein showed that the rules given by this complicated theory [i.e. Lorentz's and Fitzgerald's] for deducing the relation between the laws observed by two observers who are in relative motion are exactly the same as those given by the Principle of Relativity; these rules, of course, involve only the relative velocity of the two observers and not their 'velocity relative to the aether.' Moreover, the form in which those rules are expressed by the Principle of Relativity is much simpler and more convenient than that in which they are given by the aether theory combined with the Lorentz-Fitzgerald hypothesis.

3.c. E. Cunningham

One of the first references made in Great Britain – if not the first – to Einstein's 1905 contribution, appears in one of Ebenezer Cunningham's papers. Consequently, and although his main work will be discussed later on, I think it is important to refer here, albeit briefly, to that paper.

In the 1907 October issue of the *Philosophical Magazine*, Cunningham[30] joined the, at the time, important discussion of the *electromagnetic* mass of a moving electron. Specifically, he opposed Max Abraham's objection[31] to Lorentz's conception of an electron as having, at rest, a spherical shape, but in motion the shape of an oblate spheroid. "The present paper," Cunningham states on page 539, "reconsiders Abraham's discussion and comes to the conclusion that the objection is not valid."

The reasons why he arrived at such a conclusion were based on the fact that "it has been proved that Maxwell's equations represent equally well the sequence of electromagnetic phenomena relative to a set of axes moving relative to the aether, as relative to a set of axes fixed in the aether."

The proof in question was none other than the substitution of Galileo transformations between inertial frames of reference for what are now called Lorentz transformations, whose expression Cunningham duly reproduces in his paper, adding that the "above transformation renders the electromagnetic of a system independent of a uniform translation of the whole system through the aether."

For my purposes here, it is important to point out that, *while dealing with that electromagnetic problem*, Cunningham showed his knowledge of Einsteins's contribution, the special theory of relativity.

The transformation in question [he wrote in a footnote on p. 539] is given by Einstein in a paper in the *Annalen der Physik*. . . . It is in substance, the same as that given by Larmor in 'Aether and Matter,' chap. xi, though the correlation is only proved as far as the second power of v/c. Prof. Larmor tells me he has known for some time that it was exact. Vide also Lorentz, *Amsterdam Proceedings*, 1903–4.

However, "knowledge of the *existence* of" does not necessarily entail "knowledge of the *meaning* of." Thus, taken by itself, Cunningham's 1907 paper does not offer any evidence that he understood that Einstein's theory was different from Larmor's or Lorentz's, in the sense that his special theory of relativity transcended electromagnetism. Quite on the contrary, the terminology, together with the exclusive electromagnetic framework and the persistent use of the expression "motion through the aether" does strongly suggest that Cunningham was still "on Lorentz's side;" that is, that he believed that Lorentz's transformations were just a sort of theorem, or property, of Maxwell-Lorentz's electrodynamics.

The point is, nevertheless, that Cunningham's early relationship with special relativity provides us with another manifestation – certainly different from Schott's or Campbell's cases – of the close connection existent between the introduction of Einstein's theory in Great Britain and the development of electrodynamics.

There is another fact, related to Cunningham's paper, which must be mentioned here. In his readiness to acknowledge Einstein's 1905 contribution, Cunningham behaved in a somewhat different manner than some of his German colleagues. Actually, we find in the pages of the *Philosophical Magazine* one illustration of this fact. In its April 1907 issue the then *Privatdozent* at Bonn University, Alfred Heinrich Bucherer, published a paper entitled "On a New Principle of Relativity in Electromagnetism".[32] In that paper Bucherer, "guided by the feeling that the *form* of the Maxwellian equations must correspond somehow to the true laws of electromagnetism," attempted a new interpretation of these equations that would "harmonize with facts." Although the *spirit* (not the structure, or even the details) of Bucherer's ideas was clearly in agreement with Einstein's, he nevertheless did not mention or refer to Einstein's name. He only said that (pp. 413–414):

It is needless to dwell on the serious difficulties which the Maxwellian theory has encountered by the well established experimental fact that terrestrial optics is not influenced by the earth's motion. The endeavours of some distinguished physicists, notably of H.A. Lorentz, to modify the Maxwellian theory in such a manner as to eliminate the effects of translatory motion have admittedly failed.

It was only when some of his contentions were criticized by Cunningham, in the above-cited paper (where Einstein's name was mentioned), that Bucherer spoke of the "Lorentz-Einstein Principle."

3.d. H.Bateman

With Harry Bateman (1882–1946), a Manchester-born mathematical physicist who became fellow of Trinity College, Cambridge, and Reader in Mathematical Physics at the University of Manchester, before he settled (1910) in the U.S., we have a different "electrodynamical approach" to relativity. It is, in this sense, important to pay attention to him, inasmuch as he will help us to realize that there was a broad range of approaches to special relativity based on Maxwell's electrodynamics.

The essence of Bateman's approach lies in its geometrical character. "Recent theoretical researches in electromagnetism," he wrote in one of his papers,[33] "indicate that the science of electromagnetism is closely connected with the geometry of spheres." Taking into account this fact, we should not be surprised to learn that Bateman had Hermann Minkowski's four-dimensional ideas in great consideration (we shall consider later on the role played by Minkowski's point of view in the introduction of relativity in Great Britain). Thus, we see that in the paper just referred to, Minkowski's name appears alongside those of Lorentz, Larmor, and Einstein. The following quotation, from the same paper (p. 624) is truly Minkowskian: "The group of *Lorentzian transformations* for which the electron equations are covariant is then represented by the group of transformations of rectangular axes in the space of four dimensions."

Another aspect of Bateman's interests on this regard, are the papers he devoted to the conformal transformation. In 1909 he published a paper entitled "The Conformal Transformations of a Space of Four Dimensions and Their Applications to Geometrical Optics",[34] in which he showed that Maxwell's equations are covariant not only under the Lorentz group, but also under the more general fifteen-parameter conformal group. As a matter of fact, this discovery was made independently by Cunningham. (The title of the paper which Cunningham published in 1910, "The Principle of Relativity in Electrodynamics and an Extension Thereof",[35] is significant in as much as it shows that Bateman's and Cunningham's approaches could contribute to the development of special relativity along routes not necessarily connected with, for instance, the ether concept.)

What has been said about Bateman should not be taken as if he were completely foreign to the problems aroused by the ether concept. A few quotations from the book, *The Mathematical Analysis of Electrical and Optical Wave-Motion*, that Bateman published in 1915 (Cambridge University Press) will serve our purposes in that sense:

Some of the modern writers on the theory of relativity [we read on p. 2] maintain that the introduction of the idea of an aether is unnecessary and misleading. Their criticisms are directed chiefly against the popular conception of the aether as a kind of fluid or elastic solid which can be regarded as practically stationary while material and electrified particles move through it. This idea has been very helpful as it presents us with a vivid picture of the processes which may be supposed to take place, it also has the advantage that with its aid we can attach a meaning to the term absolute motion, but herein lies its weakness. Larmor, Lorentz and Einstein have shown, in fact, that the differential equations of the electron theory admit of a group of transformations which can be interpreted to mean that there is no such thing as absolute motion.

If this be admitted, the popular idea of the aether must be regarded as incorrect, and so if we wish to retain the idea of a continuous medium to explain action at a distance we must frankly acknowledge that the simplest description we can give of the properties of our medium is that embodied in the differential equations [rot $\mathbf{H} = (1/c)\, \partial E/\partial t$, rot $\mathbf{E} = -(1/c)\, \partial H/\partial t$].

This last sentence is, in some sense, not too different from Hertz's famous dictum: Maxwell's theory is Maxwell's system of equations." It can be taken as implying that Bateman did not care much about the ether; a conclusion which is reinforced if one reads this book, a truly mathematical-physics monograph not much preoccupied with physical concepts. Still, the content of the only pages (pp. 2–4) of *The Mathematical Analysis of Electrical and Optical Wave-Motion* explicitly devoted to a discussion of the concept of interaction is rather inclined to favour explanations based on the ether concept, although, at the same time, leaving open other possibilities:

If we abandon the idea of a continuous medium in the usual sense only two ways of explaining action at a distance readily suggest themselves. We may either think of the aether as a collection of tubes or filaments attached to the particles of matter as in the form of Faraday's theory which has been developed by Sir Joseph Thomson and N.R. Campbell; or we may suppose that some particle or entity which belonged to an active body at time t belongs to the body acted upon at a later time $t + \gamma$. From one point of view these two theories are the same, for if particles are continually emitted from an active body they will form a kind of thread attached to it. The first form of the theory is, however, more general than the second.

Bateman's mathematical outlook can be easily seen when he writes (p. 3): "At present we are unable to form a satisfactory picture of the processes that give rise to, or are represented by, the vectors E and H. We believe, however, that some points may be made clear by studying the properties of solutions of our differential equations." And his mathematical analysis of those equations led Bateman to three distinct theories of the universe, which may be described briefly as follows:

ETHER	MATTER
– Continuous medium.	– Aggregates of discrete particles.
– Discontinuous medium consisting of a collection of tubes or filaments.	– An aggregate of discrete particles attached to tubes.
– Continuous medium.	– An aggregate of discrete particles to which tubes are attached.

"The last theory," Bateman points out on (p. 3), "may be supposed to include that form of the emission theory of light in which small entities are projected from the particles of matter under certain circumstances and produce waves in the surrounding medium. This theory might be justly ascribed to Newton." However, it was to the first theory that most of Bateman's book (especially the first part) was dedicated. "The other theories," Bateman added, "have not yet received much attention but it is hoped that the analysis of Chapter VIII will lead to further developments so that a comparison can be made between the different theories. *It is likely that one theory will be enriched by the developments of another*" [emphasis mine].

3.e. J. Jeans

There is no doubt that textbooks have great importance in the way new ideas are accepted or rejected, at the same time as they reflect the state of the art when they were written. If the book has been an influential one then the historian has, by analysing its content, an excellent instrument to carry out his work.

This is the case with James H. Jeans's treatise, *The Mathematical Theory of Electricity and Magnetism*, an excellent and widely-used textbook. Even as this book adds to the interest of its author – it was written by one of the most popular and influential physicists in Britain during the first decades of our century, – the fact that it is dedicated to electromagnetism will allow us to bring this already long section to an end by studying the changes undergone in its different editions as far as concerns the theory of relativity.

The first edition (1908) of the book[36] did not contain any discussion of relativity. It was only in the second edition (1911) that Jeans added "two new chapters . . . on the Motion of Electrons and on the General Equations of the Electromagnetic Field",[37] where some problems *related to* relativity were touched. It is important to point out, however, that Jeans's approach was fully within the ambit of electrodynamics; in this sense it did not differ from, for instance, the first edition (1907) of Campbell's *Modern Electrical Theory*. So, he dealt firstly with the problem of the variation, with respect to velocity, of the kinetic energy of an electron moving with any velocity, referring to the experiments performed by Walther Kaufmann and Alfred Bucherer, as well as to their significance for the theories proposed by Abraham and Lorentz. After that, Jeans considered the question of the motion of an arbitrary system in equilibrium. The problem here was that when a material system moves, the electric field produced by its charges is different from the field when at rest; the difference between these fields must show itself in a system of forces which must act on the moving system and in some way modify its configuration. Jeans showed in which sense the equilibrium of the original system was maintained. Once this question was settled, he stated (p. 577): "Lorentz, to whom the development of this set of ideas is mainly due, and Einstein have shewn how [this result] may be extended to cover electromagnetic as well as electrostatic forces." This was the only reference to Einstein made by Jeans in 1911.

The last two problems discussed by Jeans in the second edition of his textbook were the "Lorentz-Fitzgerald contraction hypothesis" and the "Lorentz deformable electron." The beginning of the section dedicated to the Lorentz-Fitzgerald contraction is significant:

It is now natural [Jeans wrote (p. 577)] to make the conjecture, commonly spoken of as the Lorentz-Fitzgerald hypothesis, that the system *S* when set in motion with a velocity *U* assumes the configuration of the system *S'*, this latter being a configuration of equilibrium for the moving system. *Indeed, if we suppose all forces in the ether to be electrical in origin, this view is more than a conjecture; it becomes inevitable.* Put in the simplest form it asserts that any system when set in motion with uniform velocity U is contracted, relatively to its dimensions when at rest, in the ratio $(1 - U^2/c^2)^{1/2}$ in the direction of its motion (emphasis mine).

That is, Jeans, like Campbell in 1907, viewed the Lorentz-Fitzgerald hypothesis as part of the electromagnetic worldview; there is nothing in his 1911 presentation revealing that he understood such hypothesis in the way Einstein did in his seminal paper of 1905.[38] In fact, the conclusions of the last section of the book, dedicated to the Lorentz deformable electron, reinforced Jeans's belief in the electromagnetic view of nature. Commenting on the agreement between Bucherer's experiments and Lorentz's formulae, Jeans stated (p. 579) that Bucherer's experiments seemed to lead to the conclusion (among two other possible ones) that "the mass of the electron is purely electromagnetic in its nature." This is the last sentence of Jeans's book.

The real changes in Jeans's textbook, as far as special relativity is concerned, arrived with the fourth edition (1920). In the preface, dated December 1919 (recall that 1919 is the year of the British expedition, led by Dyson and Eddington, to observe the solar eclipse of May 29), Jeans made clear which were the novelties: "It will be found that the main changes in the fourth edition consist in a rearrangement of the later chapters and the addition of a whole new chapter on the Theory of Relativity."

Looking at the actual content of this new edition, we find that, although Jeans was quite slow in opening the pages of his book to Einstein's theory, he, at last, understood it now quite well. Thus, the manner in which he introduces the "relativity-condition" is clearly a generalization that superseded specific interactions like the electromagnetic one.[39] Furthermore, he realized the consequences that Einstein's point of view had for the ether:

The hypothesis that there is an ether [we read on p. 619] may give a possible explanation of the phenomena, but the hypothesis that there is no ether provides an equally possible and very much simpler explanation . . . If the observed constant velocity of light is simply the constant velocity of propagation through an ethereal medium, it would seem to follow that each observer must carry a complete ether about with him. This at least robs the ether of the greater parts of its reality . . . Considerations such as we have mentioned do not prove in strictness that light cannot be propagated through an ether; what they prove is that if an ether exists, it must be something very different from the absolutely objective ether imagined by Maxwell and Faraday.

Ambiguous as the last paragraph may seem (it was not so easy to escape from one's upbringing), Jeans's stand about the ether was quite clear. He considered it as an unnecessary device. Indeed, he even gave a rather original argument against that elusive concept (p. 620):

If an ether existed it would provide a fixed set of axes relative to which all positions and velocities could be measured. To account for the result of the Michelson–Morley equipment, it would be necessary to postulate a real shrinkage of all bodies moving through the ether. This shrinkage could not be detected by mechanical means, for a measuring rod would shrink in precisely the same ratio as the body to be measured, but it could be detected by gravitational means unless every gravitational field of force shrunk in just a way as to conceal the shrinkage of matter.

At this point Jeans argued that if the gravitational field were not to shrink then a geoid, or surface of mean sea-level on the earth, might be a gravitational equipotential for some velocity through the ether, but could not remain an equipotential as the earth's velocity through the ether changes on account of the description of its orbit. That is, one would expect to observe seasonal and even daily surgings resulting from the earth's motion *through the ether*. However, no such events are observed, "whence" – Jeans continued – "it seems natural to suppose that gravitation also must conform to the relativity-condition," a conclusion that, he notes, was supported by the fact that "Einstein's relativity theory of gravitation has received confirmation." Consequently, it was almost inescapable to conclude that (p. 621):

If, then, we continue to believe in the existence of an ether, we have also to believe not only that all electromagnetic phenomena are in a conspiracy to conceal from us the spread of our motion through the ether, but also that gravitational phenomena, which as far as is known have nothing to do with the ether, are parties to the same conspiracy. The simpler view seems to be that there is no ether. If we accept this view, there is no conspiracy of concealment for the simple reason that there is no longer anything to conceal.

4. ANOTHER TYPE OF APPROACH: A.A. ROBB

It has been my purpose in the preceeding sections to show that there existed a rather large number of different approaches in Great Britain towards the sort of problems that characterize special relativity. It is true, however, that all the approaches reviewed so far were, in one way or another, closely related to electrodynamics. The case of Alfred A. Robb (1873–1936), a former student in J.J. Thomson's Cavendish Laboratory, and F.R.S. since 1921, is completely different. In fact, Robb's contributions inaugurated a new phase in the understanding of relativity.[40]

Of course, the historical significance of any individual does not depend only on the importance of his ideas; it is also connected with his position and social relations within the cultural context at the time he was alive. In this last sense there is still much historical research to be done as regards Robb. He certainly was not an unknown personage. According to A. Vibert Douglas, Eddington's biographer, "[L.A.] Pars recall[ed] frequently seeing [Eddington] strolling along Jesus Lane after the service to call on A.A. Robb, a kindred spirit both because of his deep grasp of the problems of time and of cosmic order, and by reason of his gift for whimsical rhyming".[41] On the other hand, the Dutchman Adriaan D. Fokker, a former student of H.A. Lorentz, recalled[42] that during 1913–1914 he "was much struck by the papers of Robb, who wrote about the theory of Einstein. In fact, he wrote an axiomatic theory for Einstein, avoiding the word relativity." Actually, during his stay in England (1914) as a postdoctoral student, he went to Cambridge only to make the acquaintance of Dirac and Robb. Later on, Fokker would write, in 1929, the first book dedicated to relativity to appear in Dutch.

It is also perhaps more than a coincidence that at the time that Robb was completing the first version of his theory, to appear as *Optical Geometry of Motion: a New View of the Theory of Relativity* (1911),[43] Whitehead and Russell were publishing (1910–1913) their famous work, *Principia Mathematica*, that aimed to reduce mathematics – arithmetic in particular – to the principles of logic, also a landmark in what could be termed the "axiomatical approach" (to mathematics in this case).

As I pointed out before, Robb was also elected fellow of the Royal Society. Nevertheless, the range of his influence within the British ambit is still to be determined. His researches contributed, as early as 1911, to the widening of the modes of facing – or of understanding – relativity, a general movement in Great Britain as well as in all the other countries. Stated in a few words, it can be said that Robb was the first to work out the idea that the geometry of the (Minkowski) space-time of special relativity can be developed on a purely causal basis. Thus, in his main work, the book *A Theory of Time and Space* (1914),[44] he derived the entire geometry of Minkowski space-time from an axiom system with only two primitives (undefined predicates), namely "x is an *element*," and "the element x is *after* the element y".[45] Their intended meaning being, respectively, that x is a world-point and that y is different from x and causally precedes x.

The origin and development of Robb's ideas is explained by himself in the preface of his book *Geometry of Time and Space*[46] – essentially a second edition of his 1914 treatise *A Theory of Time and Space* – that appeared the year of his death (1936). It is, consequently, appropriate to quote from it (p. v):

So far as I am aware the book, in its original form [that is, *A Theory of Time and Space*], was the first of its kind to be written, and a brief account of its origin may be of interest. At the meeting of the British Association held at Belfast in 1902 Lord Rayleigh gave a paper entitled: *Does Motion through the Ether cause double Refraction?* in which he described certain experiments which he had carried out with the object of testing this matter, and which seemed to indicate that the answer was in the negative.

I remember that he inquired of Professor Larmor, who was present on this occasion, whether, from his theory, he would expect double refraction to be produced in this way. Professor Larmor replied that he would not, and, in the discussion which followed considerable surprise was expressed that, in any attempt to detect motion through the aether, things seemed to conspire together so as to give null results. The impression which this discussion made upon me was, that in order properly to understand the matter, it would be necessary to make some sort of analysis of one's ideas concerning equality of lengths, etc., and I decided that, at some future time, I should attempt to carry this out. I am not quite certain that I had not some idea of the sort prior to this meeting, but, in any case, the inspiration came from Professor Larmor, either then, or on some previous occasion while attending his lectures.

Some years later I attempted to carry out this scheme, and, while doing so, I heard for the first time of Einstein's work.

I may say that, from the first, I felt dissatisfied with his approach to the subject, and I decided to continue my own efforts to find a suitable basis for a theory.

Some of the former statements are characteristic of Robb. He referred more to Larmor, and also to Lorentz, than to Einstein and Minkowski. However, it is the ideas of these last two, especially the idea of space-time as a four-dimensional physical continuum, which dominate his work, especially his great treatise, *A Theory of Time and Space*.

The first work which I published on the subject [Robb continues] was a pamphlet which appeared in 1911 entitled: *Optical Geometry of Motion: A New View of the Theory of Relativity*.

This pamphlet was of an exploratory character and did not profess to give a complete logical analysis of the subject; but nevertheless, although bearing a very different aspect, it contained some of the germs of my later work. It was, in fact, an attempt to describe Time-Space relations without making any assumptions as to the simultaneity of events at different places. Later on, the idea of *Conical Order* occured to me, in which instants at different places are regarded as definitely distinct; so that there is no such simultaneity.

Probably, Robb's most fundamental idea was that the axioms of geometry are the formal expression of certain optical facts which are at the basis of special relativity, a theory in which the velocity of light plays a most important role. Of course, it was natural for someone willing to set up a theory consistent with such an idea to choose the axiomatic approach. In fact, Robb has been nicknamed "the Euclid of Relativity." His 21 axioms and 206 theorems form what is probably the most complete and rigorous exposition of what the special theory of relativity has to say about space and time, as expressed in terms of a single basic concept, that of "conic order."

Robb's attachment to the axiomatic approach can be seen in the preface to *Geometry of Time and Space*, that we have been citing:

The working out of a scientific theory in the form of a sequence of propositions, such as was done by Euclid, Newton and others, seems largely to have gone out of vogue in these latter days and I consider that this is rather regrettable.

No doubt, in doing exploratory work, other methods are permissible and necessary, but I think that the incorporation of the more fundamental parts of a theory in a sequence of propositions should always be kept in view, since, in this way, one is able to see much more readily what are our primary assumptions, and one is able to fall back upon these in cases of difficulty.

Such were some of the ideas of Alfred Robb, whose work opened, when the "relativistic debate" had hardly begun, a new way to consider special relativity.

5. THE FIRST MONOGRAPHS DEDICATED TO STANDARD RELATIVITY: L. SILBERSTEIN AND E. CUNNINGHAM

The year of 1914 was, as far as the introduction of relativity in Britain is concerned, a remarkable year. Because it was then that two books dedicated to the special theory of relativity were published: Ludwik Silberstein's *The Theory of Relativity*,[47] and Ebenezer Cunningham's *The Principle of Relativity*.[48] As they circulated widely among British readers, it is compulsory to say at least a few words about them. (As we have seen, before 1914 discussions of Einstein's special theory appeared in some texts – the second edition [1913] of Campbell's *Modern Electrical Theory*, for instance, – but such books were not dedicated exclusively to relativity, and when they were – the case of Robb – the approach was far from the standard one.)

I shall deal briefly with Silberstein's book, because although published in Britain and based on a course of lectures delivered at University College, London, during the academic year 1912–1913, Silberstein was not a British "product." A native of Warsaw, educated in Cracow,

Heidelberg and Berlin, he was a lecturer, successively, in Lemberg, Bologna and Rome, during the period 1895–1920. However, during 1912–1920 he also held in addition to his lectureship in Rome, positions in the Research Department of Adam Hilger, Ltd., London (a firm, founded in 1874, known for the high quality of its optical work), as well as in the University of London. Finally, in 1920 he moved to the Eastman Kodak Co. in Rochester, U.S., where he worked until his retirement in 1929, remaining in the U.S. until his death in 1948. Silberstein, therefore, could hardly be considered as a representative of the British scientific *milieu*. However, as I said before, his book was widely circulated in Britain, and this fact justifies that we mention some of its main traits.

First, and foremost, let us note that Silberstein had a very fine understanding of the real meaning of special relativity. The following quotation, taken from Silberstein's book, is, by itself, an exceptionally good summary of Einstein's theory:

In the meantime, [Silberstein is referring here – on p. 87 – to the period between Lorentz's contributions and Poincaré's 1906 paper in the *Rendiconti del Circulo Matematico di Palermo*], 1905, Einstein published his paper on 'the electrodynamics of moving bodies' which has since become classical, in which, aiming at a perfect reciprocity or equivalence of the above pair of systems, S, S', and denying any claims for primacy to either, he has investigated the whole problem from the bottom. Asking himself questions of such a fundamental nature as what is to be understood by 'simultaneous' events in a pair of distant places, and dismissing altogether the idea of an aether, and in fact of any unique framework of reference, he has succeeded in giving a plausible support to and at the same time a striking interpretation of, Lorentz's extended theory. Einstein's fundamental ideas on physical time and space, opening the way to modern Relativity, will occupy our attention.

The second important point is that Silberstein fully understood (as, for example, Campbell also did) that Einstein's special relativity breaks down with the notion of an ether. Thus, no reference to this concept is contained in his formulation (in p. 99) of the two basic principles of the theory. This procedure contrasts with Cunningham's or even with Eddington's presentation in his famous *Report on the Relativity Theory of Gravitation*.[49] Commenting on the two principles, Silberstein wrote (p. 99):

Fresnel claimed this property [i.e., the constancy] of light propagation only for a certain, unique system of reference, namely the aether or a system fixed in the aether, while Einstein, by accepting [the two basic axioms], postulates it for any one out of an infinity (∞^3) of systems moving uniformly with respect to one another. With regard to this property the [inertial] systems S', S'', etc., are perfectly equivalent to the system S . . . and this is the reason why the mere notion of an 'ether' breaks down.

Finally, I want to mention that Silberstein also realized the kinematical, universal, nature of special relativity. In this sense he stated (p. 116), for instance:

This is not to say, of course, that mechanical and all other phenomena must be 'ultimately' electromagnetic, i.e. that everything must be explained by, or reduced to, electromagnetism. The theory of relativity is not concerned at all with such reduction of one class of phenomena to another. It does not force upon us an electromagnetic view of the world any more than a mechanical view. Quite the contrary; it opens before us a wide field of possibilities of asserting that even the mass of a free electron, say a β-particle, must not be entirely electromagnetic.

That is, with Silberstein's book, British students and scientists had an excellent tool to learn special relativity. However, one thing was Silberstein's clear ideas and teachings, and a completely different one the actual expectations and understandings of the theory by some of the British audience. An anonymous review of *The Theory of Relativity*, which appeared in the *Philosophical Magazine*,[50] gives an idea of the terms under which British scientists took Silberstein's explanations.

The fact is [the reviewer wrote] that without the result of Michelson and Morley's experiment there would be no justification for the theory at all. It is because it gives the most direct explanation of their null result and is at the same time not at variance with any other experimental fact, that the theory may claim serious consideration.

Thinking that what he had just said were in fact Silberstein's ideas and not his own, the same commentator went on to state: "So much the reviewer felt this to be true that he would go further, and declare that it will only be when further experimental data of a crucial kind are obtained that the theory will run much chance of becoming definitely accepted as scientific knowledge."

One of the most important events in the introduction of relativity in Britain was the publication of Ebenezer Cunningham's *The Principle of Relativity*, the first book written by an Englishman and published in Britain, dedicated to the "old restricted principle," as its author once called it.[51] Cunningham, Senior Wrangler in the 1902 Mathematical Tripos, fellow and lecturer of St. John's, Cambridge, when he wrote this book, is, certainly, not a major figure in physics, but there are several aspects of his career – his work in relativity being but one – which make him quite an interesting personality. However, here I shall restrict myself to a discussion of Cunningham's views as contained in *The Principle of Relativity*, a book that constituted the summit of his previous original researches.[52]

Cunningham's understanding of Einstein's 1905 principle of relativity shows two, *in principle*, rather contradictory characteristics; namely, he realized that the special principle of relativity is a universal one, something that not everybody understood at that epoch (in Britain as elsewhere), but at the same time he believed that such a universal principle did not entail that the ether was unnecessary. Let me quote in this connection the following passage of Cunningham's book (p. 155):

There are two interacting processes at work in theoretical science. There is the continual endeavour to form models which . . . shall imitate or represent, more or less closely, phenomena of which the *modus operandi* is not closely realized . . .

There is on the other hand the attempt to disentangle general principles of the widest possible application, not fully descriptive of each particular set of phenomena, but common to them all, and independent of the special mechanism which is characteristic and particular. Newton's Laws of Motion must be instanced, the principle of least action, the principle of the conservation of energy. With these we may class the Principle of Relativity, that is, the general hypothesis, suggested by experience, that whatever be the nature of the aethereal medium we are unable by any conceivable experiment to obtain an estimate of the velocities of bodies relative to it.

We see from Cunningham's definition of the principle of relativity, that he had not been able to free himself from the ether concept. However, he was "modern" enough, and valued Einstein's discovery so highly, that he tried to make "ether" and "principle of relativity" compatible (we have, thus, one example of an approach in which those two concepts were not viewed as incompatible):

. . . the critics of the Principle of Relativity are justified in saying that it does not admit of an 'objective fixed aether,' but it cannot be said that it denies the existence of an objective aether of any kind, until it is shewn that a medium cannot be conceived which renders account of electromagnetic phenomena and that at the same time has a motion which is consistent with the kinematics of the principle (p. 162).

As a matter of fact, Cunningham argued – presenting also the corresponding mathematical analysis – that it is possible to conceive a *moving ether* in which the conception of stress, transfer of momentum, velocity, and flux of energy are consistently related, and which is also conformable to Einstein's kinematics, its velocity being defined by relations which have an invariant form. The magnitude of the velocity is everywhere that of light.

One of the most interesting aspects of Cunningham's work concerns Minkowski's formalism,[53] especially as a *heuristic device* to carry through the restricted relativity programme. He thus wrote (p. 156):

The four-dimensional analysis introduced by Minkowski not only introduces a greater symmetry into the discussion of the relativity of electrodynamic phenomena. It gives us also a new point of view from which to regard mechanical quantities and enables us to go some way in finding what modifications are necessary to the usual statements of mechanical theory in order that they may be included within the scope of the principle of relativity.

That is, Cunningham's case is but another example of the sympathetic reception given in Britain to Minkowski's four-dimensional formulation. In fact, after having reviewed Cunningham's main ideas concerning the basic elements of special relativity, we can go on and consider other examples of British physicists who welcomed Minkowski's formulation.

6. THE ROLE PLAYED BY MINKOWSKI'S FOUR-DIMENSIONAL FORMULATION IN THE INTRODUCTION OF SPECIAL RELATIVITY IN BRITAIN

It is a well-known fact that Albert Einstein had some difficulties in accepting Minkowski's four-dimensional formulation of special relativity as a positive, and interesting, contribution. (He finally became convinced of its utility and basic importance during his search for a relativistic theory of gravitation, approximately in 1912). It seems, however, that Minkowski's four-dimensional view of special relativity favoured the introduction of Einstein's theory, at least it did so in some significant cases of British physicists. We already saw, for instance, that Bateman and Cunningham held Minkowski's approach in great esteem. Moreover, they also made use of the four-dimensional framework to further some of their particular investigations (in Cunningham's case it happened thus in his studies of conformal transformations, which he even considered a possible substitute of the Lorentz transformation).[54]

We find, also, the same kind of sympathetic welcome of Minkowski's formulation in Silberstein's *The Theory of Relativity*,[55] and in Eddington's *Report on the Relativity Theory of Gravitation*,[49] in which Minkowski's famous sentence, "Henceforth Space and Time in themselves vanish to shadows, and only a kind of union of the two preserves an independent existence," is reproduced.[56]

Indeed, Minkowski's ideas found such a congenial reception among the British, that it even appeared mixed up with theories other than Einstein's. Take, for example, the case of S.B. McLaren, professor of mathematics at University College, Reading. In 1913, McLaren published a paper entitled "A Theory of Gravity",[57] which contains a rather complicated theory, connected, according to McLaren, with some of Bernhard Riemann's ideas, as well as with "the most revolutionary principle Physicists have accepted since Riemann's day, Einstein's principle of the relativity of time".[58] But the point I want to emphasize here is the prominent role played by Minkowski's four-dimensional approach in McLaren's considerations: "I approach," he wrote (p. 639), "these problems in the first instance as a believer in the physical reality of a four dimension. I argue that the four-dimensional geometry of Minkowski is not a mathematical fiction, but a necessary inference from ordinary experience."

Whether or not this type of reception of Minkowski's ideas applies to other countries, is something, as far as I know, still to be looked into.

7. CONCLUSION

The "final" acceptance of Einstein's theories came with the results of the British expedition, led by Dyson and Eddington, to measure the light bending during the solar eclipse of May 29, 1919. After that, Einstein's fame was such that the extant opposition to his theories – including the special theory of relativity – was considerably reduced, in Britain as elsewhere.

In some sense, the year of 1923 would be a good date to put an end to our discussion. It was then that Eddington published his masterpiece, *The Mathematical Theory of Relativity,*[59] wherein generations of physicists and mathematicians, British or not, learned the special as well as the general theory of relativity. In fact, Eddington's treatise was but another, though eminent, example of what can be considered a general trend in the establishment of relativity in Great Britain during the 20's. Illuminating, in this sense, is the following quotation from the preface (p. vii) of the book *A Systematic Treatment of Einstein's Theory*, published precisely in 1923[60] and written by a Senior Lecturer in Physics at the University of Liverpool, J. Rice:

The demand on the part of the layman to know just what this revolution portends has been satisfied; as far as may be, by a liberal supply of popular works on Relativity. But the science undergraduate taking his normal courses in the University classroom, or reading his text-books of Physics and Mathematics, is anxious to ascertain in a more precise manner what changes this new idea is producing in the principles and content of physical science. It is primarily for him that this book has been written.

Finally, a word of caution. I have excluded from this paper a systematic discussion of all those British scientists who did not accept Einstein's special theory of relativity. There were, certainly, many of them. A.S. Eve recalled[61] that during a conversation held during 1910 with Wilhelm Wien and Ernest Rutherford, Wien manifested that "no Anglo-Saxon can understand relativity," to which Rutherford replied "No!, they have too much sense." And, independently of its possible jocular character, Rutherford's answer might be considered as representative of a large group of British physicists. The motivations and degrees of understanding of Einstein's restricted theory[62] were, of course, different among all these opponents to special relativity,[63] but their very existence and importance can hardly be denied. If I have not discussed them in this article it is because of two reasons: first, because there exists already a good treatment of them, namely Goldberg's,[64] and, second, because I wanted to point out the existence and importance, within the British *milieu*, of a great variety of scientific approaches not so uncongenial to Einstein's point of view as those defended by its opponents. It is one of my contentions, that in this manner, we obtain also a vision, more realistic, of British physics not so monochromal and dominated by single themes (for instance, the ether) as the one usually presented.

NOTES.

[1] S. Goldberg, 'In Defense of Ether: The British Response to Einstein's Theory of Relativity, 1904–1911,' *Historical Studies in the Physical Sciences* **2** (1970), 89–124.

[2] S. Goldberg, *Understanding Relativity* (Boston, Birkhauser 1984), p. 221.

[3] It would be better to say the "ethers," because as J. Illy ('Revolutions in a Revolution,' *Studies in History and Philosophy of Science* **12** (1981), 173–210) points out, that concept has been used in many different senses.

[4] *British Association Report*, 1899, pp. 615–624, reprinted in *Collected Scientific Papers by John Henry Poynting* (Cambridge, Cambridge University Press 1920), pp. 599–612.

[5] The question of the reception given to Ernst Mach's ideas in Great Britain is still, as far as I know, to be studied.

[6] Poynting, p. 600 of *Collected Scientific Papers*, (note 4, above)

[7] See Eddington's assessment ('Joseph Larmor, 1857–1942,' *Obituary Notices of Fellows of the Royal Society* **IV** (1942–44), 197–207, p. 205):

"He [i.e., Larmor] seemed a man whose heart was in the nineteenth century, with the names of Faraday, Maxwell, Kelvin, Hamilton, Stokes ever on his lips."

[8] Here is again Eddington (*op. cit.* note 7, p. 204):
"Larmor had an intense, almost mystical, devotion to the principle of least action. . . .
to Larmor it was the ultimate natural principle – the mainspring of the universe."
[9] *BAAS Report* (London, John Murray 1900), pp. 620–621.
[10] *Ibid.*, p. 622.
[11] *Philosophical Magazine* **19** (1910), 129–137.
[12] In his paper McAulay stated (p. 135):
"Sir Oliver Lodge seems to strike the correct note when he, somewhere, calls upon us to
explain everything in the physical world by pushing and pulling, or indeed by pushing and
not pulling. Here we seem to find cause for the fascination exercised by the vortex theory
of matter."
[13] Cambridge, Cambridge University Press 1907; second edition Cambridge University
Press, 1913.
[14] The book was also translated into French: *La théorie électrique moderne. Théorie
électronique*, translated by A. Corvisy (Paris, A. Hermann 1919).
[15] *Philosophical Magazine* **19** (1910), 181–191. This paper was the basis for the appendix
on the ether included in the second edition of *Modern Electrical Theory* (1913).
[16] There are no references to Einstein's theory in the first edition of *Modern Electrical
Theory*.
[17] For details of Schott's career see A.W. Conway, 'Professor G.A. Schott, 1868–1937,'
Obituary Notices of Fellows of the Royal Society **II** (1936–38), 451–454.
[18] Cambridge, Cambridge University Press 1912.
[19] The original title of the essay was 'The Radiation from Electric Systems or Ions in
Accelerated Motion and the Mechanical Reactions on their Motion which arise from it.'
[20] In a paper he published in 1907 ('On the Electron Theory of Matter and of Radiation,'
Philosophical Magazine **13**, [1907], 189–213), Schott defined what he understood by
"Electron Theory of Matter;" namely "any theory which assumes matter to consist of
electrical charges, acting upon each other with electromagnetic forces only" (p. 189).
[21] This was one of the main topics of Schott's investigations in the *Electromagnetic
Radiation*.
[22] See the list of Schott's published papers included in Conway's obituary, note 17, above.
[23] *Philosophical Magazine* **13** (1907), 657–687.
[24] See equations (350), Section 229, of the book.
[25] Actually, it seems that Schott accepted the special theory of relativity. In the obituary
already referred to (see note 17), Conway writes (p. 453): "The theory of relativity was
accepted by him, although mathematical difficulties in his latest papers [in the thirties]
caused him to use the conception of a rigid sphere." Let me add, however, that I think
that Conway's statement can be criticised on different grounds.
[26] Except for what I said in note 16.
[27] Of course, Campbell's "electromagnetic view of nature" would be rather peculiar in
the sense that it was not attached to the ether concept.
[28] This is not to say, however, that he did not see any problems with such a worldview. He
realized, for instance, that there were some difficulties connected with aberration.
[29] Cambridge, Cambridge University Press 1913.
[30] 'On the Electromagnetic Mass of a Moving Electron,' *Philosophical Magazine* **14**
(1907), 538–547.
[31] *Theorie der Elektrizität: Elektromagnetische Theorie der Strahlung* (Leipzig, Teubner

1905), p. 205. When dealing with Schott's case we also came across this book, although on that occasion Schott referred to the second edition (1908).

[32] *Philosophical Magazine* **13** (1907), 413–421.

[33] 'The Relation between Electromagnetism and Geometry,' *Philosophical Magazine* **19** (1910), 623–628, p. 623.

[34] *Proceedings of the London Mathematical Society* **7** (1909), 70.

[35] *Ibid.*, **8** (1910), 77.

[36] All the editions here referred to were published by Cambridge University Press.

[37] From the preface.

[38] Almost a decade later, Arthur Eddington felt obliged to clarify the issue at debate (see 'Discussion on Einstein's Theory of Relativity,' March 26, 1920, *Proceedings of the Physical Society of London* **32** (1920), 245–251):

"When we consider the matter carefully, it is not so surprising after all. The size and shape of the material apparatus is maintained by the forces of cohesion, which are presumably, of an electrical nature, and have their seat in the aether. It will not be a matter of indifference how the aether is streaming past, and there will be a readjustment of the molecules when the flow is altered.

But what does seem surprising is that the readjustment of size should just hide the effect we were hoping to find. It looks almost like a conspiracy. There have been three or four other experiments tried since then – of more technical kinds – all in the hope of detecting our motion through the aether; but they have all been defeated by the same kind of conspiracy. And to turn from experiment to theory, we find the same kind of conspiracy even in our mathematical equations. Perhaps you know sometimes, just as you are hoping to get an important result, the equations evade you, and, instead of telling you what x is, they announce the solemn but irritating truth that $0 = 0$. It is somewhat like this in our electromagnetic theory: the velocity through the aether appears abundantly in the formulae, but, whenever we try to run it down, it drops out, and refuses to be equal to anything in particular.

In consequence of these conspiracies, a great generalisation has been put forward, known as the restricted Principle of Relativity. *It is impossible by any conceivable experiment to detect the velocity of a system through the aether.* That is to say, the conspiracy is a general one; and, in fact, it is in the nature of things that this motion is undetectable.

We make this generalisation and build a branch of science on it, in the same way as we make the generalisation that it is impossible to construct a perpetual-motion machine, and build the science of thermodynamics upon it. The case for the Principle of Relativity seems to me as strong as the case for the denial of the possibility of perpetual-motion machines."

[39] Jeans's formulation is quite clear: "Systems of equations or natural laws which are such as to make it impossible to determine absolute motion may be said to satisfy the 'Relativity-condition'" (*The Mathematical Theory of Electricity and Magnetism*, 4th edition [1920], p. 597).

[40] Among those who followed Robb's lead, we have Rudolf Carnap, Hans Reichenbach and Roger Penrose.

[41] A. Vibert Douglas, *The Life of Arthur Stanley Eddington* (London, Thomas Nelson and Sons 1956), p.99.

[42] Interview, *Sources for the History of Quantum Physics.*

[43] Cambridge, Heffner and Sons 1911.

[44] Cambridge University Press.

[45] In *A Theory of Time and Space*, we read (p. v): "The special object here aimed at has been to show that special relations may be analyzed in terms of the time relations of *before* and *after*." Because of the predominance of the concept of "time" in Robb's construction, his theory has rightly been labeled "causal theory of time."

[46] Cambridge, Cambridge University Press.

[47] London, Macmillan and Co. 1914.

[48] Cambridge, Cambridge University Press 1918.

[49] The Physical Society of London, 1919.

[50] **29** (1915), 335–6.

[51] E. Cunningham, interview, *Sources for the History of Quantum Physics*.

[52] Among these we count: 'On the Electromagnetic Mass of a Moving Electron,' *Philosophical Magazine* **14** (1907), 538; 'On the Principle of Relativity and the Electromagnetic Mass of the Electron: A Reply to A.H. Bucherer,' *Philosophical Magazine* **16** (1908), 423; 'The Motional Effects on the Maxwell Aether-Stress,' *Proceedings of the Royal Society of London* **A83** (1909), 109; and 'The Principle of Relativity in Electrodynamics and an Extension Thereof,' *Proceedings of the London Mathematical Society* **8** (1910), 77.

[53] Chapter VIII of *The Principle of Relativity* is entitled 'Minkowski's four-dimensional calculus.'

[54] See p. 89 of *The Principle of Relativity*.

[55] See especially pp. 129–131 of this book.

[56] See p. 15 of the second edition (1920).

[57] *Phil. Mag.* **26**, 636–673 (1913).

[58] *Ibid.*, p. 636.

[59] Cambridge, Cambridge University Press 1923. The book was translated to German in 1925 (*Relativitätstheorie in mathematischer Behandlung* [Berlin, Springer 1925]) with an appendix by Einstein ('Eddingtons Theorie und Hamiltonsches Prinzip').

[60] London, Longman.

[61] A.S. Eve, *Rutherford. Being the Life and Letters of the Rt. Hon. Lord Rutherford, O. M.* (New York, MacMillan 1939), p. 193.

[62] There is, for instance, the case of Arthur Schuster who by 1908 understood perfectly well the meaning and consequences of special relativity, but, however, did not believe in its correctness; (see A. Schuster, *The Progress of Physics during 33 years (1875–1908)*, Cambridge, Cambridge University Press 1911, especially pp. 108–111).

[63] There were, of course, others who did not know the theory at all.

[64] See note 1, above.

Universidad Autónoma de Madrid

LEWIS PYENSON

THE RELATIVITY REVOLUTION IN GERMANY

In 1905, at around the time that the first of Albert Einstein's publications of the year was to appear, the young theoretical physicist wrote to admonish a close friend of his for not having remained in touch. The friend, Conrad Habicht, was along with Einstein a member of a small study group, the Olympia Academy, which had been meeting in Berne for several years. The tone of the letter is light-hearted; its prose is littered with clumsy attempts at irony and self-mockery. Einstein asked to see Habicht's doctoral dissertation, as he would be one of the few people around who could understand it; Einstein assessed Habicht a stiff increase in his study-group fees as a penalty for non-attendance. Then there is mention of Einstein's own work. Among the four works of his that would be appearing in 1905, Einstein indicated that one – the paper on the light quantum – set forth "very revolutionary" consequences.[1] He had the opportunity of identifying the theory of relativity as a revolutionary doctrine, and he declined to do so.

Einstein's private remark suggests that, in 1905, he saw his first paper on special relativity as a preliminary and tentative approach toward resolving some of the fundamental problems associated with the electron theory. He certainly had firm reasons for believing that his approach yielded results superior to those of other theories being discussed, but he thought that more work was required to settle the point. He sought to contribute to a revolution that was already underway. Events moved quickly. By 1912 Einstein was the master of the relativity revolution.

In the following pages, I reconsider the relativity revolution in Germany from the standpoint of social history of science. The technical content of Einstein's work and that of his colleagues (for which good analyses have long been available) recedes in favor of seeking broader explanations for the reception accorded to relativity.[2] The focus is placed on the *revolutionary* dimensions of what happened. I begin by affirming how, in the decades before 1914, physicists and other commentators spoke in unambiguous terms about the revolution that was then in progress. The thread of rhetoric regarding revolution, commencing before what we now identify as Einstein's revolutionary thought,

59

Thomas F. Glick (ed.), The Comparative Reception of Relativity, 59–111.
© 1987 *by D. Reidel Publishing Company.*

suggests that physicists in Central Europe were particularly disposed to seeking revolutionary solutions for the scientific problems that they faced. After having identified the revolutionary urge, I indicate its incarnation, as well as its rejection, in the various positions regarding relativity which were taken by physicists over the first two decades of the twentieth century. The inspiration for this recounting is found in the French Revolution of the late eighteenth century. Once the course of relativity's reception has been delineated, the issue of a social matrix for the events is considered, with special attention devoted to patterns of publishing and apprenticeship.

The literature surrounding scientific revolutions, especially in the generation following the publication of Thomas Kuhn's book on the topic, is enormous. Kuhn is largely silent, however, on the matter of revolutionary structure – of what happens during a revolutionary crisis and whether there are stages of development in the period falling between two paradigms. Some scientific revolutions have been studied with the object of delineating their structures, and broad remarks have been directed to comparing scientific revolutions in general with political revolutions in general.[3] Examining an unambiguous scientific revolution from the perspective of one of the best known political revolutions, however, seems not to have been undertaken recently. It should be clear at the outset that the author of any such attempt is well advised to take account of Max Weber's strictures, offered in 1909, about uncritically transferring forms of thinking from one discipline to another.[4]

The task of a writer seeking to fathom the twentieth-century revolution in exact sciences has recently been outlined by the Japanese historian of science Chikara Sasaki. He follows Kuhn in believing that science proceeds by nonlinear and drastic changes, or revolutions. During such scientific revolutions, there is a complete transformation in both the conceptual and institutional environments.[5] Sasaki believes that the core of the matter lies in social symbols, which condition the way people view nature. Because science is a product of everyday life, scientific revolutions seem to parallel social and political ones.[6] The two concurrent revolutions are, however, only synchronic; their forms are not necessarily homologous. Although Sasaki believes that changes in social reality precede changes in the image of nature, he affirms that scientific discourse cannot be a mere byproduct of social intercourse. Neither can institutional support be a starting point for scientific ideas.[7]

Factors external to scientific discourse do not determine disciplinary content. If internal history cannot be deduced from external history, external history is nevertheless more than a mere framework for internal history. The two structures are interactive and indissociable.[8] There is an organic relationship between ideas and environment.

In his massive investigation of scientific revolutions, Sasaki suggests, as much by what he writes as by what he leaves out, the difficulty of revealing such organic unity. The parameters of the ideational content of a scientific revolution in progress, especially, seem difficult to capture. Sasaki disallows the notion, often attributed to Kuhn, that a scientific revolution simply replaces one dogma by another. He argues for the primacy of historical specificity: To speak of dogma in general, he reminds us, is to engage in scholasticism. He allows on this point that something like the notion of a permanent revolution might provide a materialist response to idealist formulations.[9] For Sasaki, Marx and the French *annalistes* provide a useful corrective to Kuhn.[10] Because his is a materialist treatment, Sasaki avoids considering whether events in politics can provide an analogical or abstract structure for changes in ideas and institutions.[11]

I have introduced Sasaki's work at some length because it is one of the most suggestive of recent publications on scientific revolutions. Other treatments, notably I. Bernard Cohen's *Revolution in Science*, might have similarly been plumbed. The exercise is not without interest, but it would not bear directly on the question at hand: how to understand what was perceived then and now as a revolution in physics. For this reason, the present text, finding profit and inspiration in the work of Sasaki and others, does not refer to their work in a systematic way.

The revolution here is that of relativity in Germany, but in what follows it will be apparent that many of the figures were not German. Some, like Ludwig Boltzmann, Philipp Lenard, Menyhért Palágyi, Gustav Herglotz, and Jakob Laub, were born and raised in various parts of the Austrian Empire. Others, like Walter Ritz and Jakob Kunz, spent many of their early years as physicists in their native Switzerland. Einstein himself, from even before his years at the Patent Office in Berne until the late 1930s in Princeton, travelled as a Swiss citizen. All these people and many others published extensively in German journals or with German presses, and most held positions of one kind or another in Germany. Borders in German-speaking Europe were easily crossed before 1914. The relativity revolution in Germany owes much to foreign infiltrators and agitators, and for this reason although the present text focuses on Germany, it does not do so exclusively.

If the cast of characters is seen as large and somewhat heterogeneous, the period of time under consideration will be found indistinct at both beginning and end. Revolutionary *discourse* in physics extends back before the discovery of the electron, and uncertainty among senior and otherwise distinguished physicists about the basic tenets of special relativity pushes forward into the 1920s. Such a continuum, I would suggest, is intrinsic to any attempt at providing a nuanced social history of ideas which deals with relatively sudden innovations as well as with patterns of activity reflected in the *longue durée*. Introducing his major synthesis of renascences in Western art, Erwin Panofsky emphasized:

The very definition of a period as a phase marked by a 'change of direction' implies continuity as well as dissociation. We should, moreover, not forget that such a change of direction may come about, not only through the impact of one revolutionary achievement which may transform certain aspects of cultural activity as suddenly and thoroughly as did, for example, the Copernican system in astronomy or the theory of relativity in physics but also through the cumulative and, therefore, gradual effect of such numerous and comparatively minor, yet influential modifications as determined, for example, the evolution of the Gothic cathedral from Saint-Denis and Sens to Amiens. [12]

Panofsky, who was wont to think in terms of epochs spanning half a millenium, would naturally have seen Copernicus and Einstein as triumphing in a trice. He continued, in the passage just cited, to underscore that the stock of human ideas suffered attrition as well as replenishment:

A change of direction may even result from negative rather than positive innovations: just as more and more people may accept and develop an idea or device previously unknown, so may more and more people cease to develop and ultimately abandon an idea or device previously familiar.

For the historian of physics, this passage calls to mind the electromagnetic ether, which, into the twentieth century, had a steadily decreasing circle of admirers as it proved to be more trouble than it was worth. The fate of the ether is only one story that may be informed by Panofsky's vision. The difficulties, but also the rewards, of his approach are reflected in the following pages.

1. THE REVOLUTIONARY URGE AND THE REVOLUTIONARY IDENTIFICATION

Germans may choose from a wide range of possibilities in order to bring nuance to the English notion of revolution. There is, of course, *Revolution*, coming directly from the French term. One also has *Grundstürzung*, a revolution with destructive or radical connotation. Then come the transfigurational senses: *Umsturz*, an overthrow, downfall, or ruin; *Umwälzung*, an upheaval or overthrow; *Umgestaltung*, *Umwandlung*, *Umänderung*, a transformation or metamorphosis; *Umstellung*, a transformation or conversion; *Umschwung*, a sudden change, like a giant swing on a gymnastic high bar. In the period under study, all these words were used to describe what was happening in physics. It was a choice of vocabulary special to Central Europe. French writers did not often refer to a *révolution* in physics. British commentators, too, did not speak in revolutionary terms.[13]

Revolution is unmistakably a part of the vocabulary of physicists, mathematicians, and scientifically-minded writers who were active in Germany in the years around 1900. Sought was a new vision that would be able to synthesize the specialized knowledge that had, for generations, been issuing from the pens of the nation's learned men. The mathematician and poet Felix Hausdorff, in reviewing one of Wilhelm Ostwald's books on energeticism, pointed to the issue with signal clarity in 1902:

Whoever understands the signs of the times cannot fail to recognize that the enterprise of natural science has recently undergone an *Umgestaltung*. The period of unquestioned, unqualified specialization, the signature of the last decade, appears close to its end, while the synthetic, comprehensive, universalist spirit dares emerge from its long winter's sleep. The present yearns for a unified worldview in which the great profusion of isolated knowledge will be built into a unified, systematic whole.[14]

The researcher had to prepare himself for the possibility of completely changing his understanding of things, Paul Drude concluded in his 1894 inaugural address at the University of Leipzig. Truth was subjective, in his opinion, and one's approach to it could not be rigid. "The researcher will not despair, in striving toward the truth," Drude went on, "if a new tool requires *Umwälzungen* in familiar views."[15] In reviewing the development of pure mathematics over the preceding fifteen years, Siegfried Aronhold noted, in 1899, the *Umwälzung* in physical theories in molecular physics, acoustics, optics, thermodynamics, and electricity. These theories were "in part completely revised" due to the applications of mathematics.[16]

The preceding extracts, having been advanced in the passive mode, obscure the extent of opinion on the structure of scientific revolutions. Revolutions were, for a number of writers, the result of one man's revolutionary insight. According to the young neoidealist philosopher Ernst Cassirer in his first book-length publication of 1902, Leibniz sought "not a *Revolution* in the outcome of thought but rather in the way of thinking."[17] The young astrophysicist Karl Schwarzschild lectured the Cassel Naturforscherversammlung in 1903 on Henri Poincaré's "*revolutionierenden* law" of 1890 in celestial mechanics.[18] Schwarzschild's Göttingen colleague, mathematician Felix Klein, spoke in 1904 about Heinrich Hertz's thoroughgoing *Umänderung* of the intuitive view of mechanics, which, although keeping to the basic ideas of Helmholtz and Maxwell, replaced the notion of potential energy by that of kinetic energy of hidden motion.[19] When, in the same year, Klein pointed to the successful *Umschwung* in mathematics instruction at the secondary schools and chided his colleagues in higher education to bring about a similar *Umschwung*, he sought to claim the mantle of leadership in a pedagogical revolution.[20] Klein no doubt knew that just two years previously one of his supporters, the mathematician Paul Staeckel, had addressed the *Umschwung* in mathematics instruction which had taken place during the middle of the preceding century.[21]

Revolutions, commentators noted, produced counterrevolutionaries. In considering the relationship between mechanical engineering and science and life, Otto Kammerer emphasized in 1899 that the new technological man was the product of revolutionary Europe at the end of the preceding century: The "storm of the *Revolution* against the worn-out edifice of the state" resulted in the freeing of a profession, although the "new ideas" were stiffled by *Reaktion*.[22] The theoretical physicist Ludwig Boltzmann, in fact, identified himself, in 1899, somewhat ironically on the side of the reaction. He noted that "when I look back on all these developments and *Umwälzungen* [that had taken place since his student days], I feel like a monument to ancient scientific memories." He felt that he was a "*Reaktionär*, one who has stayed behind and remains enthusiastic for the old classical doctrines as against the men of today."[23] Returning to Austria in 1902, Boltzmann then changed the terms of revolutionary reference. In his inaugural lecture at the University of Vienna, he noted the *Reaktion* against the mechanical explanation of nature. He portrayed himself as the liberating revolutionary leader arrayed against unrepentant recidivism.[24]

Giving Boltzmann cause for alarm were the partisans of the electromagnetic view of nature – those hoping that some improvement in the laws of electrodynamics would be found to lie at the base of all physical theory.[25] The industrial and commercial applications of electricity contributed to its popular image as a Promethean liberator – the image portrayed, for example, in the official promotional illustrations for the Frankfurt electrotechnical exhibition of 1891.[26] This would reasonably have been the reference intended by writer Paul von Zech, who in 1892 referred in the *Deutsche Revue* to the "stupendous *Revolution*" of nineteenth-century electrical theory.[27] Zech's view extended to university lecture halls. Entirely familiar with laboratory experiments of the day, Hermann von Helmholtz emphasized in his lectures of 1888/89 that a "complete *Umwälzung*" was brewing in electromagnetic theory, once the contemporary crisis in the field had been resolved.[28]

In such a climate of revolutionary optimism, it would be normal that the startling results and experiments surrounding the electron theory, beginning to appear about 1899, were seen as signaling the arrival of a new vision of physics. Walter Kaufmann, one of the first experimentalists to focus all his attention on the electron, emphasized in 1901 that the history of science experienced, from time to time, major changes in long-accepted views. One such example of this kind of change was the *Umwälzung* in views regarding electrical processes which had occurred over the past decades.[29] The various theoretical formulations of electron dynamics, however, even after having taken into account Einstein's and Hermann Minkowski's seminal work, left unanswered questions regarding the basic shape and constitution of the elemental charge of negative electricity and, which is more noteworthy from our perspective, experimental verification of the electron's variation in mass as a function of velocity. Especially in view of the latter, unresolved question, the old Göttingen experimentalist Eduard Riecke, in speaking to the Naturforscher in 1909, qualified his sense of the revolution that had just occurred. He observed that future physicists would speak of the early twentieth century as the epoch of the electron theory. The theory had not fulfilled the expectation that it would be able to solve all the problems of physics, but it "concluded a movement which lifted the whole structure of our mechanics off its hinges and which threatens to *umzustürzen* deeply rooted views."[30] Partisans of the electromagnetic view of nature, in the throes of elaborating a dynamics for the electron, also expressed reserve about the new point of view. Wilhelm Wien, writing on the electron theory in 1904, cautioned: "The great *Umschwung* that physics has undergone in recent times doubly admonishes us to look to other theoretical possibilities and thus always to leave open the possibility of a change in the chosen path."[31] A Dutch physicist, Cornelis Harm Wind, writing in the *Physikalische Zeitschrift* in 1905, emphasized that the new revolution could itself be overthrown. Reflecting on the "crisis" in mathematical physics announced by Henri Poincaré at the St. Louis World's Fair in 1904, Wind wondered whether the invariability of mass with respect to velocity, recently proposed, would *umstürzen* the electron theory."[32]

We have seen that Einstein, in 1905, wrote to his friend Conrad Habicht about how his forthcoming paper on the light quantum was revolutionary – a judgment explicitly withheld from his papers on relativity and Brownian movement. The next year, he wrote ironically to Maurice Solovine that he had succumbed to stationary and sterile old age, when *revolutionäre* feelings were given over to youth. In Einstein's letter, the word revolutionary is written over a false start that begins "reak. . . ."[33] Einstein, it seems, was taken here by the same polarity that Boltzmann expressed on several occasions. Einstein's appeal to revolutionary rhetoric does indeed erode with time; it is likely that the adjective does not appear in his later scientific publications or correspondence at all. The word drops from Einstein's vocabulary, however, just before his identification in the popular literature as a scientific revolutionary for his work on special relativity.[34]

Foremost in arranging Einstein's apotheosis was Max Planck, Germany's most distinguished theoretical physicist and the senior editor of the world's most prestigious physics journal, the *Annalen der Physik*. Planck, a conservative physicist both politically and scientifically, described to a popular readership in 1910 how theoretical physics had experienced a major movement "of such a radical, *umwälzender* kind," which had given rise to such scientific controversy, that it could be compared only to the Copernican achievement. Planck expanded, later in his article, on the "*revolutionären* consequences of the principle of relativity," on the interpretation of the concept of time, whereby the exact hour and date always have a physical sense, depending on the motion of the observer.[35]

Planck's view was extended by those who had studied with him.

Ernst Lamla, in the dissertation that he wrote under Planck on a hydrodynamical formulation of the principle of relativity, emphasized in 1912:

The principle seems so simple and it describes such far-reaching *Umwälzungen* in all fields of physics that one is forced to give up or to entirely revise many views and concepts of the old physics to which one has become accustomed.[36]

In 1914, Planck's former student Hans Witte, an *Oberlehrer* at the Wolfenbüttel Gymnasium and *Privatdozent* for physics at the Brunswick Institute of Technology, compared the principle of relativity with Copernicus's achievement. It was "a complete *Umstellung* in our grasp of the world and especially of the essence of time." Witte indicated to his readers that he had previously proposed a general mechanical theory of electricity (he had surveyed the state of such mechanical theories in his 1906 dissertation under Planck's direction), but that he recanted, proclaiming Einstein, next to Hendrik Antoon Lorentz, as the discoverer of the new relativity principle. This principle, in his view, had *grundstürzenden* meaning for the essence of space and time. The greatest question of physics was how far the principle extended, whether it was really the long-sought "universal principle."[37]

Planck was not alone in his judgment of 1910. Others who were not especially productive researchers expressed similar views. Johannes Classen, professor of physics at the state physical laboratory in Hamburg, noted in a periodical reaching secondary-school teachers in 1910 that no one could fail to observe the *Umschwung* of late in certain basic views. A number of sentences further along Classen was clear to identify the origins of the *Umwälzung* in the principle of relativity.[38] A related message was echoed in the same year by Otto Berg, a young physicist writing in Göttingen philosopher Leonard Nelson's journal of neofriesian persuasion:

Certainly the principle of relativity places stronger demands on the capacity for abstraction than any other law of theoretical physics, and none other demands that one breaks so basically with traditional notions.[39]

Philosopher Max Frischeisen-Köhler, who did not especially share Nelson's outlook, elaborated on the revolutionary nature of relativity in 1912: "The *Umwälzung* in the field of the physical worldview brought about by this principle is comparable only to the introduction of the Copernican system."[40]

The sense in which the revolution depended on authority and belief came clearly to one of Planck's colleagues. Max Bernhard Weinstein, titular professor of physics and *Privatdozent* at the University of Berlin, emphasized in a general article of 1913 that many serious researchers saw in the principle of relativity a threat to the sound development of science, and for this reason they were demanding a return of earlier visions. In this view they were right, insofar as

the meaning of this principle has often been extended so far and without reflection that an intolerable intolerance of other opinions has developed to the most silly assertions, which is almost comparable to a medieval constraint of belief [*Glaubenszwang*].[41]

Weinstein, unlike Planck, saw Minkowski as the largest revolutionary figure. In his exhaustive, 1913 summary of relativity physics, Weinstein observed that, in Minkowski's formalism, complex equations belonged to reality, in contrast to the way that, previously, complex equations really had no proper association with physical objects. In his view, "that certainly means one of the greatest *Umwälzungen* in our familiar views."[42] Others concurred in this judgment. Also in 1913, mathematician David Hilbert, writing to a high administrator in the Prussian Kultusministerium, underscored that the consequences of Einstein's " 'relativity principle' were *revolutionierend* for mathematical and physical science," although in effect he attributed the revolution to Minkowski's four-dimensional "world mechanics."[43]

As the quotations from Weinstein and Hilbert will have suggested, all physicists of the time were not sympathetic with the new revolutionary spirit regarding relativity. One notable reactionary visited Germany from abroad to stake his claim as an architect of the new mechanics. When French mathematician Henri Poincaré spoke in Berlin around 1910, he ironically addressed the revolution in modern physics that seemed to have called into question the surest foundations of classical mechanics. He used the word *Revolution* in deference to current usage, he implied, for he thought that the revolution was really only a "menacing spectre, because it is quite entirely possible that in the short or long term some old, tested, Newtonian dynamical principles will emerge from this struggle as the winner." Poincaré thought that there was something positive to be gained from all this talk about revolutions, however. It provided the opportunity for a valuable lesson in conventionalist epistemology.[44] Less accomplished than Poincaré, but also having a vested interest in associating his own ideas with the current rage, was Greifswald physicist Gustav Mie, a firm believer in the electromagnetic ether. Mie emphasized in 1911 that over the recent past physics had seen the beginning of a process that in all likelihood would lead "to a complete *Umwandlung* of the basic concepts of science." The new development was "almost imperceptible to nonspecialists, but it will become important in the foreseeable future also for wider fields of learning." It concerned the concept of matter itself.[45] Mie who was working toward his own four-dimensional ether-based, gravitational theory, like Poincaré chose not to mention Einstein. Much farther down the scale was the Hungarian philosopher of science Menyhért (or Melchior) Palágyi, one of Ernst Gehrcke's fellow-travellers. Palágyi rejected Einstein's relativity because of the "indeterministic grasp of nature" that in his view resulted from it; relativity was to have broken the "causal bond" between the inner motion of a system and its collective motion. The relativity theories of Minkowski and Einstein, Palágyi wrote in 1914, "are now in a state of crisis" and were in danger of being treated as a theory of current fashion (*Modetheorie*); such a sensational theory "is raved about by young, overzealous researchers, only immediately to fall into oblivion." This was precisely "the fate of certain *revolutionärer* but unfruitful theories."[46] And the Leipzig experimental physicist Otto Wiener similarly worried about the ongoing revolution in physics. He underscored in his contribution to the major, encyclopedic collection, *Kultur der Gegenwart*, in 1915:

To the uninitiated who hear of the *grundstürzenden* new theories of physics, the impression may be given that physics forever presents new structures from philosophical theories and these destroy the old theories. That is in no way the case. The old theories are completely integrated into the new ones.[47]

In Wiener's view, the mechanical theory of light had been surpassed by the electromagnetic theory of light, but in reality nothing had changed except the names. The views of Wiener, Palágyi, Mie, and Poincaré are reactionary ones. By 1915, the revolutionists had come to dominate the Convention of physics in Germany.

2. A STRUCTURE FOR THE REVOLUTIONARY ACTION

A discussion of revolution in Wilhelmian Germany would have brought to mind events in Paris during the last decade of the eighteenth century. The French Revolution and its imperial successor intervened directly in the course of German history, both in political and in

intellectual realms: Kant interrupted his daily walk to ponder news from
Paris; the battle of Jena finds an echo in Hegel's *Phenomenology of
Mind*. Of all the features of the French upheavals, furthermore, the
notion of Jacobin excesses remained as a clear warning throughout the
nineteenth century; in the assemblies and conventions of the early
1790s, class struggle – not unlike that which threatened to erupt in
Germany – was nakedly revealed.

When physicists and mathematicians wrote about revolution, they
generally avoided analogies to events in the realm of politics. Yet given
the anxieties of the aristocracy and the bourgeoisie, one naturally comes
to wonder about allusions conjured up by calls for a revolution in
physics during the years around 1900. The question assumes a greater
intensity when it is recalled that the beginning of the twentieth century
has consistently been viewed as the time of a revolution in exact
sciences: Writers from Gaston Bachelard to Lewis Feuer have used
events and circumstances from this period – largely extracted from the
German-language scene – to inform models of scientific change.

In view of rhetoric about the relativity "revolution," then and now, it
is of more than passing interest to consider the extent to which what
happened in Germany, around 1900, may be said to parallel events in
the French Revolution. Such an enquiry, of course, should be seen as a
hazardous undertaking. It would lie beyond the bounds that Marc Bloch
established for comparative history. It would conflate bloody events in
politics with texts in the scientific literature. It would cross time, space,
and cultural frontiers.[48] The exercise retains its interest, nevertheless,
especially given recent writings in sociology of science regarding how
scientific ideas change. Some commentators, in fact, have come to assert
that scientific discourse reflects nothing other than the same sorts of
venial and selfish personal interests that determine, no doubt, the
course of political affairs.[49]

Critics have cautioned us not to expect to find many features of
political revolutions in the world of science, but it is surprising that there
has been no consideration of the extent to which any particular scientific
revolution follows or departs from one or another political revolution.
The following pages respond to this lacuna by reconsidering the relativity revolution in Germany from the point of view of the French Revolution. Developments in physics, considered on their own terms, will be
cast following the institutions and structures that dominated Paris during the 1790s. It will be seen that although a number of features of the

relativity revolution find no counterpart in Revolutionary France, the parallel can be extended to great lengths. Just as in Bloch's comparative history, heuristic rewards follow from seeking to document one set of events by reflecting them against what happened elsewhere.[50]

Political revolution, deriving as it does from perceived injustice, requires organization and technology to succeed, for flagrant oppression and systematic abuse in themselves can readily become structural features of a political regime. Revolutionaries, those who define injustice, if lacking in organization are condemned to live as oracles; revolutionary organizers without technological means are quickly liquidated. When Paris deprived the Bourbons of power, France was Europe's richest country; it was vital, both intellectually and commercially. The revolutionaries established a national political infrastructure and defended their gains through the actions of a technologically superior army.[51]

Physics in Germany at the end of the nineteenth century, under the old regime of classical mechanics, was in its own way no less vital than life in old-regime France. Magnificent new institutes broadcast the power and prestige that the subject commanded. Hundreds of young physicists, newly certified with doctorates, labored at highly specialized research in the hope of receiving a permanent position. They and their senior colleagues published their results, of enormous variety, in dozens of learned periodicals. Foreigners flocked to sit at the feet of the German masters who knew the compass of nature's laws. German researchers led the world assembly of physicists.[52]

Discontent with classical mechanics was building in the 1890s, and it gave expression to precipitous formulations like Wilhelm Ostwald's and Georg Helm's energetics. The act that symbolized a break with the past came unambiguously and suddenly: In 1897 Joseph James Thomson discovered the electron in cathode rays. Others had intimated the discovery before Thomson, but the director of the Cavendish Laboratory at Cambridge had the acuity to identify the electron as the object radiating spectral lines, the constituent element of atoms, and the smallest unit of electrical charge. Thomson's discovery was the storming of the Bastille; it provided a definitive break with the old regime of physics.[53]

In rapid sequence came theoretical formulations of electron dynamics and measurements of the ratio of electrical charge to electron mass.

Under the influence of the elegant syntheses of Hertzian electrodynamics produced by Henri Poincaré and especially Hendrik Antoon Lorentz, German researchers scurried to elaborate how electrons behaved both in their pure state and bound up in matter. Previously, under classical mechanics, electrodynamics was but a large part of physics, intended to be subsumed under the fundamental laws of mechanics. Within several years after the discovery of the electron, however, electrodynamics was cut free from its mechanical underpinnings. Proponents of a new vision, the electromagnetic view of nature, argued that Maxwell's equations or some modification of them were to replace Newtonian mechanics as the fundamental starting point for physics.[54]

The electrodynamicists – members of the Third Estate of physics in Germany – disputed the traditional powers of the other two estates: mechanics and thermodynamics. The role of Mirabeau – aristocrat and concilliator – in the early phase of the revolution is reflected in the activity of Dutch theoretician Lorentz. Fluent in German and by 1900 a senior statesman of physics, Lorentz circulated easily in Central Europe. His eloquence and consistent equanimity of spirit allowed him to speak with authority before physicists of all persuasions.

The fall of the Bastille and the agitation of revolutionary firebrands did not in themselves define a major change in law and administration. The latter was manifest by the convening, in 1791, of the Legislative Assembly, a single deliberative body with places for thinkers representing all factions and opinions. Previously, between 1787 and 1791, there had been a National Constituent Assembly that ironed out the form of a constitutional monarchy, but its accomplishments were not sufficient to forestall its demise. The Legislative Assembly, charged with defining a new course for civic life, was France's first revolutionary government. It is not without some justification, then, that one can associate the Assembly's first meetings in the Manège with discussions and debate following publication, in 1905, of Einstein's first papers on the special theory of relativity. A tour of the seats in the Legislative Assembly of German respondents to relativity will place this homology in clearer focus.

On the extreme right wing of the Assembly were the traditionalists, supporters of the old regime of classical physics. Following French-Revolution terminology, we may call these men royalists, or blacks. There were a number of prominent speakers. First came Ludwig Boltzmann, adamant in his defiance of the electron revolutionaries while at the same time inspiring, in his seminal work on statistical mechanics over the preceding decades, the early reflections of men like Einstein, Paul Ehrenfest, and Fritz Hasenöhrl. Boltzmann was La Fayette to the assembled physicists – a veteran of another major transformation and a man who, although a confirmed traditionalist, led revolutionary forces into battle. A second prominent royalist was Vilhelm Bjerknes, a Norwegian physicist with strong German ties who was a partisan of various classical, hydrodynamical, atomic models.[55] In this role, Bjerknes paralleled the Duc d'Orléans. Along with Philipp Lenard, the 1905 Nobel Laureate at Heidelberg, Bjerknes believed in the physical reality of the electromagnetic ether.[56] Their point of view was supported by other physicists, notably Arthur Korn and the talented Swiss-American researcher Jakob Kunz.[57]

Astronomy figures intimately in Einstein's general relativity, and for this reason it is significant to observe that another prominent royalist appeared in the form of Hugo von Seeliger, Germany's doyen of astronomy, who sought to explain deviations from Newton's law of gravitation by the hypothesis of dust rings in the solar system. He opposed everything about the electron theory and Einstein's relativity: "Seeliger *hates* this most recent development with all the sincerity and impulsiveness of his being," Arnold Sommerfeld exclaimed to Karl Schwarzschild in 1910.[58] Following Seeliger's lead were other astronomers, for whom the entire program of the electron revolutionaries was suspect. Egon R. von Oppolzer, professor of astronomy at Innsbruck, wrote to Schwarzschild in 1902 that he had just read about Max Abraham's electron theory in the *Physikalische Zeitschrift* reports of the Carlsbad Naturforscherversammlung. "I do not understand how one can accept such a thing," Oppolzer protested. "I hope that we will soon have this erroneous concept behind us." Oppolzer was astonished that Abraham – and by extension the other contributers to the electron theory – would have confused "mol ions and gas ions with, what the devil, lovely little particles." It was all a dead end, "sterile ashes." Oppolzer hoped to be alive in twenty years "to see what remains of all the skeletons without flesh and blood." In his view, "everything is a trivial consequence of hydrodynamical theory."[59]

Moving to the right of center, we find those who were prepared to change classical mechanics, but not to dethrone it and abandon the ether. They may be called the monarchists, or, to use the name given supporters of an eighteenth-century English-style bicameral, constitutional system in the French Revolution, the *Anglomanes*. The monarchist physicists were anglophile in following the nineteenth-century English ether theorists, while retaining a healthy respect for the mathematical methods of classical mechanics. Here sat Gustav Mie, a long-time advocate at Hertzian electrodynamics;[60] physicist Emil Kohl at Vienna; and Emil Wiechert, an important theoretical voice in the first years after the electron's discovery. Joining them was Johannes Stark, a firm believer in the ether whose sympathetic feelings toward the mechanist right later drew him into a political alliance with Lenard.[61] None of the monarchists worked on the electron theory with single-minded intensity; each published substantial contributions in other areas.

In dead center it is convenient to place a rather diffuse group of experimentalists. The men who measured the variation of electron mass with velocity, of course, approached their tasks with dispositions and prejudices that supplemented their degree of technical competence. Walter Kaufmann labored with overly soft vacuums at Göttingen to verify Max Abraham's rigid-sphere electron model; Alfred Bucherer's unreliable experiments verified not his own electron theory but that of Lorentz and Einstein; Adolph Bestelmeyer, Kurt Wolz, Erich Hupka, and Wilhelm Heil were all cautious about embracing Einsteinian relativity, even though they worked with the encouragement of relativity-supporter Max Planck at Berlin.[62] In general, the experimentalists were willing to adjust their theoretical allegiances as a function of laboratory results. Kaufmann, like Bucherer, eventually abandoned a rival theory in favor of Einstein's relativity. Heil, in his doctoral dissertation of 1909, expressed great caution in considering whether experiment spoke in favor of Einstein.[63] Polymath Jakob Laub, Einstein's early collaborator and faithful supporter, believed in 1910 that special relativity had been verified in the laboratory; yet in the same year he undertook experiments, as Lenard's less-than-loyal postdoctoral assistant, to place bounds on the density of the electromagnetic ether![64]

From the center to the left in the physicists' Assembly sat the relativist patriots, those who heralded relativity as the dawn of a new age. The largest block of patriots circulated around the University of Göttingen. These were the electron theorists. Mathematicians or mathematically-inclined physicists, they sought to see just how far the electromagnetic view of nature could be extended. In matters of physical credence they were generally agnostic, believing only in the overwhelming heuristic power of pure mathematics for calling into being new physical laws. In the years around 1905, they applied classical action principles and potential theory to the Lorentz electron theory, and they regularly commented on the physics of faster-than-light motion. None seemed especially eager to banish the electromagnetic ether to obscurity. Among the most important Göttingen-inspired electron-theorists we find Hermann Minkowski, Karl Schwarzschild, Gustav Herglotz, Arnold Sommerfeld, and Paul Hertz. As a group they may be said to have valued, above all, the techniques of classical mechanics, which they were not prepared to abandon without a struggle. Unlike the monarchists, they envisioned a secure place for the remnants of the old order. Their homologues in the French Revolution are the *Feuillants*, or the Constitutionals.

Beyond the *Feuillants* in the Legislative Assembly were those who sought, from the point of view of physical laws, a clean break with classical mechanics. Some of these Democrats, or *Jacobins*, were uncertain about the path to follow: Wilhelm Wien, a former supporter of the electromagnetic view of nature, became an early follower of Einstein without himself undertaking much innovative research in the field. The majority of the Democrats proposed one or another alternative to classical mechanics or seriously questioned whether conventional physics could be the whole story. Max Abraham, indeed, belongs here: Although he initially opposed Lorentz's electron theory and then seriously queried both Minkowski's electrodynamics and Einstein's general relativity, he wrote with a brilliantly clear vision of Einstein's special relativity as a physical theory. The same may be said of Walter Ritz, who proposed an emission theory of light in opposition to Einstein's first postulate of relativity. Here, too, sat Hans Witte, Planck's student who abandoned mechanical electrodynamics in favor of relativity, and Boltzmann's student Ehrenfest. The strongest Democrat of all was Max Planck, who set his Berlin entourage to explore relativity from all conceivable points of view. Around 1905, Planck's position was in the minority, but it was he who, more than any other physicist, provided crucial support for Einstein.

It is well to emphasize that around 1906 there was no general agreement regarding what special relativity consisted of and how it related to the electron theory. At the Stuttgart Naturforscherversammlung in that year, Bruno Strasser and Max Wien observed that acceptance of the principle of relativity, which argued against speaking of motion relative to the ether, was "in flux."[65] At the same meeting, Max Planck referred to the "Lorentz-Einstein" electron as the *Relativtheorie* – the 'relative theory' – a term by which special relativity would often be known over the next five years. Planck found it more congenial than Max Abraham's spherical electron theory, which was, in his view, purely electro-dynamical.[66] Arnold Sommerfeld injected a generational issue into the discussion: Men under forty supported the electrodynamical postulate (the electrodynamical view of nature); men over forty were in favor of the mechanical-relativistic postulate of Lorentz and Einstein.[67] Sommerfeld, having by 1906 been forced to reverse his position on a number of points in the electron theory, still saw this theory as the leading edge of research in physics.[68] For Sommerfeld, Einstein was behind the times.

After 1906, when Planck, Germany's leading theoretician, came out completely in favor of Einstein's relativity theory, the Legislative Assembly dissolved. It was simply no longer possible for the royalists and the monarchists to participate with the Democrats in the relativity revolution. Traditionalists like Mie – those holding to the fundamental principles of classical mechanics – withdrew to consider their next move. In place of the Legislative Assembly came the National Convention of the First Republic, wherein all delegates sought a new unifying vision of physics. Here began the romantic phase of the relativity revolution, where its ideas were sent to do battle against traditionalists from abroad and at the same time diffused within Germany. Just as in the French Revolution, the relativity Convention fell into three sections.

The first section, and at the beginning the dominant one, was that of the heritors of the electron-theory *Feuillants*. Hermann Minkowski's firey oratory and new four-dimensional formalism swayed dozens of uncommited physicists and above all mathematicians to relativity's cause. Minkowski's imperializing message was not lost on his supporters and followers, many of whom came to speak of the relativity revolution with little, if any, mention of Einstein. Advancing Minkowski's formalism came to be seen as a worthy, revolutionary end in itself; repeating Minkowski's rhetoric was taken as a declaration of faith. Much of the activity of Minkowski's followers led to little in the way of concrete innovation: Max Born, Gustav Herglotz, Waldemar von Ignatowsky, Wilhelm Behrens, Erich Hecke, Ludwig Föppl, Percy Daniell, and Gunnar Nordström (a Finnish physicist strongly influenced by the Göttingen school) ran into dead ends when trying, after Minkowski's death in 1909, to formulate a relativistic notion of classical rigidity;[69] the four-dimensional formalism completely seduced others, like the ageing mathematician Felix Klein, his younger colleague Fritz Noether, and Gustav Jaumann. As might be expected, the followers of Minkowski were generally unsophisticated about the physical bases of relativity theory. In fact, many continued to believe in some form of the electromagnetic ether. They constituted, by analogy with the Convention of the French Revolution, the *Gironde*. Minkowski, leader of the *Gironde*, finds a reflection in Vergniaud, the independent-minded *Girondin* orator; and perhaps, too, there is something of the *Girondin* philosopher and mathematician Condorcet in Minkowski's make-up.

Arrayed against the *Girondins*, high at the back of the seats in the Manège, sat the Mountain – the advanced revolutionaries who were more or less firmly convinced of Einstein's revolution. The Mountain fell under the sway of Planck, his students, and his associates. It was Planck, the relativist Danton, who in 1910 elevated Einstein to the status of a new Copernicus. Among his students, Planck counted Max von Laue, author in 1911 of the first definitive textbook on special relativity; Ludwik Silberstein, author of the second such textbook, published in 1912 in English; Hans Witte, the reconverted mechanist; the experimentalists Kurd von Mosengeil, Erich Hupka, and Wilhelm Heil. Among his critical charges as coeditor with Wilhelm Wien of the *Annalen der Physik* was winnowing acceptable manuscripts on relativity.

Planck's student Max Abraham also sat as a *Montagnard*. He played the role of a firebrand Marat, functioning as a safety valve for the revolutionaries. Just as Marat belonged to Danton's club, the *Cordeliers*, so Abraham was of Planck's physicalist school. Neither accepted nor well liked, Abraham spoke for many by voicing stern disapproval of the *Girondin* penchant for mathematical abstraction, and he probed, perhaps more critically than any other physicist, for logical inconsistencies in the various formulations of relativity which appeared from month to month in the scientific literature. Like Marat, Abraham was less dangerous within the Convention than outside it, and like Marat, Abraham did not survive the revolution. In 1909 he took up a chair in Milan.[70]

The third and largest section of the National Convention of relativists was the Plain, or the Marsh. Here sat followers of the revolution who watched the debates between the Göttingen *Girondins* and the Berlin *Montagnards*. They would have been attuned, initially, to the great success of the *Girondins* in dealing with threats from counterrevolutionary critics. Poincaré, in 1910, addressed Berlin physicists on the conventionality of all scientific ideas and the regularity of progress in physics; in his sardonic introductory remarks, as we have seen, he studiously avoided mentioning Einstein's name. Mie, in 1910, published a treatise on ether mechanics, and Lenard thundered forth in the same year on how the ether related to matter. Emil Wiechert, an early architect of the electron theory, serialized a small book in the pages of the *Physikalische Zeitschrift* in 1911; his aim was to recast relativity in a kind of absolute space. Energeticist Georg Helm came to Wiechert's defense. Experimentalist Ernst Gehrcke called relativity a mathematical-physical "jest"; mathematician *Oberlehrer* Alois Höfler and physicist *Oberlehrer* Friedrich Poske argued against introducing relativity – a "passing fashion"in Höfler's view – into the schools; Leo Gilbert saw relativity as "foppery"; Italian physicist Michele La Rosa sought to persuade his German colleagues to resurrect the ether.[71]

Poincaré here plays the role of La Fayette, a founder of the *Feuillants* who in 1792 went over to the revolution's enemies; Mie corresponds to La Rochejaquelein, leader of the Vendée counterrevolutionaries; Lenard and company, along with Seeliger, represent the Spanish and German armies arrayed against France; and Wiechert is Dumouriez, a revolutionary general who turned coat in 1793. The revolution suffered a backlash from traditionalist physicists just when senior *Jacobins* like Planck, Wien, Sommerfeld, and Einstein decided to devote the first Solvay Congress to the problems of quantum physics because to their mind relativity posed no fundamental problems. The assertion of victory signaled an irrevocable break with classical physics.

In his chronological tabulation of the number of publications dealing with relativity, Maurice Lecat revealed a sudden dip around 1911. This, the single exception to a monotonically rising curve, provides support for the contention that the decline of the *Girondin* electron theorists was largely independent of the advent of general relativity. The *Gironde* fell from power, in fact, because of its initial success. Its work was greatly suggestive for mathematicians, but it was finally unable to arrive at experimentally verifiable propositions. It exuded little more than ebullient joy in proposing mathematical elaborations or refinements – epicycles on Minkowski's *oeuvre* – before Einstein and Marcel Grossmann introduced differential geometry to deal with gravitation in 1912.

With the publication of Einstein and Grossmann, the Convention ended. The Mountain, under the leadership of Planck, pressured the formalist *Gironde* to retire by flogging Einstein's publications. Around

1912, Einstein, the Robespierre of the small core who determined the progress of the coming treatment of gravitation, was personally unassailable. Over the next three years he pursued his program of general relativity with constant devotion. Planck, the eloquent Danton, invited Einstein to sit in Berlin as an extra-university oracle, and Einstein accepted. There he joined Planck's former student Laue and others, like astronomer Erwin Findlay Freundlich, whom he found congenial. From Berlin he easily disposed of the alternative theories advanced by his revolutionary colleagues Nordström and Abraham. As his attention turned inward, he dismissed out of hand the counterrevolutionary work of Gustav Mie; he omitted answering other contenders, such as Jun Ishiwara, with whom he had personal contact. Einstein left Planck behind with general relativity, not completely unlike the way that Robespierre superseded Danton: Although the senior theoretician continued to evaluate manuscripts on the subject for the *Annalen der Physik*, he declined to engage differential geometry himself; he was not entirely comfortable with light bending in general relativity. Between 1912 and 1916 nothing withstood Einstein's withering gaze. When Gunnar Nordström objected to general relativity on the grounds that the velocity of light would depend, in it, on the gravitational field – in his view this marked "at least a complete *Umwälzung* in the foundation of the previous theory of relativity – Einstein easily rejected the alternative that Nordström proposed.[72] These were the years of Einstein's Terror.

Late in 1915, mathematician David Hilbert, a close associate of the old *Girondin* electron theorists, published the same covariant field equations of general relativity as Einstein, and nearly at the same time. Hilbert drew strength from Mie's rather traditionalist theory of gravitation; he proceeded from an entirely nonrelativistic point of view by postulating the ether, absolute space, and special properties of the electron. His publication was widely hailed as a Thermidor, an end to the relativity revolution's excesses. Mie predictably greeted Hilbert's work with enthusiasm while ignoring Einstein's contribution entirely. He wrote to Hilbert in 1916:

I have read your work on the foundations of physics with growing wonderment and enrapture. Although I must naturally take much trouble in doing so, still I am excited by what I have already understood. It is to me as if I had been feeling in the dark; the theory that you have given is certainly in the right direction. But I lack the mathematical force to replace these obscure feelings with clear concepts, and I didn't hope to venture that such a powerful step as you have taken could be made.[73]

Hilbert's triumph stimulated Felix Klein to take up relativity once more. Seeliger and his astronomer colleagues, notably Hermann Struve, took the occasion to strike a blow against Einstein by removing Einstein's coworker Freundlich from his position at the Berlin-Babelsberg observatory.[74] Planck remained loyal to Einstein, even if he did not completely support general relativity, by processing a dozen or so new authors of works on the topic through the *Annalen*, and foreigners like Lorentz elaborated on general relativity in an Einsteinian vein.[75] The mathematicians, however, came under the influence of the old Göttingen formalists, and they dominated the field between 1916 and 1919. Hermann Weyl, in his treatise on the unified field theory, set the tone for much of what followed in the theoretical response to general relativity in Germany.

The old Göttingen formalists paraded as *Muscadins* and *Incroyables*, while in the new revolutionary Directory sat Hilbert as Barras and Klein as Carnot. Poor Freundlich, beginning in 1911 the only German observational astronomer who worked to verify general relativity, was Babeuf. Carnot betrayed Babeuf to the Directory's counterrevolution; Klein – Freundlich's old doctoral dissertation adviser – did not intervene to save the young astronomer from Seeliger and Struve's wrath.

Viewed in the large, the Directory of Hilbert and Klein was ineffective in bringing about a consensus on relativity. It little more than sanctified a long-standing pattern, where pure mathematicians felt able to jump in and deal with one or another aspect of the electron theory. Things might have been different had the clouds not parted on Principe to permit Arthur Eddington's photographs, for in that case William Wallace Campbell might have published his results from the 1916 Goldendale eclipse – results that seemed to speak against Einstein's light-deflection prediction. As it turned out, however, British astronomers used the Principe results to vindicate not Hilbert, who had signed the notorious manifesto to the civilized world of 1914, but Einstein – the Jewish pacifist of Swiss nationality who had only moved to Berlin a short while before the war began. The announcement of the Principe results was the 18th Brumaire for Hilbert and Klein's Directory.

A new Einstein emerged around the time of the announcement that general relativity had been verified. The theoretician became reified – purged, as it were, of observationalist sentiment. He rejected modernist positivism in favor of critical idealism. He read Kant with admiration and Schopenhauer with empathy. He enjoyed Lichtenberg's aphorisms. He came to hold, along with Leibniz, that pure mathematics could prefigure the laws of the physical universe. Einstein indeed flirted with the spirit of the Göttingen Directory. In 1928, he was welcomed into the fold of its successors when he became an editor of the *Mathematische Annalen*.

It would be unkind to ask a generous reader who followed the revolution thus far to go ahead through the 1920s. An argument could be made for the extent to which the positions taken in debates over special and general relativity recur in the development and early reception of quantum mechanics; the story of Lenard and Stark's furious campaign against Einstein's 'Jewish' physics might have been limned; the tortuous progress of attempts to perform the three 'classical' tests of general relativity might have been reconsidered. These themes have received a certain amount of attention and no doubt could benefit from a closer look. In the 1920s, however, the drama of the relativity revolution had run its course among physicists in Germany.

The meaning of an analogy between relativity in Germany and the French Revolution may not be set aside so lightly. The dissymmetries have been, surely, as striking as the parallels. Both Boltzmann and Poincaré have been cast as a La Fayette at different stages in the revolution. A large and heterogeneous group of experimentalists have been placed at the center of the Manège – a conventional position for the traditional adjudicators of theoretical questions, to be sure, but one that tends to undervalue the extent of the experimentalists' theoretical prejudices and stylistic inclinations. Minkowski is cast as the leader of the electron-theorist *Girondins*; it must be remembered, though, that the leaders of the *Gironde*, Vergniaud and Condorcet among others, were liquidated by the Terror, while Minskowski died early in 1909 at the height of the electron-theorists' prestige. Casting Einstein as Robespierre risks a quarrel, for although his star rose like that of the revolutionary *Jacobin*, Einstein always had the friendliest of relations with his colleague and supporter Max Planck – relativity's Danton. Should Ernst Mach be perceived as a disconsolate forerunner of relativity, a revolutionary who later came to regret the theory's excesses?[76] And which revolutionary shall Karl Schwarzschild be – the electron theorist of 1903 who in 1913 sought to find the gravitational reshift in the solar spectrum and in 1916 produced a lovely result with Einstein's covariant field equations?[77] When one expects, by analogy with 1799, to see an unprincipled Bonaparte take charge in the aftermath of the British eclipse results, none other than Einstein – albeit somewhat transfigured – reemerges from the shadow of the Thermidor to become the Man of the Hour.

It is well to emphasize that, in setting down the analogy, the focus has been on men and their scientific thought. Counterparts to the Parisian *sans-culottes* or the provincial *chouans*, insofar as such comparisons would have any meaning, are reserved for the last section of the present text. Women actors have not been identified, although in an extended analogy one would do well to remark on the activity of women of critical social importance, for example, Käthe Jerosch Hilbert, Helena Joseph Weyl, and Elsa Einstein, and to compare them with the salon keepers Tallien, de Staël, and Recamier. Politics and war have been kept in the background in order for essential parts of the analogy to be constructed: In the political realm one can find few congruences between the conservative nationalist Max Planck and the democratic pacifist Einstein, two men who stood side by side on the issue of relativity.

Despite the apparent limitations of analogy, rephrasing the relativity revolution in Germany in terms of the French Revolution furnishes a model of considerable explanatory power. It clearly places Einstein in the revolutionary tradition of the electron theorists without in any way detracting from his unique and prodigious accomplishments. It illuminates the longstanding disagreements between Einstein and those who seriously grappled with the physical implications of relativity, on the one hand, and formalist, mathematically oriented researchers, on the other hand. It suggests that Max Abraham (a student of Planck's) was closer to Einstein's general way of thinking than was Max Born (an assistant of Minkowski's), despite the fact that Abraham persistently questioned Einstein's theories of relativity while Born acquiesced to them. The analogy preserves Einstein's own notion that special relativity was 'in the air' around 1905 whereas general relativity was his special achievement. We see that many contributors to special relativity who styled themselves as revolutionaries were rather anti-Einsteinian in their beliefs, just as the range of revolutionary sentiment among the *Jacobins* ran from bourgeois republicanism to popular democracy. The analogy suggests that Einstein's final formulation of the covariant field equations of general relativity was challenged by a counterrevolutionary Thermidor. Einstein's understanding of general relativity triumphed after the stunning and partisan coup brought about by the British astronomers, although the Einstein of 1919 was not the Einstein of 1905.

The analogy is of greatest interest because of the structure that it imposes on the relativity revolution. There is seen to be a progression of change – several transformations – not unlike the case in France at the end of the eighteenth century. To seek *which* development was the essential one, from this point of view, is to engage in a historicist enterprise. Who was, after all, the 'essential' actor of the French Revolution? Was Mirabeau more of a significant force than Condorcet, Carnot, Marat, Robespierre, Barras, Babeuf, or Napoléon? The answer – insofar as a meaningful answer exists – depends on one's ethical perspective.

The structure that has been suggested by the analogy allows for deep, religious-style conversions as well as other changes of mind within a fixed or slowly changing ensemble of prejudices and methodological inclinations. Many actors fundamentally changed their opinions without having a deciding effect on the course of the revolution. Hans Witte came to reject mechanistic explanations of electrodynamics; Walter Kaufmann came to lose confidence in his experiments verifying Max Abraham's electron theory; Max Born lost interest in attempting to extend Hermann Minkowski's formalism to account for rigid-body motion; Gunnar Nordström abandoned Minkowski's relativistic formalism in favor of Einstein's special relativity, and then rejected his own gravitational theory in favor of Einstein's general relativity; Max Abraham directed his immense talent toward finding logical inconsistencies with Einstein's work; Einstein himself oscillated on a number of basic propositions. All the foregoing researchers changed their minds without changing their basic approaches to their discipline. It is certainly the case that Max Planck, Wilhelm Wien, Max von Laue, and other early supporters of Einstein's work accepted his interpretation of special relativity after having been schooled in a contrary interpretation of things, and there is a story to be told in understanding how they did so. Biographers have long been interested in the making of great minds – the moment in his life when, for example, Danton became a democrat. From the present perspective, however, when the structure of a revolution is under study, such a question seems less relevant than asking about Danton's relationship with Marat, Robespierre, and the *Girondins*.

The structure revealed here, then, seems compatible with Thomas

Kuhn's understanding of scientific revolutions. It can accommodate his notion of a *Gestalt* switch in the mind of a scientist who has been converted to a new point of view. It also suggests the sense in which Kuhn is right in emphasizing that many scientists willingly attempt to contribute to their own understanding of a new point of view without having undergone such a change of mind. Neither conclusion should astonish us. After all, with regard to formalism, there is little to distinguish Lorentz's electron theory from Einstein's special theory of relativity and nothing to separate Hilbert's gravitational theory from Einstein's general relativity. In the revolutionary process, as Kuhn would allow, one finds a continual adapting of innovative principles – great and small – to create a new synthesis, rather than a struggle fought between two clearly demarcated forces on the plains of a scientific Armageddon. Finally, relativity, revolution is portrayed as a self-conscious process called into being by the historical actors themselves. The revolutionary call is taken up and repeated at the same time that new syntheses are being proposed and debated. The requirement of an explicitly *revolutionary* alternative, indeed, appears in the last days of the old regime, before an acceptable solution has been set down. Far from being fostered on a complacent scientific community by the unexpected appearance of a work of genius, the revolution is willed into being by communal exertion. From this perspective, both revolutionaries and reactionaries wander through a period of intense disagreement and confusion – the revolutionary break – before the establishment of a new scientific regime.

All structures and their resulting hagiography depend on the historian's choices. One may ask whether it is more than fortuitous that, in following the high road of personalities in the French Revolution, homologies have been found in the relativity revolution in Germany. Modern historiography surely takes a dim view of studies that begin and end with men at the top. It exhorts us to look for the big picture of life in society at large, to focus on actors with weak voices and slender means, to seek structures in patterns of daily activity rather than in the vagaries of parliamentary seating. Let us, then, for the moment set aside critiques and contentions emanating from the legislature sitting in a converted riding stable. Let us turn to consider the relativity revolution from below.

3. THE LONG VIEW OF REVOLUTIONARY CHANGE

Personalities, the short of history, are fixed to persons who act in surroundings that, taken in the large, are quite unrecognizable without further explanation. Whereas qualities of personality – courage, perseverance, love, honesty, loyalty – are read directly (if not always accurately) in historical lives, constraints on and motivation for action can require an appreciation of conditions special to time and place. To gauge innovation and originality, indeed, it is necessary to reach an understanding of contemporaneous, long-term strategies of rather ordinary people in a community or society. The most innovative historical questions of the past generation have concerned the dimensions of such phenomena as the revolutionary crowd in Paris and the counterrevolutionary crowd in the Vendée; village life in provincial, medieval France; joint-stock investors in seventeenth-century England and proletarians in the first, English, industrial revolution. Historians of science have not been immune to the claims of a 'bottom-up' perspective, where the King's honours list assumes less importance than the grocer's balance sheets. From this perspective, the significance of even spectacular, contingent events, such as the discovery of X-rays or the publication of Darwin's *Origin of Species*, tends to recede into the background.

In this section, I take a longer view of the relativity revolution by considering two of its structural features: the field of literature and the process of appprenticeship. These are not the only themes from social history which may inform the nature of the scientific discourse in question. They are of special interest here, however, because of the way that each helped to determine the course of relativity's reception in Germany.

a. The Field of Literature

In their degree of persistence with the task at hand, electron-theory and relativity revolutionaries were not unlike their counterparts in the French Revolution. Some, like Max Abraham, dedicated themselves exclusively to elaborating the meaning of the new vision of the world. Others, like Einstein, Planck, Laue, and Hilbert, ranged across physics to consider a number of burning issues, notably radiation and the light quantum. Still other men contributed to the early states of the revolution (Walter Kaufmann, Hendrik Antoon Lorentz) and then turned to consider less revolutionary matters. To judge, indeed, from the level of quality in publications of the time, the intense field of discussion was accessible to virtually any interested party; one was able to gratify the urge to appear in print by commenting briefly or at length in either popularizing glosses or highly technical treatises.

It will be evident that the publishing restrictions of the Ancien Régime have no real counterpart in Wilhelmian Germany, where authors heaped ridicule on scientific adversaries shamelessly and without fear of legal reprisals. The jury is out on whether knowledge circulated in eighteenth-century France more widely and among a wider readership than in other countries. Yet with what we know about the dimensions of the literary establishment and the literary underground, it seems not unreasonable to think that the reach of the printed word was longer in France than elsewhere. This long reach, from the top of society to the bottom, must figure, as Robert Darnton has emphasized, among the intellectual origins of the French Revolution.[78]

Taken as a whole, publishing outlets in Wilhelmian Germany proved rather accommodating to the community of physicists. Colorful debate appeared to an extent that, one thinks, would not be tolerated today in the world of specialist journals. Quite unlike the situations in Great Britain and France, where relativistic debate occurred – to the extent that it did occur – within the confines of a small circle of elite periodicals, significant contributions appeared in as many as a score of German journals. With one partial exception – Johannes Stark's *Jahrbuch der Radioaktivität und Elektronik* – the journals sought to appeal to a general readership of physicists, scientists, philosophers, or educated onlookers. Unlike the case in the United States and France, journals devoted to special parts of physics did not proliferate in Germany until after the First World War.

Though the field of periodical publication in Germany is polyvalent and complex, one may distinguish in it three broadly-defined kinds of journals: those controlled by learned corporations, those run by broad-based societies, and those issued by individual proprietors. It is well to consider each kind in turn.

There is no clearer indication of the extent to which science has remained an elite activity than the persistence, for three centuries, of high-quality academy proceedings. By the end of the nineteenth century, many of these offered fellows and their protégés a secure means of laying claim to an idea or to a discovery. A communication, made public by having been read in a sitting of the academy, could appear in print in the academy's *Verslagen, Sitzungsberichte*, or *Comptes-Rendus* within a matter of weeks. By virtue of exchange ententes with other academies around the world, the publication would circulate quickly among the high and mighty. Furthermore, academies were empowered to grant their authors substantial quantities of reprints, which could then be sent among less eminent, but perhaps more active, researchers. Academies at Berlin, Göttingen, Vienna, and Amsterdam, especially, published large numbers of works dealing with the electron theory and relativity. But because academies only published what passed through the hands of their fellows, their engines of publication depended on the presence of a full-time stoker. Relatively little on the electron theory and relativity appeared, for example, from sister academies at Munich, Königsberg, Heidelberg, and Haarlem, and from junior cousins like the Physical-Medical Society at Würzburg.

The second kind of periodical, that published by broad-based scientific societies, generally avoided the question of personal sponsorship. Much depended, though, on the society in question. A society like the Naturforscher or the German Physical Society would be receptive to including in its *Verhandlungen* the text of just about any communication delivered orally at one of its meetings.[79] Limitations of space naturally prevented most authors from producing a sustained argument. Other societies, such as the German Mathematicians' Society, the astronomical society, and the mathematical societies in Berlin and Hamburg, were catholic but principled in accepting contributions to their *Jahresberichte, Vierteljahrsschriften, Sitzungsberichte*, or *Mitteilungen*; the editor and his friends no doubt sought more to prevent an embarrassment from appearing than actively to seek long and innovative articles. Authors in a hurry would often knock on other doors first, for, with the exception of the astronomical society, a manuscript could wait for a long time before being printed in the society's outlet. Still other societies, such as the Silesian Society for the Civilization of the Fatherland or any of the various school-reform associations, were self-selecting in the kinds of material that would come their way. In the latter category should be placed small institutional publications, like the *Abhandlungen* of the Göttingen Fries-Schule, which existed to promote one or another philosophical or social point of view.

Important as the elite-academy and democratic-society publications were in the relativity revolution, most attention focused on the third kind of periodical, the proprietary journal. Such journals were controlled by an editor or coeditors backed up by a small circle of disciplinary colleagues. Some proprietary journals (so called because they existed to make money for the owner, or at least to break even as a result of subscriptions and advertisements) had affiliation with a disciplinary society – the most famous of these affiliations being between the *Annalen der Physik* and the German Physical Society – but in no sense was the journal a house organ. The proprietary journals solicited contributions from anyone – German or foreigner, academic or amateur – in the hope of thereby appealing to a broad community of interest. Indeed, the overwhelming majority of foreign contributions to the relativity literature in Germany appeared in proprietary journals. Different from their counterparts who had been appointed by learned corporations, editors of proprietary journals – Max Planck and Wilhelm Wien of the *Annalen der Physik* and Emil Bose and Felix Krüger of the *Physikalische Zeitschrift* – would not infrequently revise contributors' prose or impose other conditions on having a manuscript accepted. Editors and publisher developed a clear idea about the profile of prospective readers, and they wanted above all to retain the special character of their journal. Some proprietary journals, like the philosophically-oriented *Kantstudien* and *Annalen der Naturphilosophie*, no doubt received few submissions that required much deliberation. Others, like the *Physikalische Zeitschrift* and the *Zeitschrift für Mathematik und Physik*, to judge from their content, seemed to publish nearly anything that came their way. Johannes Stark's *Jahrbuch der Radioaktivität und Elektronik* published timely reviews of the literature which the editor usually solicited. *Himmel und Erde* sought material vaguely related to astronomy. *Die Umschau* sought to present scientific questions before intelligent laymen.

One of the valuable functions of proprietary journals was to keep researchers informed about the latest monographic treatments in their field. From the very beginning, the electron theory generated vast, book-length treatises devoted to one or another point of view. Some of these, like Max Bernhard Weinstein's summaries of relativistic physics, were tedious exercises in algebra. Others, like Emil Cohn's, Alfred H. Bucherer's, and Gustav Mie's books, were attempts to set forth a general philosophy for dealing with electromagnetic phenomena. Still others – Johannes Classen's or Joseph Petzoldt's lectures, for example – were philosophical elaborations of recent scientific debates. A number of monographic treatments, however, became classics that a serious researcher ignored at his peril. Into this class fall Lorentz's contributions in 1903, 1904, and 1909 on electromagnetism and the electron theory; the two editions, in 1904 and 1909, of Abraham's book on the electron theory; Laue's book on special relativity in 1911; Silberstein's recapitulation, in English, of 1914; and Pauli's encyclopedic treatment of 1919. No one less than Einstein looked to monographs as authoritative summaries. As might be expected, references to books sometimes appeared in journal articles.

Editors manipulated their own publishing outlets featuring works on relativity for special ends. *Oberlehrer* Friedrich Poske used his *Zeitschrift für den physikalischen und chemischen Unterricht*, and unabashedly so, as a forum for promoting science instruction in the schools in particular and a vision of a humanistic-scientific worldview in general. Mathematician Felix Klein and his colleagues used the *Nachrichten* of the mathematical-physical section of their academy, the Royal Scientific Society at Göttingen, as a rapid means of publishing hastily composed, interim reports and mathematical speculations about the electron theory; they usually excluded reflective reviews of the literature and general overviews. Of greatest interest here, however, is the editorial policy of the *Annalen der Physik*, without doubt the premier physics journal of the period.[80]

Beginning in 1877, with the appointment of Gustav Heinrich Wiedemann as its editor, the *Annalen der Physik und Chemie* enjoyed financial support from the Berlin Physical Society. The Society exacted a measure of control over the proprietary journal, published by the firm of Johann Ambrosius Barth, as the price of its support: It delegated Germany's most distinguished physicist, Hermann von Helmholtz, as its factotum in the editorial office, and it made sure that Helmholtz's name appeared right under Wiedemann's, in smaller type, on the title page. When Helmholtz died in 1893, Wiedemann named his son as a coeditor, but the Berlin Society (from 1898, the *German* Physical Society) still wanted to have an overseer; in 1895 it designated Max Planck to succeed Helmholtz in this role. Wiedemann died in 1899 and his son declined to continue alone. The Physical Society turned to Paul Drude, professor of physics at the small University of Giessen and someone at ease with both experiment and theory, as the new editor. This time the Society named four experimentalists, however, to join Planck as an editorial council, and it dropped the word 'chemistry' from the journal's title.

After five years of editing the *Annalen*, Drude was called to become professor of experimental physics at the University of Berlin. It was the most prestigious post in the discipline. Drude lasted one year. Overwhelmed by teaching, administrative, and editorial responsibilities, he shot himself in 1906.

Into Drude's editorial chair stepped Max Planck. In an attempt to distribute some of the burden, Planck asked Wilhelm Wien to become coeditor. Wien, professor of physics at Würzburg, was like Drude both an experimentalist and a theoretician. In the sharing of tasks, though, Planck remained the senior partner. Either editor could, it appears, accept a paper without setting conditions on the material, but the coeditors agreed to contact each other about submissions to be revised or rejected. Over the next decade or so, the two men seem to have accepted at least around 80% of what crossed their desks.

With regard to papers on special and general relativity, the coeditors were generous, firm, and principled: generous to first-time authors with something new or innovative to present; firm about rejecting tedious summaries of previously published work or long, formalistic digressions offering no new results; principled in seeking to avoid, at all costs, personal invective and bad feelings. In rejecting a manuscript of Austrian physicist Emil Kohl's, Planck wrote to Wien that among all Kohl's equations there was not "a single one in which a new relationship between measurable quantities is provided." Urging rejection of a submission by industrial physicist Emil Arnold Budde, Planck wrote that Budde's work contained "no original thought that is not already found in the scientific literature, and done better there." On a paper of F. Grünbaum's, eventually rejected, Planck remarked that the argument was "correct, but it includes nothing really new and its physical interest is only very indirect." And let us listen in on Planck's evaluation of a manuscript by a young Polish physicist, Felix Joachim Wiśniewski: "The author defines every last thing in a formal way and assumes that, behind it all, these definitions have physical meaning. But nothing new comes from it."[81]

Planck's generosity, however, extended to more serious and sophisticated submissions. He published contributions by such authors as Walter Ritz, Philipp Frank, Ferencz Jüttner, and Gunnar Nordström, even though he was not convinced that their formulations of relativity would turn out to be correct. He was glad to handle the submission of secondary-school teacher Ernst Reichenbächer, in 1916, on the interrelationship between electricity and gravitation. Planck reacted to Reichenbächer's theory with skepticism. "The value of such a theory," he felt, "lay in what it finally delivers." Of greatest importance in any theory were "simplicity and intuitiveness and above all whether it has such characteristic consequences that can be tested by experiment."[82] Reichenbächer's work failed to meet these criteria. Planck returned the manuscript to the author. Reichenbächer submitted a revision. Planck urged Reichenbächer to contact Einstein. A meeting took place, and Planck soon published Reichenbächer's paper as the first unified field theory following Einstein's covariant field equations of general relativity. Planck, the senior statesman of physics in Germany and the relativity revolution's Danton, sought to accommodate any serious and flexible author.

b. *Patterning and the Schools*

Between 1890 and 1910 a major transformation occurred in pedagogy in German-speaking Europe. Before this time, graduates of the *Gymnasien* (the classical schools offering Greek and Latin) retained many official privileges not extended to graduates of the *Realgymnasien* (the semi-classical schools offering Latin but not Greek) and the *Oberrealschulen* (the non-classical schools offering neither Greek nor Latin). In place of Greek, the *Realgymnasien* offered modern languages and a bit more mathematics and natural science, a substitution pushed farther by the *Oberrealschulen*. In nineteenth-century Germany, however, passage through the *Gymnasium* was held to be necessary for anyone with broader cultural ambitions – and certainly for anyone who sought to contribute to the world of science and scholarship. As a result, although in the early years of the German Empire, the Prussian *Realgymnasien* were awarded the right to send students to the universities to study modern languages and natural sciences and in 1882 the same privilege was extended to the *Oberrealschulen*, until 1900 most university programs and many governmental positions remained closed to students trained in these *Realanstalten*. Only in 1901 did the Prussian educational authorities grant the *Realanstalten* virtual parity with the *Gymnasien*, a policy taken up as well by the other German states. A similar liberation of spirit occurred in 1899, when the German institutes of technology were given the right to award a doctorate of engineering. German educational authorities acknowledged the modernist basis of their Em-

pire's technological strength, then, more than a generation after the
advent of the second industrial revolution.[83]

The transformation in pedagogy may be called a revolution. It involved more than simply breaking the monopoly on higher culture that
had been held for nearly one hundred years by neohumanist-minded
educators in the classical *Gymnasien*. It involved, as well, as restructur-
ing of the curriculum offered in all secondary schools – both in terms of
content and in terms of spirit. Beginning around 1900, elements of
differential and integral calculus were increasingly taught in secondary-
school mathematics, especially in the *Oberrealschulen*. Many modernist
schools introduced pupils to natural history, chemistry, and electrotech-
nology, while at the same time devoting increasing attention to foreign
languages like French and English. Pedagogical technique adapted to
the new ethic of usefulness and practicality. In place of rote memorizing
and systematic regurgitation of assorted facts came an emphasis on
organic learning, where pupils were encouraged to discover meaning for
themselves in their studies. Secondary-school instructors, the *Oberlehrer*, shed their nineteenth-century image of scholarship and idealism –
an image savagely caricatured as nothing more than pedantry; they
became guides and seers, essential interpretors of a complicated modern
world who attempted to mold the personalities of their charges into
harmonious wholes.

The revolutionary changes followed much agitation from three consti-
tuencies in the academic world. The first comprised science and mathematics *Oberlehrer*, who increasingly saw themselves as the purveyors of
a new, scientific humanism that was appropriate for the technological
age. A second constituency came from practising engineers and their
Dozenten in the polytechnics and institutes of technology, who saw
themselves as holding the keys to mastery of the universe. Third and
last to come out in favor of educational changes were mathematics and
science professors at the universities. The *Oberlehrer* and the engineers
organized pressure groups, associations, and house organs to demand
that *Realen* – practical arts and useful skills – be granted equality in
education with *Idealen* – abstract and ennobling studies unrelated to the
world of industry and commerce. The German universities, based as
they were on the notion of pure learning, were not eager to abandon
their claim that education be at its base *allgemein* and *wissenschaftlich* –
general and scholarly or scientific – rather than *fachlich* and *angewandt* –
specialized and applied. And, as may be expected in entrenched institu-

tions with special privileges, even professors who were intimate with the needs of industry and technology, men like Helmholtz and Bunsen, were reluctant to see science students entering their classes without proper grounding in Latin and Greek.

The most visible proponent of *Realen*, among university professors, was mathematician Felix Klein. Since coming to Göttingen in 1887, he labored to implement a vision of mathematical universalism, where pure mathematicians worked together with scientists to solve the pressing questions of physical theory and industry. Klein used his considerable influence with Friedrich Althoff, the Prussian authority in charge of universities, to bring together at Göttingen a large circle of sympathetic and brilliant colleagues and to obtain funding for a handful of new physical institutes. A portion, by no means the largest portion, of the cost of this new development was supplied by Klein's industrialist friends. At the same time that he was setting his Göttingen empire in place, Klein presented himself to reform-minded *Oberlehrer* and engineers as a supporter of their cause. Perceptive engineers immediately realized that Klein sought to bring them into his orb and domesticate them, essentially reserving for the universities the privilege of carrying out most of the fundamental research and instruction related to applied problems. Major confrontations ensued between Klein and engineers like Alois Riedler. The *Oberlehrer*, though, were generally charmed by Klein's attentions, and most of them did not realize that their university patron was striving to revise the secondary-school curriculum in order the better to place, as mathematics and science teachers, young university graduates.

Klein's imperialist vision, designed to preserve the dominant position of the universities in the new materialistic age, also reserved a special place for pure mathematics, his cherished specialty. Pure mathematics, a discipline flourishing in the late nineteenth century and one identified with the *Wissenschaftlichkeit* of the neoclassicists, was along with other forms of *Idealen* threatened by the demands of practical-oriented reformers. Reformist schoolteachers and engineers viewed mathematics as a tool for understanding and transforming the world; they were not sympathetic to the classical vision of pure mathematics as a schooling of the intellect. Klein conceived of a way for pure mathematicians to respond to the demands of scientists and engineers without compromising the integrity of their 'abstract' discipline. Pure mathematicians were in Klein's view to become the general staff of an industrial army, condescending to draw up tactics and plan strategy for dealing with problems in the real world. Under no circumstances were pure mathematicians to abandon their calling, however, and flock to the new field of applied mathematics.

The organizing skill of Klein and his fellow reformers combined with institutional inertia to guarantee a secure place in the secondary schools for graduates of university programs in both pure and applied mathematics. The curriculum in both the schools and the universities shifted to respond to the new interest in mathematization. The first decade of the twentieth century saw an overture – an openmindedness – toward applying the structures of pure mathematics in physical situations. It was as if pure mathematics extended the promise of resolving longstanding scientific questions by the force of deductive reasoning alone. In the schools, especially in the *Oberrealschulen*, calculus made strong inroads; in the universities, group theory, integral equations, differential geometry, and finite difference equations were turned to problems in the electron theory, geophysics, and spectroscopy.

Given the variation in individual trajectories both on the secondary-school and university levels, the effects on physicists of this transformation in mathematics instruction are difficult to assess. We know that whereas the class structure of young physics graduates remained unchanged between 1900 and 1914, the percentage of physicists educated in *Realanstalten* increased from around 20% to around 60%, an increase greater than that for other fields of study, although only a fraction of this young physics community engaged relativity theory in print.[84] An attempt can be made, nevertheless, to correlate positions taken during the relativity revolution with the secondary instruction that physicists and mathematicians received.

The special relationship between secondary-school education and positions taken from the benches of the revolutionary Convention may be expressed best in a negative way. There is *in the large* no association between having attended a *Realanstalt* and supporting or rejecting Einstein's theories of relativity: After all, such diverse intellects as Ritz, Debye, Lenard, Wiechert, Bucherer, and Einstein himself received the *Abitur* from a non-classical secondary school. Nor can one observe, across the board, a marked proclivity for *Realanstalt* graduates, distinct from *Gymnasium* graduates, to engage in complicated mathematical techniques when grappling with relativity: Mie, Witte, Herglotz, Sommerfeld, Wien, Silberstein, Laub, and Ehrenfest all graduated, for example, from *Gymnasien*, and their attitudes toward relativity diverged in a number of ways. The principal distinction with regard to secondary education, however, seems to lie in views regarding *Minkowski's* approach to the electron theory. There seems to be no *Realanstalt* graduate who wholeheartedly followed Minkowski's lead and joined the *Feuillantiste* and then the *Girondin* faction of the revolutionary assemblies.

The process at work here is that of displacement. Minkowski's call centered around the notion of a preestablished harmony between pure mathematics and the physical world.[85] A *Gymnasium* education was predicated on the notion that abstract and ideal values would as a matter of course find application in daily affairs. In the classical *Gymnasium*, the accent was placed on values central to Roman and especially Greek authors. It required little effort, then, for a young mathematician or physicist to find in pure mathematics a replacement for Greek and Latin. Such a displacement did not become equally manifest in all young students. One must bear in mind that a number of *Realanstalt* graduates – Wiechert, Ritz, Lamla, Blaschke, and also Einstein – used highly sophisticated mathematics in extending or in criticizing the electron theory or relativity; it is reasonable to identify these researchers as having been exposed to the new pedagogy in mathematics. The new pedagogy extended to many *Gymnasium* graduates who were partisan to Einstein's way of viewing relativity, or at least informed in their analysis of it: To mind come Erich Kretschmann, Jakob Laub, Kurd von Mosengeil, Wolfgang Pauli, Walter Schottky, Karl Schwarzschild, Ludwik Silberstein, Wilhelm Wien, Paul Ehrenfest, and Erwin Finlay Freundlich. A number of these men in fact passed through Klein's Göttingen: Schwarzschild was a professor and Laub, Ehrenfest, and Freundlich were students there.

Pedagogical styles and academic trajectories exhibited much variation in late Wilhelmian Germany, and any thesis of educational determinism would be of limited significance in explaining the development of scientific ideas. One is nevertheless struck by what may be called the 'physicalist' thought apparent in the approach of a number of researchers educated outside Germany in modernist secondary schools. The centrality of an intuitive grasp of physical laws is seen in the work of both Einstein and Philipp Lenard, two graduates of foreign *Realanstalten* who squared off against each other as early as 1910.

Just as Einstein was marked by the lessons of his mathematics and

science instructors in Munich and Aarau, so Lenard found much to
admire in his secondary-school instructor of physics, Virgil Klatt, at the
Pressburg *Realschule*. Having taught himself calculus and spectroscopy,
Lenard came to spend time in Klatt's physical cabinet, "a rather small
room with 3 windows," two stories up in the *Realschule* building. There
teacher and pupil would spend their Sunday mornings together. Later,
on vacation from his university studies at Heidelberg, Lenard returned
to Klatt's laboratory where he tried out what he had learned from
Robert Wilhelm Bunsen. In his autobiography, Lenard devoted many
pages to Klatt; he never mentioned his doctoral dissertation or even the
name of his dissertation adviser. Lenard took pains to describe, however, how his progress in German universities was impeded by not
having had formal instruction in Latin. He had to pass a special examination on translating Livius in order to receive a doctorate from Heidelberg. When he wanted to receive the *venia legendi* as Heinrich Hertz's
assistant at Bonn, however, the university demanded that he satisfy
their own Latin expert. Hertz asked the faculty to be permitted to test
Lenard in all *Gymnasium* subjects, but the senior philologist objected:
"'Hertz is a young man; he cannot judge,'"[86] The same as Einstein,
Lenard saw little use in learning Latin. The pattern here seems similar
to that found in other 'physicalist' foreigners – notably Debye, Ritz, and
Frederick Alexander Lindemann – who attended modernist secondary
schools.

c. The Impact of Doctoral Apprenticing

It will be apparent that, just as events from outside the world of school
can have great impact on the formation of a scientist's worldview, so, for
many people, the circumstances of advanced training may override the
prejudices retailed in secondary school. Elaborating the process of
advanced certification in Germany, indeed, has loomed large in historiography. Masters of specialized learning produced disciples to fill
university institutes, industrial laboratories, and governmental bureaucracies. University education in Germany, however, admitted of variations, especially in the matter of receiving a doctorate of philosophy.
For this reason, it is well to reconsider some features of the German *Dr.
phil.*, or Ph.D.[87]

The doctorate in philosophy was awarded by the faculty of a university for an original research essay expected to be published at the

author's expense. The requirements for defending a dissertation were defined by the individual faculties. Usually, but not always, the only prerequisites were possession of a *Gymnasium* leaving certificate (the *Abitur*) and about six semesters of study at a German university. In some universities, the student had to sit for state examinations in his field of study. Originality of research, length of dissertation, and residence requirements were all set by the faculties or, in practice, the student's faculty adviser, his 'promoter.' It appears that throughout the nineteenth century, a Ph.D. could have been obtained by almost any determined *Gymnasium* graduate. Deserving foreigners, Jews, and other minorities were seldom if ever refused a Ph.D. Indeed, for those foreigners who had difficulty with German, standards were often relaxed. Questionable practices resulted in occasional scandals, as in the 1870s, when charges of awarding doctorates in absentia were brought against Göttingen, yet this practice continued at various universities into the twentieth century. Lenient standards were to everyone's advantage. Professors received fees for processing candidates, universities received publicity through their alumni, and the young doctor and his wife bore on their greeting card a title reserved for *Kulturträger*, or bearers of civilization. The system allowed professors to supplement their incomes and disseminate their good name as well as to introduce students to scientific research. Yet, unless one were seeking a position as laboratory assistant or unless one thought to write a second thesis in order to become *Privatdozent*, the Ph.D. was of little practical use. It paled beside membership in the officer corps. Government and private employers could require evidence of a *Gymnasium* education, but only implicitly did they ask for a Ph.D. Far more important was mastery of a specialty, an attainment verified by success in state-administered examinations. The examination taken by many future scientists was the one that allowed entry into the lists of secondary-school teachers. It was seen, throughout the nineteenth century, as a difficult hurdle to pass.

A comparative picture is available for Ph.D. production, in physics and a number of other fields, in the years around 1900. In such a picture, physics and the other physical sciences are seen to have attracted students from higher social strata than was the case in other fields. Into the twentieth century, however, the *Gymnasien* ceased to furnish the majority of future physicists and mathematicians. In fact, the *Realanstalten* sent a torrent of students to take university doctorates in the physical sciences by 1913. This change in the secondary-school preparation of physicists is a demographic movement on a large and relatively sudden scale.[88]

Having remarked on the Ph.D. in physics in general terms, I now turn to consider one case of special interest: the students processed through the University of Berlin by Max Planck during the decade before the First World War. It was usual at Berlin for a student to have a principal adviser – the *Doktorvater* or promoter – and a supplemental adviser, an expert whose charge would have been to verify that a particular area of expertise was authentically represented in the dissertation. Beginning in 1904/05, these arrangements are specified in the *Jahres-Verzeichnis* of German doctoral dissertations. According to this source, between 1904/05 and 1913/14 Planck was principal adviser for 16 dissertations and supplemental adviser for 29 more; on the average, he sat on between four and five examining committees per year. In his role as supplemental adviser, he, the theoretician, surveyed for the most part experimental dissertations. His principal faculty collaborator was Heinrich Rubens, who in 1906 succeeded Paul Drude as Berlin's chief experimentalist and who, across his long career, provided essential information to an entire generation of theorists working to extend first Maxwellian electrodynamics and then the quantum theory. In the decade under consideration, Planck coadvised 12 of Rubens's students, and he invited Rubens to sit as coadviser of several of his own, theoretician students. Next in order of collaborative frequency came physical chemist Walther Nernst; Planck coadvised seven of Nernst's students. After Nernst came experimental physicist Arthur Wehnelt, inventor of the interruptor induction-coil that bears his name and discoverer, in 1903/04, of a method for producing slowly moving electrons and soft canal rays; four of Wehnelt's students benefitted from Planck's expertise. Planck also occasionally coadvised students of meteorologists Wilhelm von Bezold and Gustav Hellmann and mineralogist Theodor Liebisch. Planck's relationship with Rubens reflected a continuing committment to interaction with the experimentalists. Planck advised two of Paul Drude's students during Drude's unhappy year in Berlin, and in 1904/05 the theoretician advised two students of Drude's predecessor, Emil Warburg.

A substantial fraction of Planck's research students received secondary-school training in *Realanstalten* (Table I). It is clear that Planck's interests in directing students ran overwhelmingly in the areas of the electron theory and the special theory of relativity. Eleven among

TABLE I

| | Kind of Secondary School | | | | |
	Gymnasium	Realgymnasium	Oberrealschule	American	Total
Major adviser	8	6	2	0	16
Minor adviser	14	5	6	4	29

Secondary-school background of Max Planck's research students, 1904/05 to 1913/14, as provided in the *Jahres-Verzeichnis der an den deutschen Universitäten erschienenen Schriften*. Indicated are dissertations for which Planck served as the major adviser or minor adviser, as specified by the order of advisers' names in the *Jahres-Verzeichnis*. *Major adviser*: Günther Falckenberg, Walther Meissner, Kurd von Mosengeil, Hermann Bönke, Ernst Lamla, Walter Schottky, August Gehrts, Abraham Esau, Wilhelm Heil, Adolf Semiller, Hans Witte, Karl Körner, Walther Bothe, Erich Kretschmann, Erich Henschke, Fritz Reiche; *minor adviser*: Egon Alberti, Hans Schneider, Alfred Schülze, Heinrich Bahr, Ernst Genzken, Gustav Hertz, Artur Lehmann, Frederick Alexander Lindemann, Fritz (Alfred) Schulze, Walter Brückmann, Frederick Russell Gorton, Julius Edgar Lilienfeld, Erich Regener, Franz Ritter, Hans Alterthum, Wilhelm Burmeister, Fritz Eckert, Rudolf Lehnhardt, Theodor J. Meyer, Gustav Weitzel, Luise Wolff, Hermann Behnken, Hans Mortiz Cassel, Willy Drägert, Isolde Ganswindt, John Torrence Tate, Arthur Quincy Tool, Hans Vogel, George E. Washburn, Gustav Zickner. Dieter Hermann provides only a partial list of Planck's students in his article, "Max Planck als akademischer Lehrer," in Horst Kant, ed., *Berliner wissenschaftshistorische Kolloquien VIII: Die Entwicklung Berlins als Wissenschaftszentrum (1870–1930); Die Entwicklung der Physik in Berlin* (Berlin, 1984), pp. 55–71.

the fifteen students for whom he served as a major adviser wrote a dissertation – generally a theoretical one – in these areas.

The historian is not recommended to attempt a determinist argument that would distinguish research style merely on the basis of secondary education, especially in such a polyvalent environment as Wilhelmian Germany. The varieties of biographical experience doom such attempts to failure. This point can be elaborated by considering three doctoral dissertations directed by Max Planck in 1912, those of Walter Schottky, Ernst Lamla, and Erich Henschke.[89] Each was an attempt to arrive at a synthetic formulation of the special theory of relativity from a particular point of view. Schottky (who alternately used the terms *Relativtheorie* and *Relativitätstheorie*) sought to synthesize the current state of knowledge on relativistic thermodynamics; Lamla studied the question of compressibility in relativistic hydrodynamics; Henschke rederived recent relativistic results by reconstructing relativistic electrodynamics according to the principle of least action. The three young theoreticians, each of whom must be said to share Planck's approach to relativity, provide indication of a common bibliographic orientation. They began with Einstein's first paper on special relativity and then reconsidered succeeding treatments by Einstein, Minkowski, Abraham, Planck, Born, Ehrenfest, Sommerfeld, Herglotz, Fritz Noether, Tullio Levi-Città, Gilbert Newton Lewis, and Waldemar von Ignatowski; they also considered classical works by Lorentz, Schwarzschild, Lorentz, Poincaré, Helmholtz, and Emil Cohn. Planck's inspiration is much in evidence in all three texts; he is often thanked for having suggested one or another point of departure.

What attracts our eye from the present perspective, however, is the fact that the three went through different secondary schooling: Schottky, the son of a university professor of mathematics, graduated from the Steglitz *Gymnasium*; Lamla (who did not provide the occupation of his deceased father in the dissertation *Lebenslauf* even though he dedicated the text to his father's memory) graduated from the Friedrichs *Realgymnasium* in Berlin; and Henschke, son of a late fabric manufacturer, graduated from the Cottbus *Oberrealschule*. If their understanding of Einstein's relativity was largely comparable – all, guided by Planck's strong hand, rejected Minkowski's approach – their partiality toward the use of mathematics showed certain clear variations. Schottky, the gymnasiast, sought, almost defiantly, to retain a physical point of view, one mirroring that of Planck and Einstein. He formulated his propositions and motivations by long digressions where no equations entered at all. The second part of his dissertation began with a "thought experiment," surely with the model before him of Einstein's thought experiment on the principle of equivalence. Lamla, the *Realgymnasium* graduate, was more comfortable with mathematical formalism, in his case Hamiltonian integrals; he used his basic equations to reexamine notions of rigidity and rotation which had previously been explored by other theoreticians. Henschke, the *Oberrealschule* graduate, was most at home with mathematics. The first section of his dissertation was devoted exclusively to a review of four-dimensional vector analysis. In later sections he formulated electrodynamics for the vacuum and for moving, material bodies.

The dissertations of Schottky, Lamla, and Henschke illustrate the difficulty of attempting to arrive at a causal connection between fine distinctions in physical discourse and gross educational experience. From general arguments it might be concluded that Henschke's affinity for mathematical elaboration derived from his passage through an

Oberrealschule – certainly Planck would not have been the source of such a proclivity. Yet general arguments cannot have revealed the extent to which *Realanstalt* graduates were unimpressed by Minkowski's approach to relativity. The displacement phenomenon, reasonable in retrospect, does not seem to have been predictable from first principles; it is, after all, an artifice, not unlike Sigmund Freud's notion of the same name. One must not hold such a circumstance against the notion of displacement in the present context: It remains valuable because of the distinction, independently formulated, between Einstein's and Minkowski's ways of viewing relativity.

Displacement is general enough to serve in other historical situations. The notion seems to explain what happened in Germany after 1933, when National Socialist mathematicians were attracted to L.E.J. Brouwer's intuitionist philosophy of mathematics. In Brouwer's view, the primitive notions of mathematics were intuitively posited, and logical development meant finite constructibility of objects to be talked about. Intuitionism seems not inherently more 'intuitive' than formalism, the mathematical doctrine, elaborated by David Hilbert, in reaction to which Brouwer conceived his approach. Resentment against establishment logic and learning led mathematicians sympathetic to the Nazi revolutionaries to look favorably on an alternative approach to the foundations of their discipline.[90] Here, as in the case of gymnasiast Minkowskians, the displaced allegiance cannot be identified as a necessary development.

Although many historians still "engender concepts from disincarnated minds which live their lives beyond time and space," to use Lucien Febvre's phrase of 1928, historian of education Roger Chartier recently emphasized that the past generation has witnessed a fundamental reexamination of literary products. Categories of analysis previously accepted without question have been subjected to intense scrutiny. Chartier identified three of these categories: First there is the traditional distinction between popular and learned activity; then there is the division between producing ideas and consuming them; finally there is the abyss separating hard reality from imagined representations of it.[91]

The preceding pages reflect the *rapprochement* between Chartier's old polar essences. Einstein's relativity has been diffracted through the prism of popular interest in the ongoing scientific revolution. Educational patterns have been probed for their contribution to a susceptibility for one or another approach to the mathematization of nature. The requirements of publication have turned out to shape the form taken by scientific ideas. Relativity in Germany has been rephrased in terms transposed from another historical setting, leading to a portrayal of the physics as, strictly speaking, it never was.

This way of viewing relativity in Germany has placed short- and long-term phenomena under equal scrutiny; it has spoken to a change in historical periods, or epochs, in the small and in the large. The approach would not have appealed, in his later years, to the most important historical actor under consideration – Albert Einstein. The "great revolution" in modern physics, Einstein wrote in 1931, took place with the work of Faraday, Maxwell, and Heinrich Hertz, with Maxwell having "the lion's share in this revolution."[92] The only significant appeal to revolutionary change in Einstein and Leopold Infeld's *Evolution of Physics*, indeed, is to the French *Revolution*; classical physics and quantum physics are, in that book, merely held to "unterscheiden sich radikal."[93] The root of the mature Einstein's views on revolution in science seems to be identical to that of his views on political revolution. He wrote to a Canadian communist in 1935:

I maintain that, in the long run, the masses cannot be helped unless they achieve political maturity. While you consider the political maturity of the masses a prerequisite to revolution, I suggest that, once the goal of maturity is realized, a revolution will no longer be necessary.[94]

Mature thinkers – physicists and proletarians – would as a matter of course direct themselves to act in a progressive way. The notion of progress here seems to be essentially one dating from the nineteenth century.

In his perceptive essay of 1975 on Einstein and scientific revolutions, Martin Klein has written that Einstein viewed revolutions in science as complete, fundamental, and permanent changes. The twentieth century, he felt, had not witnessed such changes: "When Einstein reacted skeptically to claims that this or that new discovery or new theory had revolutionized physics, he did so in the name of True Revolution."[95] Here, as elsewhere in his work, Einstein resisted what he thought to be passing fashion in the world of physics. In some cases, furthermore, observation could be held to be related to fashion. Beginning at least around the eve of the First World War, Einstein believed that the essential truth of a theory could not be keyed directly to the outcome of an apparent experimental confirmation. He wrote to Infeld in 1949 to emphasize his "strict adherence to logical simplicity and [his] lack of confidence in the merit of ever impressive confirmations of theories, whenever questions of principle are involved."[96] The man whose ideas, more than those of any other figure, revolutionized twentieth-century physics, was chary of grandiose proposals for transforming natural laws. He is reported to have affirmed before around 1935: "I have an unbounded mistrust of nature."[97]

"Les révolutions se font malgré les révolutionnaires," Ernest Labrousse noted in 1948.[98] His is the viewpoint adopted today by most theorists of revolutionary change in science. Scientists are seen as not fully conscious of their social roles, and in their comportment they respond to social imperatives. Scientists may hold that they deal only with reasonable arguments advanced in the world around them, but they not infrequently appear in the light of history as if they were, like Hamlet, taking cues from an unseen ghost of their prejudices and emotions. It is on the level of delineating intangible motivation that one may fruitfully employ an analogy between cognitive and social processes. As Chikara Sasaki and others have emphasized, scientific revolutions are social revolutions within the set of scientists under consideration. Cognitive cleavages can and should be defined not merely in social terms, but in socially dynamic ones. For this, bold analogy is an effective vehicle. The material for reconsidering relativity from the perspective of the French Revolution has long been available to historians. It would be of great interest to see whether, in their analogical structures, other scientific revolutions may resemble the flow of circumstance in the world of politics.

ACKNOWLEDGEMENTS

I am grateful to John Heilbron and Thomas F. Glick for criticism. Kuniko Kawachi Mitrovic assisted me in studying Chikara Sasaki's text. My research has been supported in part by the generosity of the Vice-décanat à la recherche, Faculté des Arts et Sciences, Université de Montréal.

NOTES

[1] Einstein to Habicht, 1905, in Carl Seelig, *Albert Einstein: A Documentary Biography*, trans. Mervyn Savill (London, 1956), p. 74; Martin J. Klein, 'Thermodynamics in Einstein's Thought,' *Science*, **157** (1967), 509–516, on p. 511.

[2] Systematic, technical analyses of relativistic physics and related questions during the first two decades of the twentieth century may be found in a large number of works. An exhaustive listing would serve little purpose, as the 'relativity literature' spans a broad spectrum of epistemological, ideological, and historical approaches. There seems to be little agreement about how to make sense out of the historical record. The reader's attention is nevertheless drawn to the following texts: Jun Ishiwara, 'Bericht über die Relativitätstheorie,' *Jarhbuch der Radioaktivität und Elektronik*, **9** (1912), 560–648; Ludwik Silberstein, *The Theory of Relativity* (London, 1914); Wolfgang Pauli, *Theory of Relativity*, trans. Gerard Field (London, 1958); Jean Chazy, *La Théorie de la relativité et la mécanique céleste*, 2 vols (Paris, 1928–1930); Eugene Guth, 'Contribution to the History of Einstein's Geometry as a Branch of Physics,' in Moshe Carmeli, *et al.*, eds, *Relativity* (New York, 1970), pp. 161–207; Vladimir P. Vizgin's two related volumes, *Relyativists-kaya Teoriya Tyagoteniya (Istori i Formirovanie. 1900–1915)* (Moscow, 1981) and *Yedinye Teorii Polia v Pervoi Treti XX Veka* (Moscow, 1985); Arthur I. Miller, *Albert Einstein's Special Theory of Relativity: Emergence (1905) and Early Interpretation (1905–1911)* (Reading, Mass., 1981); Abraham Pais, *Subtle is the Lord: The Science and the Life of Albert Einstein* (New York, 1982); Stanley Goldberg, *Understanding Relativity: Origin and Impact of a Scientific Revolution* (Boston, 1984); Lewis Pyenson, *The Young Einstein: The Advent of Relativity* (Bristol, 1985). Some of the issues that I treat in the following pages have been raised by József Illy in 'Revolutions in a Revolution,' *Studies in the History and Philosophy of Science*, **12** (1981), 173–210, a text that brings to mind Régis Debray's study of armed struggle and political revolution in Latin America; I have been unable to follow all of Illy's arguments in detail. In the present text, I generally do not document primary, published sources that may easily be retrieved from standard works, such as the various editions of *J.C. Poggendorff's biographischliterarisch Handwörterbuch* and the *Dictionary of Scientific Biography*, 16 vols (New York, 1970–1976), edited by Charles C. Gillispie; complete references are provided, however, for unusual sources and archival material.

[3] Thomas S. Kuhn, *The Structure of Scientific Revolutions* (Chicago, 1962; 1970). Some of the philosophical response to Kuhn's work is catalogued in the bibliography to *Scientific Revolutions*, ed. Ian Hacking (Oxford, 1981). An illuminating discussion of the history of the concept of a scientific revolution, and of how scientific revolutions differ from political revolutions, may be found in Lewis Feuer, *Einstein and the Generations of Science* (New York, 1974); the matter is considered at length in I. Bernard Cohen, 'The Eighteenth-Century Origins of the Concept of Scientific Revolution,' *Journal of the History of Ideas*, **37** (1976), 257–288. Neither Goldberg, *Understanding Relativity* (ref. 2) nor Illy, 'Revolutions in a Revolution' (ref. 2) is primarily interested in delineating the structure of the relativity revolution. Among a number of studies that reexamine Kuhn's notion of a scientific revolution from a historical perspective, Karl Hufbauer's recent book is of signal clarity: *The Formation of the German Chemical Community (1720–1795)* (Berkeley, 1982). Noteworthy is the philosophical essay by Paul Scheurer, *Révolutions de la science et permanence du réel* (Paris, 1979).

Since the time that I completed this study, I. Bernard Cohen's large book, *Revolution in Science* (Cambridge, Mass., 1985), has come to hand. The book considers how many distinguished scientists, and also historians of science, have made use of the notion of a scientific revolution. According to Cohen (pp. 28–29), actual scientific revolutions proceed in four stages. First comes the apperception or discovery of a radical new solution to outstanding problems. Then there is a commitment to the new way of doing things. Commitment is followed by a dissemination of the new approach ("a revolution in paper," in publications). Finally one finds a large number of scientists carrying out their own research in the new way. In the present text, I depart substantially from Cohen's application of the four-stage model to the reception of the theories of relativity (pp. 36–37); the relativity revolution in Germany seems more complicated than he indicates (pp. 406–419), especially in the period before 1919.

[4] Max Weber, "'Energetische' Kulturtheorien (1909)," in Weber, *Gesammelte Aufsätze zur Wissenschaftslehre*, ed. J. Winckelmann (Tübingen 1973), pp. 400–426, on p. 424. Weber's article is an essay review of Wilhelm Ostwald's *Energetische Grundlagen der Kulturwissenschaft* (Leipzig, 1909). In Weber's view, Ostwald was guilty of attempting to reify abstractions from natural science, trying to transfer forms of thinking from one to another discipline in an uncritical way, trying to brand events and circumstances with an "energeticist" tag, and transforming the *Weltbild* of a discipline into a *Weltanschauung*; Weber thought that Ernst Mach made many of the same mistakes.

[5] Chikara Sasaki, *The Historical Structure of Scientific Revolutions* (Tokyo, 1985), p. v. The book is in Japanese.

[6] *Ibid.*, pp. v–vi.

[7] *Ibid.*, p. 10.

[8] *Ibid.*, p. 552.

[9] *Ibid.*, p. 49.

[10] *Ibid.*, introduction (pp. 1–56) and chapter 6 (pp. 599–628).

[11] *Ibid.*, p. 279.

[12] Erwin Panofsky, *Renaissance and Renascences in Western Art* (Stockholm, 1960; New York, 1969), pp. 3–4.

[13] I noted and commented on this difference in an earlier publication: 'La Place des sciences exactes en Allemagne à l'époque de Guillaume II,' *Europa: A Journal of Interdisciplinary Studies* (Montreal), **4**, no. 2 (1981), 187–217, on pp. 199–200. Compare Cohen, *Revolution* (ref. 3), pp. 254–261, where the changing language of revolution in Germany is discussed.

[14] Felix Hausdorff, in *Zeitschrift für mathematische und naturwissenschaftliche Unterricht*, **33** (1902), 190–193, on p. 190.

[15] Paul Drude, *Die Theorie in der Physik: Antrittsvorlesung gehalten am 5. December 1894 an der Universität Leipzig* (Leipzig, 1895), p. 15.

[16] Siegfried Aronhold, 'Die reine Mathematik in den Jahren 1884–1899,' in E. Lampe, ed., *Die reine Mathematik in den Jahren 1884–1899 nebst Actenstüken zum Leben von Siegfried Aronhold* (Berlin, 1899), pp. 7–31, on p. 31.

[17] Ernst Cassirer, *Leibnitz' System in seinen wissenschaftlichen Grundlagen* (Marburg, 1902), p. viii.

[18] Karl Schwarzschild, "Ueber Himmelsmechanik," *Physikalische Zeitschrift*, **4** (1903), 765–773, on p. 771.

[19] Felix Klein, 'Mathematik, Physik, Astronomie an den deutschen Universitäten in den Jahren 1893–1903,' *Physikalische Zeitschrift*, **5** (1904), 764–775, on pp. 766–767.

[20] Felix Klein, 'Bemerkungen zum mathematischen und physikalischen Unterricht,' *Physikalische Zeitschrift*, **5** (1904), 710–717, on pp. 712, 713.

[21] Paul Staeckel, 'Bericht über die Entwicklung des Unterrichtsbetriebes in der angewandten Mathematik an den deutschen Universitäten,' *Physikalische Zeitschrift*, **3** (1902), 92–97, on p. 92.

[22] Otto Kammerer, 'Ueber den Zusammenhang der Maschinentechnik mit Wissenschaft und Leben,' *Physikalische Zeitschrift*, **1** (1899), 186–190, on p. 189.

[23] Ludwig Boltzmann, 'On the Development of the Methods of Theoretical Physics in Recent Times,' in Boltzmann, *Theoretical Physics and Philosophical Problems*, ed. Brian McGuiness, trans. P. Foulkes (Dordrecht, 1964), pp. 77–100, on p. 82. Boltzmann's text appeared in the *Physikalische Zeitschrift*, **1** (1899/1900), and the extract is located on p. 78.

[24] Ludwig Boltzmann, 'Antrittsvorlesung, gehalten in Wien im Oktober 1902,' *Physikalische Zeitschrift*, **4** (1903), 274–277, on p. 276.

[25] The electromagnetic view of nature is analyzed in a series of brilliant publications by Russell McCormmach: 'H.A. Lorentz and the Electromagnetic View of Nature,' *Isis*, **61** (1970), 457–497; 'Lorentz, Hendrik Antoon,' in the *Dictionary of Scientific Biography* (ref. 2), *s. v.*; 'Einstein, Lorentz, and the Electron Theory,' *Historical Studies in the Physical Sciences*, **2** (1970), 41–87.

[26] Pyenson, *Young Einstein* (ref. 2), pp. 42–43.

[27] Paul von Zech, 'Die Physik vor hundert Jahren und heute,' *Deutsche Revue*, **17**. I (1892), 188–198, on p. 197.

[28] O. Krigar-Menzel and M. von Laue, eds, *Vorlesungen über Elektrodynamik und Theorie des Magnetismus von H.v. Helmholtz* (Leipzig, 1907), pp. 3–4. The lectures were reworked from Helmholtz's notes.

[29] Walter Kaufmann, 'Die Entwicklung des Elektronenbegriffs,' *Physikalische Zeitschrift*, **3** (1901/02), 9–15, on p. 9.

[30] Eduard Riecke, 'Die jetztigen Anschauungen über das Wesen des metallischen Zustandes,' *Physikalische Zeitschrift*, **10** (1909), 508–519, on p. 508. The Naturforscher, or Gesellschaft deutscher Naturforscher und Aerzte, held annual meetings that were attended by a large proportion of Germany's practising scientists.

[31] Wilhelm Wien, 'Zur Elektronentheorie,' *Physikalische Zeitschrift*, **5** (1904), 294–295, on p. 295.

[32] Cornelis Harm Wind, 'Elektronen und Materie,' trans. Alfred Gradenwitz, *Physikalische Zeitschrift*, **6** (1905), 485–494, on p. 493.

[33] Einstein to Solovine, 3 May 1906, in Albert Einstein, *Lettres à Maurice Solovine* (Paris, 1956), pp. 2–7.

[34] Revealing in this regard is Max Born's recollection that Einstein spoke to the Salzburg Naturforscherversammlung in 1909 on "die neueren Umwandlungen, welche unsere Anschauungen über die Natur des Lichtes erfahren haben." Born, 'Physics and Relativity (1955),' in Born, *Physics in My Generation* (New York, 1969), pp. 100–115, on p. 107. Einstein published his paper, however, with the title: 'Ueber die Entwicklung unserer Anschauungen über das Wesen und die Konstitution der Strahlung,' *Physikalische Zeitschrift*, **10** (1909), 817–825, the same title that is provided in the program for Einstein's session which was published in the *Verhandlungen* of the German Physical Society. At the beginning of his paper, Einstein emphasized that in the future "a deep-going *Aenderung* of our views of the essence and of the constitution of light is indispensable." Here Einstein chose not to speak about an *Umwandlung* or *Revolution*. For Born, nevertheless, the import of Einstein's message was in the direction of a revolutionary change.

[35] Max Planck, 'Die Stellung der neueren Physik zur mechanischen Naturanschauung, *Physikalische Zeitschrift*, **11** (1910), 922–932, on p. 923.

[36] Ernst Lamla, *Ueber die Hydrodynamik des Relativitätsprinzips* (diss., Univ. of Berlin, 1912), p. 7.

[37] Hans Witte, *Raum und Zeit im Lichte der neueren Physik* (Brunswick, 1914), quotations on pp. 2, 59, 83, 77.

[38] Johannes Classen, 'Ueber das Relativitätsprinzip in der modernen Physik,' *Zeitschrift für den physikalischen und chemischen Unterricht*, **23** (1910), 257–267, on pp. 257–258.

[39] Otto Berg, 'Das Relativitätsprinzip der Elektrodynamik,' *Abhandlungen der Fries'-schen Schule*, **3** (1910), 333–382, on p. 381.

[40] Max Frischeisen-Köhler, *Wissenschaft und Wirklichkeit* (Leipzig, 1912), p. 324.

[41] Max Bernhard Weinstein, 'Die Relativitätslehre und die Anschauung von der Welt,' *Himmel und Erde*, **26** (1913), 1–14, on p. 1.

[42] Max Bernhard Weinstein, *Die Physik der bewegten Materie und die Relativitätstheorie* (Leipzig, 1913), p. 307.

[43] David Hilbert to Hugo Andres Krüss, 1913. Göttingen, Niedersächsische Staats- und Universtätsbibliothek. Hilbert Nachlass 494.

[44] Henri Poincaré, 'Die neue Mechanik,' trans. P. Schwahn, *Himmel und Erde*, **23** (1910/11), 97–116, on p. 97.

[45] Gustav Mie, 'Ionen und Elektronen,' in *Fortschritte der naturwissenschaftlichen Forschung*, **2** (1911) [ed. E. Abderhalden], 163–192, on p. 163.

[46] Melchior Palágyi, *Die Relativitätstheorie in der modernen Physik; Vortrag gehalten auf dem 85. Naturforschertag in Wien* (Berlin, 1914), quotations on pp. 32, 44–45, and 5, in sequence.

[47] Otto Wiener, 'Entwicklung der Wellenlehre des Lichtes,' in Emil Warburg, ed., *Physik*, (Leipzig, 1915), pp. 517–574, on p. 572. The general editor of the series *Kultur der Gegenwart* was Paul Hinneberg.

[48] The main lines of current thought regarding comparative history may be retrieved from recent articles and debates in the pages of the *American Historical Review*, **85** (1980),753–857, 1055–1166; **87** (1982), 123–143, and also in *Comparative Studies in Society and History*, **22** (1980), 143–221. Featured are contributions by twenty-four scholars. Comparing the relativity revolution in Germany with the French Revolution would not be excluded in principle by Theda Skocpol and Margaret Somers, 'The Uses of Comparative History in Macrosocial Inquiry,' *Comparative Studies in Society and History*, **22** (1980), 174–197.

[49] There has been much huffing and puffing about such analyses of late. For two philosophers making a case against and for this way of doing history: Larry Lauden, 'The Pseudo-Science of Science,' *Philosophy of the Social Sciences*, **11** (1981), 173–198; David Bloor, 'The Strengths of the Strong Program,' *ibid.*, 199–213. Neither Lauden nor Bloor seems much interested in marshalling and evaluating the secondary literature in history and sociology. For a balanced and critical (if not overly optimistic) survey, see: Vojin Milić, 'Sociology of Knowledge and Sociology of Science,' *Social Science Information*, **23** (1984), 213–273.

[50] I have been guided in my understanding of the French Revolution by François Furet and Denis Richet, *La Révolution* (Paris, 1965). A translation and abridgment by Stephen Hardman appears as *The French Revolution* (London, 1970) and includes a summary bibliography. A further list of references would serve no purpose, as each reader will have

favorite surveys. It would be pleasing to think that readers might forgive here, as elsewhere in the present text, the absence of exhaustive referencing. The recent remarks of Dominick LaCapra are not unrelated to this point: 'Is Everyone a *mentalité* Case? Transference and the 'Culture' Concept,' *History and Theory*, **23** (1984), 296–311.

[51] The books of Barrington Moore, Jr., have provided much to my understanding of revolution: *Social Origins of Dictatorship and Democracy: Lord and Peasant in the Making of the Modern World* (Boston, 1966); *Injustice: The Social Bases of Obedience and Revolt* (White Plains, N.Y., 1978). In the diffuse and vast literature on revolutionary etiology, I have found Herbert Marcuse's *Counterrevolution and Revolt* (Boston, 1972) of use. I have come back a number of times to ponder the clear and synoptic presentation in Charles Tilly, *From Mobilization to Revolution* (Reading, Mass., 1978), especially regarding the matter of when revolutionary situations produce revolutionary outcomes (pp. 189–222). A sharp and short statement is found in Eugene Kamenka, 'Revolution: The History of an Idea,' in Kamenka, ed., *A World in Revolution?* (Canberra, 1970), pp. 1–14. The stage theory of Crane Brinton – one that draws heavily on the French Revolution – holds some attractions, but I have not consciously followed it here. Brinton, *The Anatomy of Revolution* (New York, 1938; 1965).

[52] Pyenson, 'La Place' (ref. 13); Paul Forman, John Heilbron, and Spencer Weart, *Physics circa 1900: Personnel, Funding, and Productivity of the Academic Establishments* (Princeton, 1975) [*Historical Studies in the Physical Sciences*, **5**]; David Cahan, 'The Institutional Revolution in German Physics, 1865–1914,' *Historical Studies in the Physical Sciences*, **15** (1985), 1–65. Cahan considers the rise of the big physical laboratories.

[53] John Heilbron, 'Lectures on the History of Atomic Physics 1900–1922,' in Charles Weiner, ed., *History of Twentieth Century Physics* (New York, 1977), pp. 40–108, on p. 40.

[54] See McCormmach's publications cited in ref. 25.

[55] Robert Marc Friedman, *Vilhelm Bjerknes and the Bergen School of Meteorology, 1918–1923: A Study of the Economic and Military Foundations for the Transformation of Atmospheric Science* (diss., Johns Hopkins Univ., 1978), pp. 17–19.

[56] Lenard publically advocated retention of the ether in 1910; evidence for his continuing belief in it until then is circumstantial and may be keyed to his high regard for Bjerknes's hydrodynamical theories, which is recorded in his letters retained in the Vilhelm Bjerknes papers, Institute for Geophysics, University of Oslo. See Andreas Kleinert and Charlotte Schönbeck, 'Lenard und Einstein: Ihr Briefwechsel und ihr Verhältnis vor der Nauheimer Diskussion von 1920,' *Gesnerus*, **35** (1978), 318–333.

[57] Arthur Korn, 'Ueber die Theorie der universellen Schwingungen mit Anwendungen auf die Theorie der Gravitation und der intramolekularen Kraefte,' in *Atti del IV Congresso internazionale dei matematici*, ed. G. Castelnuovo (Rome, 1909), **3**, 81–88; Jakob Kunz, *Theoretische Physik auf mechanischer Grundlagen* (Stuttgart, 1907).

Kunz continued to consider the ether as a construct worth mentioning when he first moved to the United States: 'On the Photoelectric Effect of Sodium–Potassium and Its Bearing on the Structure of the Ether,' *Physical Review*, **29** (1909), 212–228. In the following year he reported how the principle of relativity supported the quantum theory of radiation and rendered the ether superfluous: 'On the Initial Velocity of Electrons as a Function of the Wave-Length in the Photoelectric Effect,' *Physical Review*, **31** (1910), 536–544, on p. 536. Some of the deeper background for the mechanist reactionaries may be found in: Paul Drude, 'Ueber Fernwirkungen,' *Annalen der Physik*, **62** (1897), i–xlix.

[58] Sommerfeld to Schwarzschild, 8 December 1910. New York, American Institute of Physics, Karl Schwarzschild papers. Einstein's threat to Seeliger was clearly perceived by their contemporaries: Hugo Dingler, 'Ueber das Newtonsche Gravitationsgesetz,' *Zeitschrift für positivistische Philosophie*, 1 (1913), 220–226, on pp. 225–226. For the deep background, see Peter Lebedew, 'Die physikalischen Ursachen der Abweichungen vom Newton'schen Gravitationsgesetze,' *Vierteljahrsschriften der astronomischen Gesellschaft*, 39 (1902), 220–226.

[59] Oppolzer to Schwarzschild, 25 August 1902. New York, American Institute of Physics, Karl Schwarzschild papers.

[60] Mie's early work is clear enough in this regard. For his views around 1905: Hans Witte, 'Ueber den gegenwärtigen Stand der Frage nach einer mechanischen Erklärung der elektrischen Erscheinungen,' *Physikalische Zeitschrift*, 7 (1906), 779–785; in the discussion following the paper (presented at the Stuttgart Naturforscherversammlung), Mie was identified as requiring the ether in a mechanical theory.

[61] Arnold Sommerfeld roundly criticized Stark's giving serious credence to the ether in 1909. Sommerfeld to Stark, 10 December 1909. Berlin, Staatsbibliothek preussischer Kulturbesitz, Nachlass Stark. Early in 1910 Hans Witte thanked Stark for his "benevolent judgment" of Witte's work on the ether. Witte to Stark, 11 January 1910, *ibid.*

[62] The work of the experimentalists is convincingly analyzed in Giovanni Battimelli, 'The Electromagnetic Mass of the Electron: A Case Study of a Non-Crucial Experiment,' *Fundamenta Scientiae*, 2 (1981), 137–150.

[63] Pyenson, *Young Einstein* (ref. 2), pp. 202–203.

[64] Lewis Pyenson, *Cultural Imperialism and Exact Sciences* (New York, 1985), pp. 165–167.

[65] Bruno Strasser and Max Wien, 'Anwendung der Teleobjektivmethode auf den Dopplereffekt von Kanalstrahlen,' *Physikalische Zeitschrift*, 7 (1906), 744–746.

[66] Max Planck, 'Die Kaufmannschen Messungen der Ablenkbarkeit der ß-Strahlen in ihrer Bedeutung für die Dynamik der Elektronen,' *Physikalische Zeitschrift*, 7 (1906), 753–759.

[67] Sommerfeld, in discussion reported after *ibid.*, p. 761.

[68] Sommerfeld's work on the electron theory in the years before 1906 is considered in Pyenson, *Young Einstein* (ref. 2), pp. 120–128.

[69] Erwin Voellmy, *Die allgemeine Bewegung des Bornschen Elektrons; eine integrallose Darstellung von Parallelkurven im dreidimensionalen Raum* (diss., Univ. of Basle, 1917).

[70] Abraham's life and impact in Italy are analyzed in Judith Goodstein, 'The Italian Physicists of Relativity,' *Centaurus*, 26 (1982/83), 241–261. Beginning in 1909, Abraham was an associate professor at the Milan Institute of Technology; he accepted this call following an unhappy number of months as a professor of physics at the University of Illinois in Urbana. See also Giovanni Battimelli and Michelangelo De Maria, 'Max Abraham in Italia,' in F. Bevilacqua and A. Russo, eds, *Atti del III Congresso Nazionale di Storia della Fisica* ([Palermo], 1982), pp. 186–192.

[71] The work of Poincaré, Mie, and Wiechert is well known. Helm made his position clear in : 'Das Relativitätsprinzip in der Aetherhypothese,' *Physikalische Zeitschrift*, 13 (1912), 157–158. Gehrcke queried the relativity of time in 'Bemerkungen über die Grenzen des Relativitätsprinzips,' *Verhandlungen der deutschen physikalischen Gesellschaft*, 13 (1911), 665–669. La Rosa's book, *Der Aether* (Leipzig, 1912), announced his point of view. Höfler, Poske, Gehrcke, and La Rosa are analyzed in Joseph Petzoldt's essay, 'Die

Relativitätstheorie in der Physik,' *Zeitschrift für positivistische Philosophie,* **2** (1914), 1–56. Also useful in this regard is Moritz Schlick, 'Die philosophische Bedeutung des Relativitätsprinzips,' *Zeitschrift für Philosophie und philosophische Kritik,* **159** (1915), 129–175.

[72] Gunnar Nordström, 'Träge und schwere Masse in der Relativitätstheorie,' *Annalen der Physik,* **40** (1913), 856–878.

[73] Mie to Hilbert, 13 February 1916, in Pyenson, *Young Einstein* (ref. 2), p. 185.

[74] Pyenson, *Young Einstein* (ref. 2), p. 235.

[75] József Illy, 'Geometrical Objects in Lorentz's General Theory of Relativity of 1916' (Budapest, 1985) [Hungarian Academy of Sciences, Institute of Isotopes, preprint IZIN-6/1985].

[76] Gereon Wolters has recently argued that Mach's putative rejection of relativity, published posthumously in 1921, was in fact constructed by Mach's son, under the influence of the philosopher Hugo Dingler. Gereon Wolters, 'Ernst Mach and the Theory of Relativity,' *Philosophia Naturalis,* **21** (1984), 630–641.

[77] Otto Blumenthal to Karl Schwarzschild, 30 December 1913. New York, American Institute of Physics, Karl Schwarzschild papers. Blumenthal remarked on Schwarzschild's last letter, in which he had indicated that he was trying to verify the gravitational redshift in the sun's spectrum.

[78] Robert Darnton, *The Literary Underground of the Old Regime* (Cambridge, Mass., 1982), p. 37; Pierre Goubert and Daniel Roche, *Les Français et l'Ancien Régime,* 2: Culture et société (Paris, 1984), pp. 219–254.

[79] Consider, for example, the paper published by a certain C. Beckenhaupt of Altenstadt-Weissenburg: 'Ueber die physikalischen Verhältnisse, welche bei dem Relativitätsprinzip und der Vierdimensionalität in Betracht kommen,' *Verhandlungen der Gesellschaft deutscher Naturforscher und Aerzte,* **83** (1911), *Sitzungen,* 105–110. The ether is held to be compatible with Einstein's and Minkowski's views on relativity, and some elementary arithmetic is used to illustrate the bounds on the density of the ether. It is difficult to see how Beckenhaupt's publication could have been taken seriously, even by confirmed mechanists.

[80] The succeeding paragraphs follow Pyenson, *Young Einstein* (ref. 2), pp. 194–214.

[81] Quotations, in sequence: Planck to Wilhelm Wien, 30 November 1909; Planck to Wilhelm Wien, 30 May 1911; Planck to Wilhelm Wien, 9 February 1911; Planck to Wilhelm Wien, 29 June 1913. All located in Berlin, Staatsbibliothek preussischer Kulturbesitz, Nachlass Wien.

[82] Planck to Wilhelm Wien, 25 August 1916, *ibid.*

[83] These issues have been addressed in a number of recent works, although there exists a substantial literature from the early part of the twentieth century. See Lewis Pyenson, *Neohumanism and the Persistence of Pure Mathematics in Wilhelmian Germany* (Philadelphia, 1983) [American Philosophical Society, *Memoirs,* **150**]. The following paragraphs draw on this work.

[84] Pyenson, *Young Einstein* (ref. 2), pp. 162–163.

[85] *Ibid.,* pp. 80–100.

[86] Philipp Lenard, 'Erinnerungen eines Naturforschers der Kaiserreich, Judenherrschaft und Hitler erlebt hat,' manuscript dated September 1943. Berkeley, University of California, Office for History of Science and Technology. Quotations from pp. 13–22, 41–43.

[87] The following paragraph follows Pyenson, *Neohumanism* (ref. 83), pp. 17–18.

[88] Pyenson, *Young Einstein* (ref. 2), pp. 162–163; Lewis Pyenson and Douglas Skopp, 'Educating Physicists in Germany *circa* 1900,' *Social Studies of Science*, 7 (1977), 329–366.

[89] Walter Schottky, *Zur Relativtheoretischen Energetik und Dynamik. Abschnitt I und II* (diss., Univ. of Berlin, 1912); Ernst Lamla, *Ueber die Hydrodynamik des Relativitätsprinzips* (diss., Univ. of Berlin, 1912); Erich Henschke, *Ueber eine Form des Prinzips der kleinsten Wirkung in der Elektrodynamik des Relativitätsprinzips* (diss., Univ. of Berlin, 1912).

[90] Herbert Mehrtens, 'Anschauungswelt versus Papierwelt: Zur historischen Interpretation der Grundlagenkrise der Mathematik,' in Hans Poser and Hans-Werner Schütt, eds, *Ontologie und Wissenschaft: Philosophische und wissenschaftshistorische Untersuchungen zur Frage der Objektkonstitution* (Berlin, 1984), pp. 231–276.

[91] Roger Chartier, 'Intellectual History or Sociocultural History? The French Trajectories,' in Dominick LaCapra and Steven L. Kaplan, eds, *Modern European Intellectual History: Reappraisals and New Perspectives* (Ithaca, 1982), pp. 13–46.

[92] Cited in Pais, *Subtle* (ref. 2), p. 319.

[93] Einstein and Infeld, *Physik als Abenteuer der Erkenntnis* (Leiden, 1938), p. 147. Contrast Infeld, *Albert Einstein: His Work and Influence on Our World* (New York, 1950), where Infeld, freed from Einstein's restraining influence, used the word "revolution" in five of his six chapter titles. Infeld wrote: "What do we mean by a revolution in physics? We mean a sudden clarification of our concepts, the forming of a new picture, an unexpected resolution of contradictions and difficulties." Infeld, *Einstein*, p. 22. *The Evolution of Physics* provides a discussion of Einstein and Infeld's ideas on how science progresses. Compare Cohen, *Revolution* (ref. 3), p. xv.

[94] Otto Nathan and Heinz Norden, eds, *Einstein on Peace* (New York, 1960), p. 269.

[95] Martin J. Klein, 'Einstein on Scientific Revolutions,' *Vistas in Astronomy*, 17 (1975), 113–120. Compare Gerald Holton, 'Einstein's Search for the *Weltbild*,' *Proceedings of the American Philosophical Society*, 125 (1981), 1–15, on pp. 14–15. Also compare Cohen, *Revolution* (ref. 3), pp. 435–445.

[96] Einstein to Infeld, 20 September 1949, in Leopold Infeld, *Why I Left Canada: Reflections on Science and Politics*, trans. Helen Infeld, ed. Lewis Pyenson (Montreal, 1978), p. 142.

[97] Reported by Otto Blumenthal in 'Lebensgeschichte,' in David Hilbert, *Gesammelte Abhandlungen*, 3 (Berlin, 1935), on p. 417.

[98] Ernest Labrousse, '1848–1830–1789: Comment naissent les révolutions,' in *Actes du Congrès historique du centenaire de la Révolution de 1848* (Paris, 1949), pp. 1–20, on p. 1.

Université de Montréal

MICHEL PATY

THE SCIENTIFIC RECEPTION OF RELATIVITY IN FRANCE

1. INTRODUCTION: A MIXED RECEPTION

At the very least, the reception of relativity in France can be described as "mixed". Within the scientific community as a whole, the concepts of relativity made slow headway and initially encountered either indifference or hostility. With rare exceptions, teaching, textbooks, and university programs reflected the indifference toward the subject until the 1950s. The hostility was obvious in the position taken by important academics until the mid-1920s. A third significant tendency was represented by such eminent mathematicians as Emile Picard and Paul Painlevé, who remained skeptical or withheld their approval before finally allowing themselves to be won over.

However, the attitude of the French scientific community cannot be reduced to this essentially negative reaction, although it has been the majority opinion and for that reason deserves consideration: that response is indeed representative of a mentality conditioned by a somewhat inflexible tradition which greeted innovation with hostility. Although it was in a relatively marginal fashion, the relativistic ideas took hold within a certain scientific milieu which, despite its minority status, nevertheless played a considerable role through its own contributions and its influence upon the larger scientific community over the long term: it prepared the way at least for the tacit acceptance of the new theories, and determined what was later acknowledged to be the most important and fruitful directions of research. More significantly, Henri Poincaré and Paul Langevin discovered fundamental elements of the theory of relativity independently of Einstein when he was developing his theory of special relativity. It is interesting to note that Poincaré rejected relativity in the Einsteinian sense, although it is in his writings (in 1904) that the expression "the principle of relativity", in the sense of special relativity, was first used. Furthermore, his rejection of the new theory cannot be likened to a "non-reception" of the relativity which he pioneered. An analysis of his seemingly contradictory position will be especially useful for understanding the epistemological conditions of the

113

Thomas F. Glick (ed.), The Comparative Reception of Relativity, 113–167.
© 1987 *by D. Reidel Publishing Company.*

"reception" of relativity in all their complexity. Unlike Poincaré, Langevin fully accepted relativity as soon as he became aware of it, and went on to articulate various important and unnoticed implications of the theory; in effect, it was through his influence over an elite corps of scientists formed and informed by him that the ideas about relativity slowly gained credibility in France. As in the case of Poincaré, an examination of Langevin's original work – close in many respects to Poincaré's – will allow us to understand fundamental characteristics of the new ideas in view of the reaction to them.

The understanding resulting from these studies of the receptiveness to relativity among these pioneers will shed light on the debates among scientists at the beginning of the 1920s, in the heat of what might be called the "battle of relativity". Underlying these debates and topics is a certain kind of scientific method, a network of epistemological preconceptions which, in addition to the actual work, have taken root in a particular intellectual, cultural, and institutional soil; we will not analyze them in detail, but a brief review of these contexts will serve as our starting point.

2. THE SCIENTIFIC, INSTITUTIONAL, AND EPISTEMOLOGICAL CONTEXT

Although the way in which the theory of relativity and relativistic notions were received by French scientists differed notably from case to case, especially among those distinguished scientists who contributed to their development, the receptiveness of scientists depended upon the context and the diversity of characteristics of their milieu. In the first place, a fundamental vision of science and physics, as it was taught in universities and through textbooks, and practiced at the theoretical level or in laboratories, was at stake. Also involved was a particular kind of institutional organization which nurtured these ideas and practices. Finally, underlying all of this was a network of epistemological presuppositions which, although relatively diverse, described general tendencies constituting an easily characterizable common framework. Taken together and tied to other cultural traits of the scientific community and its societal link, these constitute what some have called an "archeological base", in which a deep permanence, a morphological stability of the milieu under consideration may be seen.[1] In our particular case, this group of characteristics created an unfavorable terrain for the cultivation of new ideas.

From the way in which research was organized, the physics community in France during the first decades of this century could be called unreceptive to changes: it was a closed world in which the deadening pedagogical influence of the *grandes écoles*, with their caste system, was widespread and entrenched. An "old boys'" network cultivated a conservative orthodoxy that insisted on orienting itself toward the past the more it recalled past achievements. The prevailing individualism and gerontocracy of the day also acted as obstacles to the introduction of new ideas.

The great avenues of research in physics were those of experimental physics, in the traditional directions of mechanics and optics; theoretical output came mainly from mathematical physics and was the work of mathematicians. From the beginning of the century until the 1930s, there was practically no theoretical physics in the true sense of the word; Duhem, in physical chemistry, and Langevin, in physics, were the exception. It is significant that it was not until later that a distinction could be drawn between theoretical physics and mathematical physics,[2] and it is true that the dividing line was not always clear, as we shall see with Poincaré, whose work in physics cannot be reduced to that of a pure mathematician. Except for the particular case of Poincaré, which is more complex and demands closer analysis, the great French tradition of mathematical physics is something of an obstacle to relativity because of its separation from physics, properly speaking, which it saw in mainly formal terms. The *de facto* separation which existed in research between a mathematical physics and a physics which was regarded as experimental prohibited any truly theoretical work requiring a formal as well as conceptual reorganization of a body of knowledge. But relativity appeared precisely in the form of such a *theoretical* reorganization of *physics*. In a terrain which was ill-prepared in this respect to receive it, the work of a Langevin will show *a contrario* in enlightening fashion how theoretical physics was put to work, between mathematical formulation and experimentation. This work filled a void for experimenters, arousing interest in the point of view of the new theory, and for mathematicians, showing how *physics* was involved, and how it was a question of something completely other than a purely formal rearrangement.

The predominant spirit in physics – with an emphasis on the experimental side, to the detriment of theory, and the mistrust of the latter[3] – helped explained the delayed reception of Maxwell's electromagnetism,[4] statistical mechanics, relativity, and quantum theory as well. But this mistrust was general and not directed only against "revolutionary" theories: in their fields, a Duhem, a Le Châtelier, who opposed these theories, had to struggle for consideration of the organization of knowledge in a rational body of theory. This characteristic of physics would be perpetuated by the educational system.[5]

As for the general conceptual framework, mechanics was all-powerful and affected a faultless orthodoxy that was spread by a system which taught physics as a rigid and solid body of doctrine from which all weak points had been eliminated or disposed of. Science was presented as a fully realized achievement, encased in certainty, organized around Newtonian categories and justified by a positivism that took on the colors of mechanism, or, in a lesser fashion, those of energetics.[6] Not unreasonably, Langevin detected in the new notions of relativity the unraveling of the positivist spirit (in the Comtian sense) which organized knowledge in a hierarchical manner around absolute and normative categories.

In truth, one would have to explore all these elements in greater detail for a full explanation of how alien the special and general theories of relativity appeared to the French scientific community. But the global characteristics of a context do not suffice to determine precisely the conditions and circumstances of receptiveness or non-receptiveness. Moreover, in most cases, individual research programs and their epistemological bases lead to different explanations. (For example, the hostile position of a Duhem has unique features that are inadequately explained by the context; they arose in particular from his vision of theoretical physics, which valued the importance of theory but restricted its scope because of his conventional outlook and his effort to minimize the role of mathematics, which had no structuring effect.[8]) For this reason, there is no substitute for an in-depth analysis of concepts as they appeared in work relating to our field of inquiry, in attitudes toward the new ideas, and in the main lines of discussion engendered by these ideas within the scientific community.

3. HENRI POINCARÉ'S CLOSE APPROXIMATION AND REJECTION OF THE THEORY OF RELATIVITY

3.1. A lot has been written about Poincaré's relationship to the theory of relativity and his silence concerning Einstein's ideas. If we were to consider the theory of relativity only in Einstein's sense, there would be little to say about the role of Poincaré, who never mentioned Einstein in this regard, at least not in his published articles and his other known writings. However, Poincaré deserves our attention since he was, in France and in the international community, one of the initiators and even one of the pioneers of the ideas of relativity, even if it was not in the same sense as that of Einstein. Moreover, although his own program in the field has not been subsequently much considered and did not give rise to any school, it was influential. The way in which the analytic results achieved by Poincaré in 1905 on the basis of Lorentz' electron theory (statement of the principle of relativity, group of Lorentz and Poincaré, notation of space-time in four dimensions that Minkowski would take up in 1908) would be incorporated into the subsequent formalism of the theory of relativity is well-known. But the teaching and writings of Poincaré also had a considerable influence. Because of him, there were many who were prepared for, if not forewarned of, the "new mechanics", whose essential features he described with remarkable insight, although it had been developed beyond the borders which he had traced. We shall see how Paul Langevin, who had taken his courses and had accompanied him to the Congress of Saint Louis in 1904,[9] took up Lorentz' and Poincaré's program and modified it so that it corresponded to Einstein's; we shall show later, during the debates of the 1920s, the lingering influence of Poincaré among mathematicians, which no doubt delayed, then facilitated, the acceptance of relativistic concepts of space, time, and physical theory. We shall only undertake a quick review of the significant elements of Poincaré's ideas and contributions to the theory of electrodynamics and to relativity, with an emphasis on their consequences and their differences from those of Einstein, as we attempt to shed some light on what made Poincaré reject Einstein's point of view.

3.2. The theoretical program of Poincaré in the field of physics which concerns us was oriented toward the search for a general theory based on the electromagnetic theory of electrons, matter, and radiation. Lorentz had undertaken the same program, around a representation of electrical particles in motion in the stationary ether. Among the theories of the day, Lorentz' appeared to Poincaré as the most satisfactory in fulfilling his requirements for a theory of the electrodynamics of bodies in motion: to account for the partial drag of the ether (that is to say, Fizeau's experiment), for the equation of continuity of the electric charge and of that of the lines of magnetic force, and to be compatible with the principle of action and reaction.[10] Moreover, Poincaré observed that no experiment could demonstrate the absolute motion of ponderable matter with respect to the ether, and that this seemed true to higher orders in v/c, as attested to by Michelson's experiment. It was in some ways the statement of a principle of relativity (he would call it that several years later, in 1904) that Lorentz' theory would have to respect as well. Furthermore, influenced in part by Mach's analyses in the *Mechanics*, Poincaré's critical analyses of the concepts and principles of classical mechanics stressed the relative nature of space and time, as well as the possibility of reconsidering the principles of classical mechanics since it was an experimental science and did not imply any absolute *a priori* assumptions.[11] In *La Science et l'hypothèse* he stated: "There is no absolute space; we conceive of only relative movements"; "there is no absolute time", and "we cannot grasp the simultaneity of two events".[12] This last remark was substantiated by the operational analysis of simultaneity which he proposed as early as 1898 and which was very close to Einstein's in the latter's paper of 1905;[13] continuing this analysis in 1904, Poincaré studied the problem of the synchronicity of clocks which were set in motion by an inertial movement with respect to the ether, but in a relative state of rest with respect to each other by using the exchange of light signals. He considered, however, the existence of a true time, different from Lorentz' local time; but since their difference was slight and imperceptible, this distinction was of no importance.[14]

Poincaré also accepted the idea of a reconsideration of Euclidean space: "Our Euclidean geometry is itself a kind of linguistic convention; we could state mechanical facts by using a non-Euclidean frame of reference which would be less convenient but no less legitimate than our conventional space".[15] As for the ether, Poincaré felt that it was simply a question of a convenient hypothesis for the explanation of phenomena, but this "convenience" was not so facultative: he viewed it as being of the same order as postulating the existence of material objects; not entirely, however, for he drew a distinction between the two, since the existence of material objects would always have to be a convenient hypothesis, "while the day will no doubt come when the ether will be rejected as useless".[16] In his view, all these concepts were open to criticism, for their origin and justification were purely experimental; nevertheless, in his subsequent research and notably in his article of 1905 on "La Dynamique de l'électron",[17] he had to make an effort to preserve absolute time and Euclidean geometry (it was probably with an end to keeping absolute time that he proposed a deformable ether instead of accepting the physical contraction of Lorentz[18]). This voluntary support is probably the clearest difference between the theories of Poincaré and Einstein at the fundamental level.

Henceforth, Poincaré's program was established: to complete the theory of Lorentz while simplifying it and correcting it on minor points; to make it more rigorous; and to make it general enough to account for all of the properties of matter. From that point on, this was the ideal which he pursued, and he never claimed to have done anything more (in 1904, when he spoke of the *principle of relativity*, he attributed it to Lorentz). In fact, he went beyond Lorentz' theory and came as close as was possible, given his program (which could be described *a posteriori* as reformist, and not revolutionary), to the results required by what he would later call the "new mechanics". In particular, his introduction of the notion of quantity of motion of the electromagnetic wave was a decisive step in developing the idea of electromagnetic inertia and of the variation of mass with velocity. From this, Poincaré was to keep the idea that particles owe all their inertia to their electromagnetic nature. After he stated, at the Congress of Saint Louis, the necessity of a unified theory of electromagnetic phenomena, it was with the intent of clarifying its essential features that he dedicated his theoretical and formal work of synthesis in 1905, which was published the following year in *Rendiconti* of Palermo.

The breadth of this work is considerable. Following up on Lorentz' 1904 electron theory (which was an update of the 1895 theory), he proposed modifications which represented, in fact, a very elaborate stage of theoretical formulation. He gave an entirely symmetrical form to formulas for transforming coordinate systems in relative inertial motion, thus systematizing the principle of relativity. He also determined the formula for relativistic composition of velocities,[21] and showed that Lorentz' equations formed a group. His familiarity with group theory made him seek out the invariants of transformations which permitted laws of physics to be stated independently of the referential system, and allowed him to pose the problem of the form of these laws so that they conformed to the principle of relativity (what is termed covariance). The simplest invariant is the quadratic form $x^2 + y^2 + z^2 - c^2t^2$ written in this fashion for the first time using the notation ict for the coordinate of time. Other invariants were pressure (including

Poincaré's) and the integral of action which allowed a general principle of minimum action, capable of representing the laws of dynamics, to be expressed. Finally, he attempted to extend the principle of relativity to gravitational forces, showing that they had to be propagated at finite velocity (the speed of light), in gradual increments, and that consequently it was necessary to modify Newton's law for higher speeds, "in exactly the same manner as the laws of Electrostatics for electricity in motion";[22] in this regard he formulated a principle of correspondence by which Newton's theory was reformulated at the approximation of the speed of infinite propagation. He suggested that "the effect will be most obvious in the motion of Mercury because this planet has the greatest velocity", and calculated the resulting deviation for the perihelion (it was partial, on the order of six seconds of arc rather than the 38 seconds that had been observed, but he noted that this correction went in the right direction).

3.3. It can be concluded that Poincaré explored the deepest implications of Lorentz' relativistic theory of the electron. For all of that, can we use the term *theory of relativity* and say that he co-discovered it with Einstein, that indeed Poincaré anticipated or even went farther than Einstein during this period (around 1905)?[23] Of course, the question under consideration is not the respective value of the works of Poincaré and Einstein, which were both immense, even if we were to limit ourselves to electrodynamics and the special theory of relativity; the real issue is knowing what relationships existed between the programs that reflected these two theories. In hindsight, such ties certainly exist. Characteristically, however, Poincaré and Einstein were silent with respect to each other's work.

It is a known fact that Einstein pondered a lot and read relatively little (he was rather isolated from the scientific community and its publications, as illustrated by his experience in Berne during the gestation period of the special theory.) If Einstein had any knowledge of Lorentz' work of 1895, he was unaware of the 1904 work (not in general circulation)[24] which contained the formulas for transformations; if he had read *La Science et l'hypothèse*, which certainly contained critical considerations of space, time, and classical concepts, he probably did not know of Poincaré's analysis of simultaneity, nor of his 1904 works (nor obviously the article in *Rendiconti*, 1906, written at the same time as his own[25]).

As for Poincaré, one might be surprised by his silence concerning

Einstein's works in the field of relativity; he never cited Einstein's name in this regard, and it was obviously not for petty reasons of priority. However, Einstein's theory fulfilled the criteria of elegance and simplicity which were not altogether present in Lorentz' theory, for which Poincaré had managed to formulate general properties of invariance in a systematic manner. It was Einstein's program in physical theory and the nature of the special theory itself which he could not accept, perhaps because the theory represented too radical a break with the past, or because it did not meet all of his requirements of a program for a physical theory of matter: if Einstein proposed in his research a comprehensive program for the properties of matter, his method was something else, for he dealt separately with problems as a function of their specificity. Moreover, others besides Poincaré misjudged Einstein's methods in his earliest research, believing him to have been interested primarily in the immediate and practical aspects of the questions under consideration, led initially by a heuristic, if not empiric approach, and guided by a somewhat uncontrollable imagination.

It is easy to imagine how, in Poincaré's eyes, Einstein's theory of relativity treated a complex web of problems in a rather cavalier way, claiming to reduce everything to two principles (whose probable validity Poincaré himself could certainly accept), without bothering about other aspects of the properties of matter, which Einstein himself considered independently in other works, such as the quanta of energy, which also provoked Poincaré's skepticism.[26] In a letter of recommendation addressed to Pierre Weiss in November, 1911, Poincaré praised the value and scientific originality of Einstein; he noted the latter's facility in "adapting to new ideas and drawing out all their implications", his ability to grasp the multiple possibilities of a problem in physics and to translate them in "anticipation of new phenomena which could one day be verified experimentally". He added: "I do not mean that all these insights will hold up under experimental conditions at a time when experiments can be devised. As he wanders about in many directions, it is to be expected that most of the paths will turn out to be dead ends; but at the same time, it must be hoped that one of the trails blazed by him will be the right one; that is enough. That is really the way to go about things. The role of mathematical physics is to ask the right questions, and for experience to answer them".[27] Several weeks before, Einstein and Poincaré had met at the first Solvay Congress, where Einstein had presented his paper on specific heats. Poincaré was impressed[28] by his scientific ability, but remained skeptical of Einstein's ideas on relativity, which were not discussed in session but were implicit in certain problems under consideration and were discussed in private besides the Congress sessions. Einstein gave this testimony: "Poincaré was in general simply antagonistic [against the theory of relativity], and, for all his acuity, showed little understanding of the situation".[29]

3.4. The contributions of Poincaré to what was to become the theory of relativity are of distinct importance. The question has been asked as to why it was not Poincaré but Einstein, a newcomer to electrodynamics, a field in which Poincaré was much better informed about ongoing work, who fully developed relativistic concepts and the theory of relativity. What interests us here is not so much getting a definite and complete response as far as Poincaré's reasons are concerned, but discovering, according to the manner in which he conceived of relativity (accessible through his program of work) and which differed from what would later be accepted (Einstein's), the signs of a theoretical and epistemological sensitivity determining certain important lines of inquiry in the intellectual terrain where relativity, having been worked out, was proposing itself for reception.

Poincaré took the idea of relativity as far as possible within the limits of a more traditional and conventional conceptual framework. He himself attested to his desire to maintain this framework; for example, in *La Science et l'hypothèse*, he wrote: "The old theories are based on a large number of numerical coincidences that cannot be attributed to chance, so we cannot split up what has been joined together. Instead of breaking the frames, we must bend them".[30]

It is precisely the maintaining of the conceptual framework of mechanics which distinguishes the physical meanings of his theory and Einstein's. Although the transformation formulas point to the same kinematics, Poincaré holds on to absolute time and ether through his explanation (dynamic in nature), through the pressure of the ether on the electron, of contraction in the sense of motion. The result is that if it can be said that Poincaré's equations for the motion of the electron give rise to a kinematics (formulas of transformation of coordinates, rules of addition of velocities), it is on the basis of a dynamics of the ether, whose effect is to maintain the old framework, that is to say, the fundamental character of the old kinematics (which henceforth only has to be *adapted*). The interpretation (in the physical sense) of the two theories is not equivalent, and relativity in the sense of Poincaré is of the type of a *dynamics*. Poincaré's program, oriented toward access to a general (electromagnetic) theory of atomic matter, prevented him from separating the kinematic and dynamic aspects of the motion of bodies. It was precisely Einstein's own point of view which permitted immediate access to a kinematics of bodies in motion independently of any consideration of dynamics, whether it was for these bodies or for an ether possessing properties of this kind. This insistence upon kinematics – that is, the structure of space-time – while economizing on any other superfluous consideration, is effective in establishing the framework for a new mechanics and in drastically modifying Newton.

In a look back at his work of 1905 and its antecedents, Einstein put his finger on the real breakthrough in his memoir: "It was to have uncovered the fact that the scope of Lorentz' transformation transcended its connection with the equations of Maxwell and questioned the nature of space and time in general. What was equally new was that Lorentz' invariance was a general condition for any physical theory".[31]

For Poincaré, Lorentz' invariance was a general condition insofar as physical theory was ultimately a theory of the *electromagnetic* properties of matter. Thus, it is the essence of Poincaré's program which is under scrutiny here, with respect to the generality of the relativity principle, and to its character of foundation with respect to the theory. If Poincaré chose his program of a totally electromagnetic (and solidly relativistic) theory of matter, it was because it was based on experiment and because the theory, in his eyes, had to be inductively derived from the base or the frame of knowledge that had been acquired up to that point. To be sure, this induction did not have the characteristic of absolute constraint (by virtue of its conventionalism, which was linked to an empiricism): for that reason, it was possible to modify the initial theory and, at the same time, to preserve the important elements of classical physics.[32] Einstein's affirmation positing the principle of relativity and the constancy of the speed of light as first principles of any physical theory must have appeared to him as an arbitrary "coup de force" intended to replace the system of Newtonian mechanics by an equally rigid system (and here Poincaré's unformulated criticism would likely have met with that of Mach, who considered Einstein's theory of relativity dogmatic).

4. PAUL LANGEVIN'S APPROACH AND HIS ROLE IN THE TEACHING AND DIFFUSION OF RELATIVITY

4.1. Paul Langevin was one of the first physicists to understand the full significance of Einstein's work on the special theory, and on the general theory, whose development he followed closely.[33] He took on the role of spokesman and propagandist within the scientific community and among philosophers, students, and the general public. His influence in France in spreading the new ideas was considerable, since it was mainly through him, and notably his courses at the Collège de France, which were taken by an elite group of physicists and mathematicians, that a whole generation was exposed to ideas which were presented, analyzed, clarified, and whose yet unperceived physical implications were developed.[34] The most renowned scientists in the world as well as French physicists have attested to Langevin's importance. According to Louis de Broglie, he was "the grand master of Theoretical Physics in France for a number of years",[35] and Einstein himself wrote in 1940: "Professor Langevin is beyond doubt one of the most distinguished scientists of our day. Among his universally known original works are his statistical theories of magnetism and ionization phenomena in the atmosphere and Brownian movement. He was but a young man when he was named Professor at the Collège de France. His courses at this famous institution had the greatest influence on the development of the young generation of French physicists. There are few today who possess his mastery over the whole spectrum of modern physics".[36] Einstein and Langevin had struck up a deep and enduring friendship since the time of the first Solvay Congress in 1911. Their personal ties were strengthened by the similarity of their views on most important scientific, ethical, and political problems. On the occasion of Langevin's death, Einstein wrote the following words, which summarize a part of what we wish to develop here: "Langevin was endowed with unusual clarity and agility in scientific thought, together with a sure intuitive vision for the essential points. It was as a result of these qualities that his lectures exerted a crucial influence on more than one generation of French theoretical physicists. But Langevin also knew a great deal about experimental technique and his criticism and constructive suggestions always carried a fruitful effect. His own original researches, moreover, decisively influenced the development of science, mainly in the fields of magnetism and ion theory. Yet the burden of responsibility which he was always ready to assume circumscribed his own research work, so that the fruits of his labors emerge in the publications of other scientists to a greater extent than in his own. It appears to me as a foregone conclusion that he would have developed a Special Theory of Relativity, had that not been done elsewhere; for he had clearly perceived its essential aspects."[37]

Einstein's testimony is of the greatest interest, above all the last statement. There were very few physicists of whom Einstein could have said, as he did of Langevin, that he was as close to Einstein's ideas on relativity at the very moment of their formulation. Working on electrodynamics like Lorentz, Poincaré, and Einstein himself, Langevin immediately recognized the originality and superiority of Einstein's viewpoint. This immediate conversion to Einstein's theory of relativity is all the more surprising since he was following a program that was closer to that of Lorentz and Poincaré – and because he succeeded, where they had not, in deriving a generalization of the inertia of energy, expressing the famous formula $m = E/c^2$ – and since he had been the student of Poincaré, whose work on the subject he knew well and appreciated. Let us skip over the ingrained intellectual honesty and selflessness which made him set aside the originality of his own approach in order to adopt enthusiastically Einstein's theory and to become its spokesman. Beyond the appropriately historical interest of Langevin's discovery of essential elements of this theory, thereby confirming and substantiating Einstein's observation that around 1905, the ideas of the special theory were ripening in several places, his orginal position allows us to understand how it was possible to go from a program which was initially very close to that of Lorentz and Poincaré, to a full comprehension of relativity in the sense of Einstein, and in doing so, to bridge the gap between the two conceptualizations. We shall also see how this position, which was theoretical, yet involved at the same time with experimental physics, had its own richness, as a variety of heretofore unnoticed but important implications unfolded, thanks to a very "concrete" physical sense.

4.2. Langevin had begun his research work in 1895 with Jean Perrin on the discharge of bodies that had been bombarded with X-rays; he spent a year in Cambridge at the laboratory of J.J. Thomson in 1897–1898, when Thomson discovered the electron. He continued his research on ionized, gases, and then on the theory of magnetism; following these leads, he found himself studying different aspects of the theory of electrons. He was called upon to present an overview of the subject at the Congress of Saint Louis, in the United States, in September, 1904, where he and Henri Poincaré were in the French delegation. "La Physique des électrons" is a long and remarkable synthesis of the experimental and theoretical problems relating to the whole of electrodynamic phenomena.[38] Langevin presented the essential elements of ongoing research, with which he was quite familiar and which included his own contributions. He tried to show their underlying unity through a systematic analysis and hoped to renew contemporary knowledge of matter. The fundamental new fact which compelled him to such a change of perspective was the mounting evidence of the granular structure of the electric charge and, in a general way, of the atomic structure of matter: the "physics of electrons" was called upon to extend its range to all of physics, and thereby "to shed light upon the basic concepts of Newtonian Mechanics itself".[39] This was the design of the program for a new physics, based on the notions of electron and atom. In his 1904 paper, Langevin described the properties of atomic and electrical matter, viewed from the experimental as well as theoretical angle; the description seemed to lead inevitably to an atomic and electronic theory of matter, with a dual base: the electromagnetic ether and the granular structure of electricity.

It was thus a question of a Lorentzian perspective, but as we shall see, there were differences which would allow him to go beyond the program of Lorentz and Poincaré. In effect, Langevin placed the desired synthesis of the physical properties of matter *outside* of Newtonian mechanics, with a shift in emphasis to electromagnetic concepts.[40] In the 1904 paper on the physics of electrons, the fundamental notion of *electrons* conceived of as charged cores (whose existence was experimentally established), "mobile with respect to a fixed *ether* as defined by Hertz' equations", was, for Langevin, the axis around which the representation of the properties of matter was organized: both inertia and dynamics were deduced from the electron and the ether.[41] In particular, the inertia of electromagnetic energy guaranteed the principle of inertia and relativity (a term not yet used by Langevin): from that point on, their formulation was arrived at beyond the limits and independently of mechanics, since up to then its involvement had not been necessary.

It is true that Lorentz' transformation and contraction in the direction of motion were required in order to ensure independence with respect to movements of translation, but Langevin did not go into detail, merely considering that the respective definition of systems in relative motion was reduced to a simple change of variables which preserved the form of the equations. Langevin did not see the need for a material representation since that would have meant postulating mechanical properties of the ether, something that he had promised himself not to do when he began his work, and which he subsequently confirmed: obviously, it was not a question of demonstrating the non-mechanicity of the ether, but of a program about it. This program, containing Langevin's innovation, was distinct from that of Lorentz and Poincaré: the electromagnetic point of view did not have to be reconciled with the mechanical perspective; it had priority over it, and mechanical notions appeared to be desirable only under certain conditions of approximation. Even Poincaré's 1906 work did not go this far. Electrodynamics showed, in fact, the limits of mechanics and its concepts (through its substantial modification of the form of its concepts, particularly the variation of inertial mass with velocity); "It is *really the ether* that has to be regarded *as fundamental*, and it is altogether natural that we should initially define it according to those properties which are known to us, i.e., the electrical and magnetic fields; these are considered to be fundamental and attainable, . . . without even having to accept the knowledge of the laws of Dynamics, the notions of mass and of force in their everyday form. We shall find the latter again as derivative and secondary ideas".[42] It must be pointed out that the ether (an expression which would linger for a rather long time in Langevin's vocabulary, until 1913) was no longer defined except by electric and magnetic fields, without any other property. For Langevin, the physics of the ether was a physics in which the fundamental concepts were those of the electric and magnetic fields.

Hence, the proposed reversed viewpoint: "To consider the analogy indicated by Maxwell between the equations of electromagnetism and those of Lagrange's dynamics as being a greater justification for the possibility of an electromagnetic representation of principles and concepts of ordinary, material Mechanics, as opposed to the inverse possibility".[43] This was what he proposed for electrical and magnetic energies, by formulating an analog of Hamilton's principle, in order to derive the equations of motion, the dynamics of electrons.[44] In essence, this was the "electronic theory of matter", capable of being applied to other known phenomena (and in particular, Langevin developed his theory of magnetism based on it). It was no longer a question of trying to adapt the old theory and to maintain the old conceptual frame: "This notion [of the electron, a link between the ether and matter] captured the imagination in a very few years, broke the frames of the old Physics, and upset the traditional order of concepts and laws, culminating in a simple, harmonious, and rich structure".[45]

4.3. In 1905, Langevin made two contributions concerning relativity which clarified or developed in a significant way certain aspects of the paper presented at the Congress of Saint Louis. The first, rather brief, was entitled "Sur l'Impossibilité physique de mettre en évidence le mouvement de translation de la terre". Langevin showed how Lorentz' recent theory of the deformable electron, with its model of longitudinal contraction,[46] could explain the negative result of Trouton and Noble's experiment on the torsion of a charged condenser in relation to the direction of the motion of the earth. The second focused "on the origin of radiation and electromagnetic inertia".[48] His primary interest was in showing the way toward a systematization of the relationship between inertia and electromagnetic energy, which Langevin would extend soon after to all forms of energy. Langevin set up the problem of the nature of inertia for matter in general, with negative electrons being a particular case. He wrote: "It is tempting in order not to have to look for two different explanations for one phenomenon" (i.e., an electromagnetic explanation for electrons and another – mechanical, perhaps – for other material particles), "to extend this result to all matter by treating its inertia as the total electromagnetic inertia of the positive and negative electrons which make it up".[49] It was by looking for a way of characterizing the inertial energy of either free or accelerated electrons (or more generally, material particles), based on the form of electrical and magnetic fields and of their variation, that Langevin proposed to clear up this question, which he knew to be connected to "the limits of validity for mechanical laws".[50]

He defined "wake", "wave of velocity", and "wave of acceleration" (used after him by Poincaré in his article of 1905), allowing him to describe electromagnetic matter whose form indicated that "the equations of Mechanics have to be modified in two distinct ways". The first related to the variation of mass as a function of velocity, which made him "believe in the electromagnetic origin of inertia and in the impossibility of establishing a convincing Mechanics without accepting electrical concepts as fundamental". The second had to do with the very form of the dynamic equations (that the variability of mass taken alone still respected): Langevin showed the formal necessity of modifying them while considering the energy of change (due to the action of exterior forces) supplied by mechanics, and radiant energy, small compared to the former (thus negligeable at first guess), but which mechanics did not supply. Such considerations, concluded Langevin, "seemed to shed some light on the inner workings of the phenomena of inertia and radiation".[51] In fact, they showed the way to a fundamental modification of the laws of bodies in motion and their dynamics. Langevin himself did not accomplish this modification. He managed to clarify some of its features, through a more extended generalization of the form of inertia, which he presented in his course at the Collège de France in 1905–1906; he did not publish it, but we have an account of it from Edmond Bauer, his assistant at the time.[52] It was not until Einstein's 1905 work was brought to his attention a short time after that Langevin saw the completion of the program which he had outlined but had not yet fully achieved on his own.

4.5. A number of questions remained unanswered in Langevin's synthesis. If the supremacy of electromagnetic concepts over those of mechanics appeared to him to favor decisively the point of view of the electromagnetic ether, it represented a programmatic position: in particular, Lorentz' contraction was accepted without the support (as in Einstein's case) of a complete view of the spatiotemporal frame and the relations of the corresponding kinematics. Because of its open (and incomplete) nature, what constituted a weakness (notably with regard to the requirements of a theory relative to the entire set of properties of matter) was also a strength with regard to a full understanding of the new point of view. Langevin left the question of a global theory in the background, unlike Poincaré, who was doubly concerned with finding a coherent theory for all phenomena while maintaining the old conceptual frame.

Although he was guided by the concern for a "synthesis" which appeared to be the inevitable outcome of these new concepts, Langevin did not feel that it was possible to accomplish this while keeping the framework of mechanics, and in particular, while insisting on a material representation of the ether. It was in the "electromagnetic ether" that the essence of the new point of view was to be found, i.e., in the fundamental notions of electrical and magnetic fields which could spread and interact with matter. But the frame for these concepts was not yet established: it was a program that had to be implemented.

A necessary condition for the realization of the program – and herein lay the originality and strength of Langevin's position – was to not give a mechanical representation of the ether. Not only did he not not feel obliged to revert to such a representation (and that would have been enough to leave his synthesis unfinished, open), he also refused to consider its possibility (thereby opening up his concepts to the possibility of another conclusion) because his conception did not shy away from radically opposing mechanical and electromagnetic notions. His attitude as a theoretician close to the data of experience did not force him to aim for such an achieved theory right away out of an immediate preoccupation with form. That is why he went a bit further than Poincaré but not quite as far as Einstein in the direction that ultimately prevailed. It is not surprising that once he was aware of Einstein's work, Langevin immediately realized that it was the totally satisfying perspective toward which he himself had been oriented; it even succeeded in combining a most striking presentation with simplicity of form, and summarized the new concepts of inertia and energy. But the concern for form was always in his thoughts, since it was the concern which made him see in Einstein's work the long-awaited synthesis, and since, moreover, he began to work within the frame of formalized expression of the theory of relativity in order to enumerate its implications at the conceptual level as well as at the level of physical phenomena.

4.6. From that point on, Langevin became the spokesman for the relativistic viewpoint which corresponded so perfectly to what he himself had proclaimed. His own work on electrodynamic problems had made him receptive to the influence of relativistic ideas and made him take up the new point of view completely (that of a dynamics of a continuous field); it also made him predisposed to grasp the concepts and formulation of relativity in an original way, as a function of what constituted his own physical and epistemological "program", henceforth firmly nestled within the sought-after theory. As a theoretician and an experimentalist, Langevin, who was sensitive to the formal perfection of the theory, did not forget, for all that, his "phenomenological" interests, and more than anyone else, could perceive certain consequences of the theory of relativity regarding the most general properties of matter and motion.

His considerations in 1913 of the implications of the inertia of energy, as regards the elementary structure of matter, were in a direct line to his own research on inertia for which the relativistic viewpoint furnished a full generalization.[54] By considering only the general nature of the inertia of energy, and without introducing any supplementary model or theory relative to the make-up of matter, Langevin was able to infer a fundamental property of matter which could be summarized by the notion of binding energy or packing loss, which would later turn out to be important.

Referring to the form of energy taken by particles, on one hand, and radiation, on the other, Langevin stated: "Any variation of internal energy of a material system through emission or absorption of radiation is accompanied by a proportional variation of its inertia".[55] Going on to question the mechanism of this variation (whatever it could be in nature), he examined the practical consequences of this law by reviewing the different fields of physics where variations of energy could be found: thermodynamics (variations of the mass of a body with temperature), chemical reactions (modifications of the mass of compounds), radioactive transformations (here the variations became measurable), deviations from Proust's law (atomic weights were not simply whole multiples of a same quantity).

In the last instance, Langevin indicated that the inertia of energy allowed the unity of matter to be restored: "*The differences originate from the fact that the formation of atoms from primary elements* (by disintegration, as we see in radioactivity, or by a yet unobserved inverse process which would result in heavy atoms) *would be accompanied by variations of internal energy through emission or absorption of radiation.* The sum of the weights of the formed atoms would differ from the sum of transformed atoms by a quantity equal to the quotient of the variation of energy by the square of the speed of light. And the deviations are such that the energies thus called into play would be of entirely the same order as those actually observed in the course of radioactive transformations".[56] Langevin ended his work on the implications of the inertia of energy[57] with a prophetic glimpse of "matter, reservoir of energy" and of the fact that what really characterized a material system was no longer its mass, but the "number and structure of elements, atoms or molecules" of which it consisted, and which alone "would remain invariable through all the changes undergone by matter, and could serve to define it". Very much ahead of its time, it showed the way to nuclear physics and elementary particles.[58]

4.7. Langevin proposed an original analysis of the new conceptual framework of relativity, even before he derived all the physical consequences of the inertia of energy. Here again, it was the intersection of his own program and Einstein's theory which allowed him to envisage – as the result of deep personal reflection – heretofore unnoticed features of the theory as a conceptual framework: they involved problems of space, time, and causality. Langevin developed them in 1911 for papers to be presented to philosophers rather than scientists: one, for the Congress of Philosophy of Bologna, drew the attention of the philosophical community to the theory of relativity;[59] the other, for the Société Française de Philosophie, primed the debates in France over these questions, anticipating the session thirteen years later in which Einstein would participate, at the side of Langevin.[60] The interest of these texts went far beyond that of a primer for non-scientists; Langevin developed physical as well as epistemological considerations which have since been integrated into the corpus itself of the theory of relativity.

In these texts, Langevin was intent on making the difference clear between the basic concepts of the viewpoints of mechanics and electromagnetism, i.e., their respective conceptual frames; he showed the way in which the notions of mechanics had to be replaced, and how, in the first place, the supposedly well-entrenched ideas of space and time were transformed.

In his considerations, Langevin did not look to replace an intuitive grasp of those ideas by another, which would be adapted to the new phenomena. From the start (and this is a characteristic of his method in physics as well as of his epistemology), he assumed the point of view of formal theory, whether it was classical or relativistic in nature, in order to state the profound difference between the two concerning the concepts of space and time, and to show thereby the limits of their intuitive acceptance. Then, after formally stating the difference between the old and new concepts, he made the physical sense clear, i.e., their phenomenal implications, which could be imagined in *de facto* situations (the consideration of these would be appropriate for evoking a new intuitive acceptance by adapting thought to facts, characteristic in his view of the evolutionary progress of human, and particularly scientific thought).[61]

In the Bologna paper on space and time as well as in the Paris lecture, Langevin started with the "experimental fact" of the identity of equations representing physical laws in systems of reference in inertial motion, i.e., with the principle of relativity (Einstein also saw it as a *fact*), and with its mathematical translation in terms of invariance under a transformation group. Each of the proposed groups (Galileo's for mechanics, Lorentz' for electromagnetism) "differed markedly" from the other "in whatever concerned transformations of space and time". Consequently, "a choice had to be made: if we want to preserve an absolute value in the equations of rational Mechanics, in the mechanism, as well as in the corresponding space and time, we have to treat those of Electromagnetism as false and abandon the elegant synthesis. . . . On the other hand, if we want to keep Electromagnetism, we have to open our minds to the new concepts that it requires for space and time. . . . Only Electromagnetism or laws of Mechanics which allow the same group of transformations would permit us to go further[62] and could assume the dominant role that mechanism assigned to rational Mechanics".[63]

Using Minkowski's universe representation, Langevin indicated how the new concepts permitted the symmetry of intervals of time and space to be re-established. He then stated the problem of the relationship between spatial and temporal distances in terms of regions of space-time determined by the cone of light (although he did not use the exact expression). We shall not describe this now classic consideration. It is Langevin who, after the introduction by Minkowski, in 1908, of space-like and time-like vectors, developed the notion of "pairs [of events] in space" and "pairs in time", which pointed to the possibility of a causal relation between them[64] and which corresponded to what was afterwards called "interval of type space" and "interval of type time". From then on, the spatiotemporal symmetry was re-established: in the *region space* (where the order of succession in time has no absolute sense), the distance of two events "passes by a minimum precisely for systems of reference with respect to which the two events are simultaneous"; in the *region time* (where the order of temporal succession is preserved, in an absolute fashion), the interval of time "passes by a minimum, precisely for the system of reference with respect to which the two events coincide in space".[65] The definition of proper time (linked to a material system) came out of this second statement: "It will be shorter" (in the midst of the material system) "than for observers who would have stayed linked to the system of reference in uniform motion" (with respect to it).[66] Once again, he translated into physical terms what had been mathematically formulated by Minkowski (the proper time as the integral along the world line of the time differential d*t*). Langevin gave two concrete examples which have since become famous: that of the lifetime of a sample of radioactive matter, and that of the traveler (later referred to as *Langevin's traveler*) who returns to earth after two years in space only to find that it is two centuries later. "The most solidly established experimental facts of physics allow us to state with conviction that this would be so," declared Langevin,[67] although these experiments were obviously still a long way from being implementable: it was because these predictions resulted from the coherence of electromagnetic laws capped by the (special) theory of relativity.

By giving these seemingly paradoxical phenomena the most detailed explanation from a practical standpoint, demonstrating that it was all possible because it conformed to a well-established theory (and the coherence of details indicated how reasonable it was), Langevin anticipated and defused any criticism of the new concepts in the name of common sense. He was to say later that it was this common sense which had to follow the evolution of ideas and be revived. His description of a space traveler was just such an example of a renewed common sense.

4.8. Langevin continued to follow closely the progress of developments in other aspects of the theory of relativity, as it can be seen in the notices pertaining to works previously mentioned, or again in an evaluation of Einstein he formulated in 1913; after reviewing the young scientist's other works, Langevin concluded with Einstein's refinement of general relativity: "Finally, Mr. Einstein is currently looking into the problem of gravitation, in which respect no progress has been made since Newton. For the first time, we can hope, through a generalization of the principle of relativity, to link gravitation to the great synthesis which already includes the phenomena of electromagnetism and optics".[68]

When he was able to resume his work and teaching after the First World War, Langevin would again devote an important part of his courses and several articles to presenting the theory of relativity, but extended to general relativity.[69] His work retained an expository clarity whose physical sense was based on thought experiments of fundamental scope – as, for example, the projectile of Jules Verne, which, incidentally, he had already envisaged in 1911,[70] and in which the passengers did not experience the force of gravity which they underwent together; in some way the acceleration canceled out gravity.[71] He also made other contributions which showed the best grasp of the theory and an interpretation which was consistent with Einstein's: the relativistic explanation of the rotating disk experiment of Sagnac, or a work on Thomas' factor reported by Sommerfeld.[72] Through Langevin's courses, French physicists had access to the theory which would be gradually accepted after a series of long controversies which we shall examine. Everyone considered Langevin the spokesman for relativity, and those who saw in it a new mystique or an attack against common sense, or were bothered by the uproar surrounding the issue, directed their criticisms against Langevin as well as against Einstein.

By its comprehensiveness in the physical representation of matter and the universe, and by its innate perfection, the theory of relativity was for Langevin the "most harmonious and well-founded (in facts) construct ever created by the human spirit".[73] In his lecture in 1919 on "The Principle of Relativity" he wrote: "The explanatory and predictive power of this theory, imposed and confirmed by the facts, is as great as its logical structure is rigorous and beautiful". In describing the theory developed "principally by Mr. Einstein", whose "admirable continuity of thought" he emphasized, he showed how it consisted of two main stages, "that of special relativity from 1905 to 1912, and since 1912, that of general relativity".[74] He reviewed its essential elements with respect to formulation and experimental verification (which included at that time intra-atomic phenomena, by the relativistic modification of the orbits of Bohr). As for general relativity's explanation of the deviation of motion of Mercury's perihelion, he wrote: "It is altogether remarkable that the general theory of relativity came up with long-sought solution without introducing any arbitrary hypothesis or constant, through the necessary development of the fundamental idea".[75] This presentation, so consistent with Einstein's concepts, would in fact be the model for all later expositions proposed by various individuals, and the features of relativity which it emphasized were the very ones which would serve to define the divisions in the ensuing debates. It must also be said that for Langevin, these features corresponded in some way to the new epistemological norms required of a physical theory, and that relativity and its elaboration constituted a tell-tale sign of the evolutionary process of scientific thought, whose "present-day outcome" is represented by this theory.[76]

5. CONTRIBUTIONS AND DEBATES DURING THE 1920s: THE ACADÉMIE DES SCIENCES AS A WITNESS

5.1. The minority view does not represent the whole of a situation, and except for Langevin's students and others who attended his lectures, most of the scientific community was unaware of the theory of relativity until the announcement of the results of the observation of the eclipse of 1919 confirming the prediction of general relativity about the curvature of light rays near the sun. It is significant that the consideration, both negative and positive, of the general and special theory of relativity by the scientific community as a whole coincided with the mass appeal of the theory. Until then, it had not made many inroads into the scientific establishment.

In this regard, the Académie des Sciences made an excellent witness. Before the end of 1919, the Académie published very few papers on relativity in its *Comptes rendus des séances hebdomadaires*; in that year, there were only eight, under the headings "astronomy", "gravitation", or "relativity", but they either ignored or deplored (e.g., Georges Sagnac's four articles) Einstein's theory. By 1920, among 15 articles on relativity or related problems, a slight but distinct change which would be more noticeable in the following volume was taking place; the number of articles on the theories of relativity (and studies which showed the influence of this theoretical trend, particularly in mathematical physics, by de Donder, Léon Bloch, A. Buhl) went up. On the experimental side, A. Pérot noted a shift in the wavelength of a spectral line of cyanogen; the measure of difference appeared compatible with the predictions of Einstein's general theory.

In 1921, 23 articles in the *Comptes rendus* were devoted to relativity; the majority of them were submitted in the second half of the year. Most of the 14 authors of these articles were favorably disposed towards the new ideas; a few of the articles dealt with special relativity and Lorentz' transformations, but a greater number had to do with general relativity and the field of gravitation; there was also one article on unitary field theories. But it was Paul Painlevé and Emile Picard who opened the door to the debate which would continue over the following months.[77]

5.2. The first six months of 1922, such as they appeared in the *Comptes rendus*, deserve particular attention; they represented the focal point of the debates and controversies over relativity and coincided (though not fortuitously) with Einstein's presence in Paris, where he met his French colleagues in the Collège de France and the Société Française de Philosophie – but neither the Académie nor the Société Française de Physique received him.[78] The reports of the sessions, week in, week out, give the impression of a stage production: a succession of players appeared, even some who were not in the Académie, but were sponsored by members whose names referred not only to individuals, but also to trends, even factions. The topics of the papers defined for some of them the stakes of a debate which was not only scientific, but also had to do with the nature and scope of physics as a theoretical science, and with its relationship to mathematics and to experience (a debate whose underpinnings were properly philosophical); for the others, they announced research developments in fields ranging from experimental physics to pure mathematics, stimulated by the theories of relativity and characteristic also of the *reception* of these theories.

In talking about the sequence of the twenty-two papers dealing with the theory of relativity which were presented to the Académie during the first half of 1922, the image of a "stage production" seems all the more meaningful starting from the second session, on January 9. At the opening of the session, Emile Picard, the permanent secretary for mathematical sciences, presented a brochure entitled "Le Principe de relativité et ses applications à l'astronomie", in which he briefly reviewed the sequence of events and focused on the problem of the measurement of time.[79] In his discussion, Picard was not hostile, merely skeptical, but the arguments which he evoked were those on which skepticism or opposition were to be based. His arguments were of two kinds, theoretical-epistemological and experimental, and were topics that had previously been touched upon by Painlevé and Picard, who had opened the debate on relativity in the Académie in 1921. From then on, the tone was set, and the discussions on relativity would evolve around these twin axes.

The first skirmishes within the Académie took place at the session of January 3. Under the heading "physics", Daniel Berthelot, one of the leading opposition spokesmen (and author of a pamphlet entitled *La Physique et la métaphysique des théories d'Einstein*, which appeared in the same year,[80] presented a note from Georges Sagnac on the theory of the mechanical ether[81] and another by E. Carvalho on "the principle of relativity in dielectrics",[82] in which the author claimed to show decisively the drag of the ether and concluded: "The consonance of electro-optics with mechanics thus appears complete, without having to upset the foundations of our knowledge according to Mr. Einstein's ideas". The physical existence of the ether and the "fragility of the bases for the theories of Mr. Einstein" was again the subject of a note presented by Berthelot, this time from E. Brylinski.[83] This continuous fire of contestations was supported by a communication from the mechanician Léon Lecornu on February 8. For the purposes of the debate, he took the side of Berthelot and G. Koenigs, the three of them leading the opposition against Einstein's concepts of relativity. Lecornu attempted to minimize the degree of conformity between Einstein's prediction and the advance of Mercury's perihelion.

It was not until the session of April 3 that the opponents of relativity took up the attack again. In the meantime, the supporters of relativity had taken over center stage, with contributions from mathematical physics and even pure mathematics, under the rubric "geometry". They were in the form of Emile Borel's presentation of notes from Elie Cartan on the geometric representation of the tensor of energy of Einstein in terms of torsional spaces or generalized congruent spaces (we shall return to this major contribution), and of a work by Enrico Bompiani on a neighboring topic. The counter-attack came in the form of a note from J. Le Roux, presented by Koenigs on April 3, which criticized Einstein's notion of the curvature of space in the name of a distinction between *mathematical* form (of metrics) and physical spaces.

During the same session, astronomy and celestial mechanics were represented by two relatively technical notes, one from Maurice Hamy, "Sur une propriété des émulsions photographiques et l'enregistrement des étoiles, pendant les éclipses totales du soleil, en vue de la vérification de l'effet Einstein"; the other, by Maurice Sanger, was about "a remarkable coincidence in the theory of relativity". Both showed a relativistic concern.

Returning to "mathematical physics" through articles by Emile Borel on issues to which we shall return, we find Borel taking up the distinction which he had made in his work on *L'Espace et le temps* between "physical hypotheses and geometric hypotheses"; Paul Painlevé proposed determining to what extent classical mechanics and Einstein's theory differred in their axioms and predictions.[84] Under "celestial mechanics", Jean Chazy, J. Trousset, M. Ferrier, and Painlevé took on the problem of astronomical predictions and their verification.[85] In the same session, S. Zaremba expressed doubts about the "assertions of the theory of relativity relative to the nature of space".[86] Other papers at the end of the first half of 1922 followed the trend to relativistic concepts: de Donder's work on the construction of an electromagnetic and gravitational field without the intervention of a mass field, May 1;[87] a note from Marcel Brillouin discussing Schwarzschild's equation not any more in the case of a homogeneous sphere, but in the case of a sphere formed out of homogeneous concentric layers of different densities or whose densities varied according to a given law, June 19;[88] finally, Edmond Bauer's work on quantum physics which mentioned the theory of relativity, presented by Brillouin, May 22.[89]

5.4 After a year – from July 1921 to June 1922 – rich in articles on or related to relativity, the intensity of the debate and the volume of contributions diminished; however, the opponents of the new concepts did not lay down their arms. The second half of 1922 saw nine notes to the Académie essentially dealing with the problem of gravitation; for the most part, they came from the opponents of the theory of relativity.

In 1923, the number of articles was still somewhat high: seventeen in the first half, including three on general relativity or unified theory, while the others had mainly doubts about experimental tests; eight in the second half, with two concerning general relativity (G. Darmois) and unified theory (de Donder), and three critical of Michelson's experiment and its interpretation. But four other notes, although dealing with quantum phenomena, referred to special relativity: one, by Edmond Bauer, Pierre Auger, and Francis Perrin, concerned the Compton scattering of x-rays where the impulse and energy of electrons had a relativistic form; three by Louis de Broglie, which are well-known, proposed the theory of matter waves. Henceforth, relativity was so to speak in action more than in question[90].

In 1924, nine articles during the first half had to do mainly with Michelson's experiment, and six, in the second half, did not renew the debate. To continue this examination of the rest of the *Comptes rendus* would be tiresome; we shall merely be content to note the decreasing interest of scientists in the subject (also reflected in other French publications) by concluding with 1925: seven papers in the first half, three in the second; only one dealt with the implications of general relativity, although it did not involve any theoretical considerations since it conceived of various corrections to the experimental determination of the advance in Mercury's perihelion.[91] The others concerned the meaning of Michelson' experiment, the relativistic variation of the mass of particles, simultaneity, and the ether.[92] The opposition was much less virulent. However, the theories of relativity only mattered to a small proportion of the scientists insofar as they were foreign to the greater part of the fields of research, which were oriented toward classical physics (mechanics, optics, spectroscopy). Essentially, it was in research in mathematical physics and in a little-known area of experimental physics, the field of radiation and radioactive phenomena, that the relativistic theories would be integrated.

6. ISSUES IN THE DEBATE OVER RELATIVITY

6.1. Certainties and Uncertainties of Experimentation

6.1.1. Although adherents and opponents saw an obvious connection between the concepts of special and general relativity, it was the question of the physical existence of the ether which was the principal backdrop for considerations of the meaning of experiments. This is not so surprising, given the solid and long-standing tradition among French physicists regarding the problem of showing movement in relation to the ether with the aid of the propagation of light. This tradition went back to Arago's experiments on refraction, which produced negative results; to Fresnel's hypothesis, in 1818, of a partial drag of the ether, which was verified by Fizeau's experiments in 1851; finally, to the work of E. Mascart and A. Potier, who concluded that any research on the effects of movement in relation to the ether which used optical methods was *a priori* doomed to failure since Fresnel's partial drag automatically cancelled out such efforts, which were of the first order. Mascart held that it was perhaps the same for all orders, and for that reason it would be impossible to demonstrate an absolute system of reference with respect to Galilean systems of reference.[93]

6.1.2. It was within this tradition (whose importance for the prehistory of relativity was considerable) that Georges Sagnac worked; his observations still dominated the experimental scene at the beginning of the 1920s. In 1910, Sagnac had conceived of and constructed a revolving interferometer which he used to observe the shifting of fringes; the effect which was obtained was of the first order in v/c. He interpreted the results in terms of an ether wind. He saw evidence against Lorentz' transformations and declared that "real relativity" (i.e., Newtonian relativity of classical mechanics) "accused Lorentz' relativity of abandoning reality for appearance".[94] Extending kinetic and statistical considerations to the field of radiation, he developed a "new mechanics" which was characterized on the one hand by an absolute movement of the source (whose speed was measured by the rotating disk experiment), and on the other by a propagation, at different speed, of the energy of radiation considered in its entirety.

He used these as the bases for a theory of electrodynamics which preserved the traditional ideas of space and time, and refuted special and general relativity at the same time. He also proposed an interpretation of the red shift of double stars in terms of his theory of an ether wind rather than attribute it to an effect of general relativity.[95]

Sagnac's experiments on the rotating disk were used in France as a powerful counter-argument to the theory of relativity.[96] But in 1921, and later in 1937, when Dufour and Prunier published the results of their experiment which was based on the same principle as Sagnac's, Paul Langevin showed that there was no real contradiction between the experimental results of the shifting of interference fringes and the theories of special and general relativity.[97] For Langevin, the experiment could be explained in terms of the conceptual framework of classical mechanics as well as in terms of relativity; the phenomenon was only of the first order in v/c and the shift was only the result of the difference in optical paths. Sagnac's experiment was nothing more than the optical analog of Foucault's pendulum, as Langevin showed,[98] by invoking the point of view of general relativity, which gave an especially simple explanation of the phenomenon.[99]

After Langevin's demonstration, only the most doctrinaire opponents would use Sagnac's experiments against relativity. Skeptical mathematicians such as Picard and Painlevé considered them compatible with Einstein's theory.

6.1.3 For nearly all physicists, the experimental base for relativity was Michelson's experiment, which showed the absence of second order effects, in v^2/c^2, in conformity with Lorentz' formulas.[100] Unlike Einstein, they had a generally inductivist outlook, and to them the special theory of relativity was directly inferred from the negative result of

Michelson and Morley's experiment on the movement of the earth in relation to the ether.[101] Consequently, these results and the significance of their interpretation had great importance in the debates over relativity. Eventually, the possibility of effects which had been previously neglected arose, such as the effect of the earth's mass on the portion of dragged ether in its vicinity,[102] as a way of suggesting that Michelson's results could only have been negative; or else the precision of the measuring instruments was questioned, in order to create a similar sense of suspicion. Despite the great number of repetitions of Michelson's experiment and the great precision which was obtained (to a hundredth of a fringe), the issue was re-examined after Miller announced a positive result, but this was subsequently found to be untrue. Numerous articles on the subject appeared regularly from 1919 to 1925, and even afterwards (notably in the *Comptes rendus* of the Académie des Sciences, but they were of uneven interest and often repetitious).

The other experimental evidence in favor of special relativity was supplied by the measurement of the relativistic variation of the mass of electrons, and its significance was rarely discussed by the opponents of relativity.[103]

6.1.4. The discussions about the three tests of general relativity are of greater interest, even if their intent was to maintain the classical concepts; they posed, in a positive way, the problem of knowing what predictive differences resulted from the general theory of relativity, on one hand, and a theory modified for Newtonian gravitation, on the other. It must be noted that the very first measurements of red shift had been made by Buisson and Fabry (iron spectral lines) as well as by A. Pérot (cyanogen and magnesium); though still imprecise, they agreed with Einstein's predictions.

First of all, the precision of the measured effects was discussed. In the case of the deviation of light rays, interference effects of deviations which could simulate the "Einstein" effect were proposed. The experiment of the 1919 eclipse, repeated by Campbell and Turner during the eclipse of 1922, raised doubts (certainly valid, in light of the imprecision of the measurements); in this regard, at the beginning of 1924, Ernest Esclangon felt that such observations "neither confirm nor deny Einstein's law of deviation", and cited "great difficulties, perhaps insurmountable difficulties, in perceiving the effect of relativity". He called for new and numerous observations as a way of arriving at a decisive result; such a demand was perfectly reasonable, and represented a kind of move toward the possible validation of the theory of relativity or involved the hypothesis of a sufficiently rigorous convergence of tests of its predictions.[104] The red shift caused even more difficulties, given the great imprecision of available measurements and the uncertainty of determining the role of other factors (e.g., the shifting of spectral lines through pressure).

6.1.5. It is interesting to note that the tests which were considered the most decisive for the theory of relativity were those of general relativity, although the criticism of hard-line opponents centered mainly on the results and interpretations of Michelson's experiment. This first observation reveals a misunderstanding over the experimental bases of the special theory. The proponents of the immobile ether and absolute space and time held this base: these were its "little effects of second order", opposed to the weighty evidence of classical mechanics. For relativists following in Langevin's wake, this base was nothing else than the collection of the phenomena of electromagnetism combined into one theory by Maxwell, Lorentz, and Einstein: the special theory of relativity represented the summit and the synthesis of electromagnetic theory. In this regard, if Michelson's experiment constituted a decisive criterion, it was not especially privileged; another test of no less importance would be the increase in the mass of electrons as a function of velocity. Curiously enough, opponents speculated very little about this. Other tests, such as the contraction of lengths and the dilation of duration – or the extremely precise measurement of the constancy of the speed of light – could not be carried out at the time, and could only be *thought experiments* (such as Langevin's traveler).

In this context, the really predictive theory was general relativity, which depended on three main tests whose precision (except perhaps for Mercury's perihelion) was not ironclad. The limited nature of these three tests (although they were spectacular) and their relative imprecision left open the possibility of finding explanations in a reworking of the Newtonian theory of gravitation. For its partisans, the theory of relativity (special and general) presented itself as a synthesis in which the old physics found its place in an approximate way (at low velocities and in Euclidean space), but justified itself through the generality and formal simplicity of its nature. The tests were compatible, and that was sufficient until the advent of other decisive predictions. For opponents or skeptics, the ideas of relativity were too far removed from common sense, with which classical Newtonian mechanics agreed; it would be better to avoid such a break with ordinary modes of thought. Accordingly, it was up to relativity to prove itself, and to submit to the continuous fire of experimental, theoretical, and logical criticism, and not vice versa. The theory was an intrusion: at best, it was regarded with curiosity, as an object with strange properties; at worst, as a monstrosity (thus accounting for the aggressive opposition of a Sagnac, for example). However, the relationship of the two theories in the face of tests under consideration was not symmetrical. The criticisms and demands for precision were addressed to relativity, whereas it was from relativity that the predictions about these effects flowed directly, in a logical and deductive manner. Inversely, it was proposed that Newtonian mechanics be refitted with supplementary models, to make it compatible with observations that would have remained foreign to it without relativity. In the view of Einstein and his supporters, this dissymmetry between logical simplicity on one side and artificiality (through the superimposition of specific hypotheses or models) stood out.[105] Between these two clearly drawn positions were those of people like Emile Picard, skeptical but not hostile, who awaited experimental results; in his pamphlet on the *Annuaire des longitudes* of 1922, he wrote: "In order for physics to travel the trail blazed by the theory of relativity, numerous experiments of a positive nature will have to be carried out in laboratories".[106]

The rarity of tests of the general theory of relativity would account for the weakening of interest in the theory among physicists (and not only in France).

Thus we have to turn to examining arguments (and above all objections) of a theoretical and epistemological nature rather than those of an experimental nature since it is the former which formed the foundation for the debate over relativity.

6.2. *Theoretical and Epistemological Issues*

6.2.1. In the minds of its detractors and supporters, relativity formed a whole with respect to theoretical arguments as well as experimental considerations. Commentators always referred to *the* theory of relativity as a solid and unique body of doctrine. The theory had two parts, the special and the generalized versions.[107] This logical connection, which conformed to Einstein's ideas, was certainly attributable to Langevin's faithful presentation of relativity in his courses and his articles – it is likely that few commentators read Einstein's memoirs directly. The reason for the logical unity of this theoretical construct varied according to the interpretation. Emile Picard declared in a Duhemian way that "the essential part of a theory [is] above all the analytic frame in which it surrounds things"; "what constitutes a theory of relativity is the quadratic form which corresponds to it".[108] For others, it was the organizing principles of the theory (physical principles to relativists, metaphysical principles to its detractors). But for all, it had to do with an *edifice*, an *architecture* marked by the same theoretical logic, the same conceptual innovativeness (in terms of space, time, invariant element), by the same type of connection to experimentation.

6.2.2 The issue of the existence of an ether was at the heart of the argumentation of opponents at the theoretical as well as experimental level. The obvious reason for it was the significance of the notions of Newtonian mechanics and, underlying them, the conviction that there could not be any physical reality without this material support. If the ether was debatable, it was not so much in terms of its existence but of its properties or make-up. For someone like E. Brylinski, to dispute the existence of the ether meant excluding the possibility of what we saw with our own eyes: that there were luminous atoms coming from the stars which constantly streaked through space in all directions.[109] To deny this kind of evidence seemed absurd: it meant reducing physical reality to an abstract and mathematical plane. This was the "fragility of the bases" of Einstein's theory (the expression was often used, even by those who were merely skeptical). Even if the requirement of an absolute ether was not always so clearly expressed – with such "naïveté" in the demand for "realism" – it was, however, the essential constituent of a mental universe and subjacent to most of the questions about the merits of the theory.

The other traditional demand was for an absolute time, i.e., in particular, absolutely separated from space. If skeptics (with regard to relativity) could accept, if they had to, the notion of abandoning the ether and absolute space, it was not the same for time, whose existence apart from any other notion appeared to be an indispensable requirement. More than any other concept, it was time, behavior in time, which determined the *physical* character of phenomena. "The modification in the fundamental notions of humanity relative to space and time" that was introduced by the theory of relativity seemed to many to be contrary to common sense, and to lead to paradoxes.

The innovation that this connection between time and space represented was often brought up when Minkowski's striking phrase was cited, whereby "space independent of time and time independent of space are no more than vain shadows; a sort of union of the two must continue to exist".[110] This union is obviously the element of metrics, the ds^2 which expresses it, and the fact that the measurements of space and time were definitely carried out by light signals. The result was a tendency – among Picard and Painlevé, notably – to exaggerate the measurement of these quantities in relation to the whole of relativistic concepts.

As for the form of ds^2 where time and space are inseparably linked – and it was this that constituted the fundamental unity of the two parts of the theory of relativity – this form drew the strongest criticism in the end. In particular, Painlevé's criticisms were based on it: the alternative Newtonian models which he proposed preserved a separate time, and his reservations in the face of the implications of the theory of relativity in this regard revealed a position rather similar to Lorentz'. Referring to the form ds^2, Painlevé contested whether the conclusions of general relativity concerning a curvature of space or a slowdown in clocks could be deduced.[111] In the case of special relativity, he felt that the result of the form was a paradox, a logical contradiction: the non-reciprocity of formulas for the dilation of time in the case of the coming and going of a train in a station (Painlevé thought to show the impossibility of the conclusion of Langevin's traveler with this example). Painlevé presented this argument in discussions during the spring of 1922 at the Collège de France, and its author agreed with Einstein's refutation of it.[112]

Whatever reservations a Picard or a Painlevé had in abandoning the traditional notions of space and a separate time, their argumentation always sought to be *physical*, even if they never succeeded in eliminating their presuppositions. This was not an *a priori* refusal to accept relativistic concepts which appeared so radical, and moreover, they were disposed to dismiss their reservations if the relativistic arguments seemed well-founded – such was the attitude of Emile Borel, Emile Picard, Painlevé, even if they did not become zealous apostles of the new concepts. At the opposite end of the spectrum was the intransigent and unyielding attitude of those opponents such as Henri Bouasse for whom certain ideas of common sense could not be challenged.[113]

6.2.3. Let us dwell for a moment on arguments intended to show that it was still possible to avoid overturning traditional categories which were so well-entrenched because of the success of Newtonian physics and their ability to account for the surrounding universe. In particular, such was the sense of the questions asked by Painlevé in 1921 and 1922. In the autumn of 1921, the questions were of two kinds.[114] First, Painlevé felt that general relativity reintroduced in disguise a Newtonian system of axes by privileged spatial coordinates and of separate time. Secondly, he thought that there were a number of possible formulas for the element of geodesic ds^2, whose results coincided nearly perfectly with the relativistic predictions, thus raising the possibility of keeping the Newtonian postulate.

Einstein answered Painlevé's arguments in a letter, at the end of 1921,[115] which was enough to convince the mathematician that his initial criticism was groundless. The rest of Painlevé's objection to special relativity and Einstein's response (the example of the train) are familiar. A little later, Painlevé took the offensive one last time by proposing new considerations based on the non-uniqueness of the form of the element ds^2.[116] These considerations had to do with the difference between the predictions of general relativity and those of an adapted Newtonian theory of gravitation, and with how far the modified Newtonian theory could go in order to make its predictions conform to the results of the theory of relativity. The question was epistemologically useful: it was a matter of staying within classical mechanics by at least keeping Euclidean geometry, as Poincaré wished, and seeing to what extent the two theories differed in their axioms and their predictions. It was, in fact, a "semi-Einsteinian theory of gravitation" that Painlevé proposed for consideration. According to Painlevé, if, under such a theory, the postulate of Euclidean geometry were removed and replaced by the idea of the distance between two points being conceived of as a geodesic of space defined by the metric ds^2, the result would be imperceptibly different propositions, undecidable according to existing tests of general relativity.

This eventuality was of great interest to Painlevé: the proposed model coincided – with the exception of the minimal and undetectable predictive differences, which particularly concerned the shift of spectral lines (a phenomenon that was not conclusively observed) – with the results of Einstein's theory but without modifications in the notion of time (conceived independently of space) and without special relativity. Shortly afterwards, it would appear that Painlevé subscribed to Einstein's notions. In 1933, he stated: "Among living scientists, Einstein is perhaps the most capable of unraveling the mysteries of the physical universe in all its manifestations".

6.2.4. The other issues which constantly arose were those of the coherence of the theory in relation to the physical nature of its predictions. Several writers believed that they had uncovered in special relativity violations of causality leading to paradoxes, if not to absurdities. Others, including Painlevé, insisted on what they thought were paradoxes or paralogisms of the notion of relativistic time. They examined the content of formulas[117] and the physical sense of quantities in certain limit conditions.

In reality, these topics were related more to the problem of the truly physical nature of a highly mathematized theory than to the problem of the formal solidity of the edifice of relativity, which was on the whole accepted. At the intuitive stage, this was the sense of Bergsons's objections to relativistic time – objections which, in a different form, were those of Painlevé and others who wished to maintain a time, which, if not absolute, was at least special and separate from spatial transformations – should not physical time be unique.[118] In his objections, Painlevé himself felt that the mere statement of invariance through relativity was a truism, that in no way did it give rise to physical law, and that in this respect, the principle of general relativity was insufficient.

Underlying this interpretation (which Einstein showed to be inaccurate) was evidence for the idea that in order to correspond to a physical thought, a mathematical formalism had to incorporate models and hypotheses which were closer to concrete intuition (as regards physics, these mathematicians were ardent empiricists: they were indeed Poincaré's disciples).

It was the difficulty of firmly grasping the exact nature of the relationship between mathematical propositions – and in this case, with general relativity it was a question of geometry – and properties of the physical universe which explained the skepticism of an Emile Picard, who wondered "if seeking to reduce Physics to Geometry is a sign of progress".[119] An extreme conventionalist version of the use of mathematics to represent physical quantities was proposed by J. Le Roux, who believed that metrics had no physical sense: there was a mathematical form and a physical form of metrics and the theory of relativity wrongly identified them in its definition of the curvature of surfaces. The quadratic form which defined metrics allowed the calculation of invariants, but it was arbitrarily selected and was not in correspondence with physical *space*. According to Le Roux, no one had the right "to make a property of space correspond to these invariants".[120]

In relation to this, Emile Borel attempted a compromise which hardly avoided the flaw of eclecticism.[121] Returning to a distinction which he made in his work *L'Espace et le temps*, he saw, in the hypotheses of physics, the summary of results of numerous experiments, whereas those of geometry were largely arbitrary ("as Poincaré asserted", Borel emphasized). In his reasoning, Le Roux depended upon this arbitrariness, whereas Einstein, to the contrary, "assumes a concrete point of view, and sees geometry as a physical science". According to Borel, this was a case of two different viewpoints which did not intersect and were thus not contradictory: there was a "ds^2 deduced from experience" and an arbitrary geometric ds^2. Borel admitted that the distinction was perhaps artificial: "It is only when our knowledge about the physical structure of the universe, in particular, in the infinitely small and infinitely large, will be more advanced that we will be able to eliminate this alternative". (In other words, more decisive tests than those of general relativity would be required, e.g., at the cosmological level). Thus, according to Borel, there were two points of view, abstract and concrete, on the statements of the theory of relativity; however, it had to be admitted that the first made a solely mathematical schema of this theory, a point of view which Einstein himself excluded.[122]

6.2.5. The question of the relationship of relativistic concepts to metaphysics was a topic of discussion that should be given some attention. In an intellectual context that was somewhat influenced by positivism, for the partisans of the new theory, relativity was obviously the most antimetaphysical of the theories because it rejected attributes which were inaccessible from physical reality, such as the ether, absolute space and time; for others, who were viscerally opposed, it was not a physical theory but a metaphysical doctrine. This goes beyond the scientific debate, which certainly had important philosophical echoes; but the philosophical implications of relativistic conceptions such as they emerged at the same period in the community of philosophers deserve to be the subject of a more detailed analysis.

This aspect was clearly present in the discussions of scientists, but it was simplified and caricatured, and it was often a case of outrageous pronouncements. Bouasse is a good example: the reason for its success "is that Einstein's theory does not return to the framework of physical theories: it is a metaphysical hypothesis" (not to mention incomprehensible); "accepting or not accepting it does not mean admitting or refusing an explanation in the sense of physicists, it means returning to the topic of the theory of knowledge and its limits".[123] Daniel Berthelot, going over the negation of the ancestral idea of time which stood out in a concept as astonishing as the twin paradox (Langevin's traveler), the fourth dimension, the finite or non-finite nature of the universe, wrote this: "Thus all the problems of metaphysics are posed, one after the other"; furthermore, he mixed in the "unsolvable enigmas of life and destiny" such as aging and death with the physical problems of relativity. He concluded that we were witnessing "the birth of a new religion. Its form is simple. Its language is abstract. Its mysteries are profound".[124]

In any event, it is undeniable that in the manner of its reception by the scientific community, relativity was perceived – and rightly so, as a matter of principle – as responsible for questioning certain traditional norms or concepts which had been previously accepted without criticism; in the country of Descartes, at the top of the list was "common sense", often confused with intuitive representations. In this regard, by the questions of an epistemological sort or having to do with the philosophy of knowledge which it was naturally led to ask, the theory of relativity was a unique meeting place for scientists and philosophers at a time when science and philosophy were neither conceived of nor experienced as totally separate.

7. RELATIVITY IN ACTION

7.1. The debates and controversy over the meaning, scope, and eventual difficulties of the new theories were characteristic of the period of assimilation (and rejection) of these theories; but the most enduring feature of their *reception* in the scientific sector was their implementation in research. This occurred in France and more generally in the world (with less of a furor than the debates, but with ultimately more effect) in such a way that the theory of relativity finally received full recognition in the fields of research which concerned it (in its special or general form); these fields would soon encompass a large portion of physics and astrophysics.

In this respect, the difference between France and other European countries or the United States is that in France, until the end of the Second World War, these fields represented only a small fraction of the research activity in laboratories and in a university system still influenced by a traditional mindset, and which were set up badly for the new directions and methods of research. For the moment, let us examine briefly some of the directions of research in physics in which the use or development of the ideas of the theory of relativity proved particularly fruitful.

Some had to do with mathematical physics; others were concerned with theoretical or experimental physics. Since it would be impossible to go into greater detail and since the period under consideration is much more important, we shall focus on the most outstanding cases.

7.2. The debates over relativity showed how mathematicians, especially those interested in mathematical physics, were more prepared than most physicists to accept relativistic concepts. Lewis Pyenson's remark, whereby "general relativity was most favorably received where physics was taught as an emerging field rather than where mathematical physics was emphasized", was certainly applicable for Leyden and Göttingen,[125] but not for France; this is the case because the representatives of mathematical physics in France fulfilled the role which was taken up by theoretical physicists elsewhere. They fulfilled it in a different manner than the latter, but in spite of everything, their idea of mathematical physics was closer to theoretical physics than axiomatic formulations. They had been taught by Poincaré, and as regards the theory of relativity, by Poincaré as seen through the ideas of or revisited by Langevin. Furthermore, they assumed their roles later than their Dutch or German colleagues; except for Langevin, who was a theoretical physicist, those among them who took on problems presented by general relativity started only at the beginning of the 1920s. But they were interested in the formal rather than physical aspects of the theory; in this sense, they shared the interest of German mathematicians, although from a different perspective, not axiomatic as with Hilbert. The examples of Hadamard and Brillouin, as well as the important contributions of Elie Cartan, give a good idea of the kind of problems that French mathematicians were interested in.

It was at the time of Einstein's lectures at the Collège de France in the spring of 1922 that Jacques Hadamard brought up the problem of singularity in Schwarzchild's equation: it had to do with a formal difficulty with physical implications, since one of the terms of which the invariant element ds^2 was the sum could become infinite, when the distance r off a point gravitating around a mass located at the origin became equal to a certain length a which was present in the equation.[126] Hadamard asked what the physical sense of the formula could be, if it could become infinite. According to an account, there was an initial moment of confusion on the part of Einstein, who christened the problem "Hadamard's catastrophe". For all the physicists who were present, if the term could become infinite through the cancellation of $r -$ a, "it would be" – according to Einstein – "a terrible tragedy for the theory; and it is very difficult to say *a priori* what could happen physically, for then the formula ceases to be applicable".

The solution was sought by numerous participants – especially Einstein, who, in the following session, brought the result of a calculation showing how the formulas of pressure and time prevented the occurrence of the "Hadamard catastrophe".[126] The logic was deemed satisfactory, notably by Hadamard, and amounted to preventing the mathematical difficulty from encountering a physical occurrence: the formal coherence of the theory was assured by estimating that the sphere of Schwarzschild was impenetrable, that the case in question did not exist in nature. The problem of what would later become the "gravitational collapse" was next taken up by de Donder and Brillouin, who proposed a more systematic study (Brillouin, in particular, proposed "to undertake a general study of the singular points characteristic of non-Euclidean spaces which were defined by equations with partial derivatives").[127]

The thing that interested mathematical physicists was preserving the coherence of the theory from the formal point of view and for its possible application in the case of physical situations. The remarkable architecture of the edifice, especially at the level of general relativity, and the unprecedented formalization which characterized it, were tempting, once the eventual conceptual difficulties attested to by the attitudes of Painlevé and Picard were overcome. In this respect, Elie Cartan's work is an exemplary case of a major contribution in mathematical physics, strictly speaking, which was directly inspired by general relativity.

7.3. It was as early as 1921 that Elie Cartan thought about the problem of a generalization of the spaces of Riemann, Weyl, and Eddington, by adopting a different perspective from theirs, that of the structure of continuous groups on which he had been working for a long time and in a pioneering way, extending and amplifying Lie's theory of groups. In fact, it was the intersection with Einstein's theory of gravitation that led him to develop a new concept and to sift out the truly geometric notions capable of accounting for Einstein's tensor. This new concept was "space with Euclidean connection" from which the notion of "distant parallelism"[128] (of a Riemannian space) was derived. Einstein would use this idea in 1929, and his explicitation was the subject of an ongoing correspondence and a close collaboration between him and Cartan, from 1929 to 1932; moreover, Einstein got Cartan to publish the observations in their correspondence and conversations which Cartan had developed for the former's use.[129] He would finally abandon this direction of research when he saw that he could not succeed in this way by writing equations for the motion of fields: "It seems", he wrote to Cartan, "that this structure has nothing to do with the true character of space".[130]

At the Institut Henri Poincaré, additional research in mathematical physics on the relativistic theory of gravitation and unified field theory was carried out, particularly after 1930; for the most part, it was a question of work of a formal nature. This direction of research was of particular interest to mathematicians, at a time when the theory of general relativity was regarded by physicists – not only in France – as a curiosity with remote experimental interest.

7.4. As we have seen in our study of the *Comptes rendus*, several researchers took up the problem of taking relativistic kinematics into account in the study of the quantum orbits of atomic electrons, following up on the work of Arnold Sommerfeld. But the most obvious link between special relativity and the development of quantum concepts was to be found in Louis de Broglie's earliest work, which saw the birth of wave mechanics with the generalization to the elements of matter of the wave-particle duality considered – since Einstein – for electromagnetic radiation.[131] The inertia of energy and the relativistic relation of frequencies such as they were evaluated in a system at rest or in motion relative to the material body in question were used from the time of the first memoir of 1923, which confirmed the existence of a periodic phenomenon internal to the body, a phenomenon which always stayed in phase synchronization to the course of the motion. "The demonstration of this important result", wrote the author, "rests solely on the principle of special relativity and the exactitude of the relation of the quanta for the fixed observer as well as for the observer who is in motion".[132] In the third memoir and in the thesis of 1924, the principles of Fermat and Maupertuis were reconciled through the relativistic form of impulse and energy,[133] thus fully illuminating "the fundamental link which unites the two great principles of Geometrical Optics and Dynamics". Louis de Broglie himself insisted on numerous occasions on "the close and profound bond which exists between Einstein's two great discoveries: Relativity and light quanta".[134] He and his school, fundamentally interested in a spatiotemporal representation of the phenomena of quantum physics, would always place special relativity at the head of their concerns.

7.5. Langevin was the first to have observed early on that the inertia of energy had important consequences in chemistry and atomic physics since it could explain atomic packing loss of chemical elements by the binding energy of the constituent elements.[135] In 1920, this observation was only a program for future research. Referring to it in 1922 in his book on the theory of relativity, Edmond Bauer wrote: "Perhaps one day it will be possible to calculate all the atomic weights of elements from the atomic weight of hydrogen, a prime element, if we know the amount of energy accompanying their formation".[136] In fact, it was in this direction that the theory of relativity, in its special form, was applied in the most widespread fashion in experimental research: it was not so much a question of verifying it as of putting it into practice, since without its concepts, atomic physics and radioactivity would not have been able to analyze as well as they did the make-up of the nucleus of the atom, thus opening up the new field – since the 1930s – of nuclear physics. It was mainly the laboratories of Marie Curie, Jean Perrin, Maurice de Broglie, and Paul Langevin which would represent the move in this direction of research in France; their laboratories, and the laboratories of their students and their successors would spawn the work of a new generation of researchers during the 1930s on artificial radioactivity (Frédéric Joliot and Irène Joliot-Curie), nuclear physics, and the physics of cosmic rays.

As for the other experimental aspects of the special and general theory of relativity (above all, the latter), they would be the subject of rare research; this is natural, given the then limited character of its predictive proofs, before the revival of astrophysics. The general theory of relativity was eclipsed (so to speak), and quickly left center stage,[137] whereas special relativity was integrated into the theoretical framework of atomic physics. But even it became secondary with respect to the new theory of quantum mechanics and its implications in the theory of the composition of matter. Quantum mechanics and atomic and nuclear physics have dethroned relativity in modern times, but this is certainly not specific to the French context.

NOTES

The bibliographic references follow the notes, which list the author's name, the year of publication, and when appropriate, a lower-case letter denoting the order of occurrence (e.g., Langevin, 1913 (b)). In cases of ambiguity, the italicized date is correct.

[1] For a detailed analysis of the French context, see Pestre, 1984.

[2] On this state of affairs regarding the scientific traditions of different countries see Pyenson, 1975. Unlike de Broglie and others (e.g., de Broglie, 1940), Poincaré did not make any distinction. On the chairs of theoretical physics in French universities, see Pestre, 1984.

[3] In 1932, Langevin emphasized in this connection the lingering influence of the "spirit of Regnault" over French physics (Langevin, 1932 (a)).

[4] This delay was especially evident in courses and textbooks. On the reception of Maxwell's electromagnetism in France, see Abrantes, 1985. In fact, French physicists took Maxwell's theory into consideration rather quickly, insofar as it was important in optics. Moreover, it was under this aspect that Poincaré introduced it in 1888 (Poincaré, 1888).

[5] See Pestre, 1984, for the analysis of textbooks during the period, and their empirico-historical presentation of physics, to the detriment of fundamental theoretical insights; Maxwell's theory is hardly mentioned at the end: with all the more reason, the theory of relativity is ignored.

[6] A good account of the dominant physical as well as epistemological concepts at the time when the theory of relativity was making its breakthrough is in the form of the contributions – diverse but not altogether divergent – of Paul Painlevé, Emile Picard, Henri Bouasse, in the collective work, *De la Méthode dans les sciences*, Paris, Alcan, 1909.

[7] See his presentation of the theory of relativity in Einstein *et al.*, 1922.

[8] See, for example, Duhem, 1902, p. 206; 1906, p. 298; 1915. Cf. Paty, 1986.

[9] Langevin recalled the week that he spent with Poincaré when they were returning from the Congress of Saint Louis in 1904 in Langevin, 1913 (c). He wrote: "I had the opportunity to see the passionate interest with which Henri Poincaré followed all the

stages of the revolution taking place in our most fundamental beliefs. With some misgivings, he saw the old edifice of Newtonian dynamics being shaken by devices which he himself had wrought; it was the same structure that he had recently crowned with his admirable work on the problem of three bodies and the equilibrium form of celestial bodies. But if his enthusiasm was more reserved than mine, he was, like the rest of us, greatly taken by the spirit of entering into an entirely new world".

[10] Poincaré, 1895, 1900 (a) and (b).

[11] See, for example, Poincaré, 1902.

[12] Poincaré, 1902.

[13] Poincaré, 1898.

[14] Poincaré, 1904. See also Poincaré, 1908, pp. 249–250.

[15] Poincaré, 1902 (a).

[16] Poincaré, 1902 (a), ed. 1968, p. 215.

[17] Poincaré, 1905 (b).

[18] Poincaré, 1905 (b).

[19] Poincaré, 1904 (a).

[20] Poincaré, 1905 (b).

[21] The credit for which he also shared with Einstein.

[22] Poincaré, 1908, p. 274. See also Poincaré, 1905 (b).

[23] Zahar, 1983.

[24] See Holton, 1964, 1981.

[25] See Einstein, 1955, cited by Kahan, 1959, p. 163.

[26] See Poincaré, 1912 (b), which contains the only published reference by Poincaré to Einstein's work, on his theory of the quanta of radiation energy. On the "silence" or "muteness" of Poincaré toward Einstein, see Holton, 1964; Goldberg, 1970.

[27] Poincaré, 1911.

[28] He wrote soon afterwards: "Mr. Einstein is one of the most original thinkers I have met; despite his youth, he has already assumed a very honorable position among the outstanding scientists of his time". (Poincaré, 1911, *ibid.*)

[29] Einstein, Letter to Zangger, November 16, 1911, quoted by Kahan, 1959, and by Miller, 1981, p. 255.

[30] Poincaré, 1902, pp. 177–180.

[31] Einstein, 1955, *op. cit.*

[32] Holton, 1964, and Goldberg, 1967, refer to Poincaré's "gradualism" to describe his concern for reconciling and preserving the framework and the old theory, which required a "suppleness" and flexibility on the part of the theory in order for it to be adapted without breaking apart.

[33] Paul Langevin (1872–1946), born into a family of artisans, was a student at the Ecole de Physique et Chimie in Paris and the Ecole Normale Supérieure. In 1902, he defended a thesis on his "Recherches sur les gaz ionisés". He was a temporary professor in 1902, then a titled professor in 1909, at the Collège de France; he succeeded Pierre Curie as a professor at the Ecole de Physique et de Chimie. His work on magnetism – he formulated, on the basis of electron theory, the theory of dia- and para-magnetism – won him international renown. From 1911, he took part in the Solvay Physics Congress, whose presidency he assumed in 1928, upon the death of Lorentz. From the time of the Solvay Congress of 1911 (where he presented his theory of magnetism), he maintained a close friendship with Einstein, whose ideas of relativity he had accepted since 1906.

[34] This influence was deeper than it was widespread since the courses at the Collège de France did not reach (or barely reached) faculty and students, and because Langevin did not bother to publish his notes, whose content has yet to be edited. It was in these courses at the Collège de France that he developed his original ideas on relativity. Thus, the theory of relativity entered France through university teaching only marginally, but at a high level. From 1906, Langevin's courses on relativity were taken by Edmond Bauer, Emile Borel, Jean Becquerel, Jacques Hadamard, Elie Cartan. At the beginning of the 1920s, Louis de Broglie and Alfred Kastler were among those who were attracted to Langevin's courses.

[35] De Broglie, 1947. Langevin was, in fact, an experimentalist as well as a theoretician.

[36] Einstein, Letter to Dr. A. Johnson, September 7, 1940, in Langevin, 1972. After being jailed, Langevin, during the German Occupation, was put under house arrest in Troyes for most of the war. Einstein increased his efforts on Langevin's behalf to obtain for him an invitation to the United States. See Langevin, 1972.

[37] Letter by Einstein on the death of Langevin, in Langevin, 1972. See Einstein 1950, pp. 231–232.

[38] Langevin, 1904 (b).

[39] Langevin, 1904 (b).

[40] Langevin's work belongs to a perspective of criticism of the fundamental role of mechanics (cf. Langevin, 1904 (a), p. 440, 441).

[41] Langevin, 1904 (b), p. 16. A model of the electron must indeed be specified, but it has little influence over the first property, inertia (p. 21), since inertia results from the electrical charge (cf. *supra*).

[42] Langevin, 1904 (b), pp. 32–33.

[43] Langevin, 1904 (b), p. 33.

[44] It is here that Langevin proposed his model of the deformable electron at constant volume (*ibid.*, p. 37). See also Langevin, 1905 (a).

[45] Langevin, 1904 (b), p. 69.

[46] Lorentz, 1904.

[47] This experiment was carried out in 1903. For a description, see Miller, 1981, pp. 68–69. Curiously enough, Miller does not mention Langevin's explanation anywhere. In a general way, this author seems to reduce Langevin's contributions to relativity to the model of the deformable electron of Bücherer-Langevin. Tonnelat (1971) is equally silent about Langevin's work and his contemporaneous contributions to Einstein's seminal work.

[48] Langevin, 1905 (a). In this article, Langevin takes up results obtained two years before which were presented in his course at the Collège de France in 1903–1904. These results were published in part, independently by Liénart in 1898 and Schwarzschild in 1903 (Langevin mentioned them).

[49] Langevin, 1905 (a) in Langevin, 1950, p. 313. (All references are to this edition).

[50] *Ibid.*, p. 314.

[51] Langevin, 1905 (a), p. 328.

[52] Bauer, 1956, quoted in Langevin, 1971, p. 58 (and in Langevin, 1972, pp. 5–6).

[53] See Paty, 1982.

[54] Langevin, 1913 (b). (We are following the edition of 1923).

[55] Langevin, 1913 (b), p. 391. The formula is obviously $\Delta m_0 = \Delta E_0/c^2$.

[56] Langevin, 1913 (b), p. 399.

[57] In addition, he raised the idea of the weight of light by mentioning Einstein's ongoing research, and concluded with the general nature of the relation $E = mc^2$, for bodies at rest or in motion.

[58] *Ibid.*, p. 402.

[59] Notably according to Bergson (Bergson, 1922), who cited, moreover, all that thinkers interested in the theory of relativity owed to Langevin's work.

[60] Langevin, 1911 (b). On the session of 1922, see Paty, 1980.

[61] Langevin's epistemology was formulated at the same time as his scientific program. In my opinion, it revolves around two considerations: it is theory which gives meaning to concepts; human thought and the theories which it produces are evolutionary, distinguished by a "progressive adaptation of thought to facts".

[62] To go further than the approximate validity of mechanics, in the case of lower velocities.

[63] Langevin, 1911 (a), p. 274; see also Langevin, 1911 (b), pp. 300–304.

[64] Langevin, 1911 (a), pp. 286–287.

[65] Langevin, 1911 (a), pp. 287–288; Langevin, 1911 (b), p. 339.

[66] Langevin, 1911 (a), p. 291.

[67] *Ibid.*, p. 294.

[68] Langevin, document addressed to the general assembly of the Collège de France to invite Einstein, cited by Luce Langevin, 1972, p. 7. Einstein presented his theory of gravitation to the Scientific Congress of Vienna in 1913. That same year, in a letter to Michele Besso, Einstein noted the interest of Lorentz and Langevin in his theories (Letter 9, end of 1913, in Einstein Besso, 1972).

[69] In 1919–1920: course on relativity; 1920–1921: course of gravitation. In 1919, Langevin produced a minor work of synthesis on relativity in which the result of the observation of the eclipse played a part (Langevin, 1919). See also Langevin, 1922 (b) and the presentation of Langevin in Einstein *et al.*, 1922.

[70] Langevin, 1911 (a), p. 290.

[71] Langevin, 1919. Inside the projectile, since the gravitational field is cancelled by the free-fall, the universe is Euclidean and light is propagated in a straight line.

[72] On Sagnac's experiment and its interpretation by Langevin, see below, and Langevin, 1921.

[73] Langevin, 1922, p. iv. See also Bauer, 1922 (b), p. 35, which meshed well with Langevin's concepts.

[74] Langevin, 1919 (ed. 1922), pp. 5–6.

[75] *Ibid.*, p. 59.

[76] See, for example, Langevin, 1919, p. 6.

[77] Painlevé, 1921 (a) and (b); Picard, 1919. The impact of these communications on the public is attested to by the press reports during the period; see also Nordmann, 1921.

[78] On the circumstances of Einstein's trip to Paris from March 30 to April 8, 1922, see Frank, 1947; Langevin, 1979; Biezunski, 1982. On the subject of the impact of the trip on public opinion and the accounts in the major newspapers of the theory of relativity, see Biezunski, 1981, 1982, 1987. Einstein was invited to the April 3 session of the Académie, but dropped the idea in view of the hostility of the majority of academy members, who had decided to boycott the individual who represented, in their minds, "German science".

[79] *CRAS* 174, 1922, 82.

[80] Berthelot, 1922.

[81] Sagnac, 1922 (a).

[82] The note was published with the *Comptes rendus* of the January 9 session: *CRAS* 174, 1922, 106–109. It was a continuation of articles in the *CRAS* of 1921. In 1934, when he was the director of honorary studies of the Ecole Polytechnique, E. Carvalho published a 56-page pamphlet entitled *La Théorie d'Einstein démentie par l'expérience* (Carvalho, 1934), in which he harshly criticized the relativistic interpretation of Michelson's experiment, citing the results of Esclangon and Miller.

[83] Brylinski, 1922 (published January 16).

[84] Painlevé, 1922 (a).

[85] Chazy, 1922; Trousset, 1922; Painlevé, 1922 (b); Ferrier, 1922.

[86] *CRAS* 174, 1922, 1416–1418.

[87] *CRAS* 174, 1922, 1228–1229. The title of the note is "Champ électromagnétique compatible avec le champ gravifique correspondant".

[88] Brillouin, 1922.

[89] Bauer, 1922 (a).

[90] Bauer, Auger, Perrin, 1923; de Broglie, 1923 (a), (b), (c).

[91] *CRAS* 181, 1925.

[92] *CRAS*, vols, 180 and 181, 1925.

[93] Mascart, 1872, 1874; 1893, chap. XV, p. 38; Potier, 1874. See Lorentz, 1895, where these results are recalled.

[94] Sagnac, 1920 (a).

[95] Sagnac, 1919 (a), (b), (c), (d); 1920 (a), (b); 1923.

[96] With the intention of weakening the theory of relativity, A. Dufour and F. Prunier repeated the experiment of the rotating disk in 1937 (Dufour and Prunier, 1937).

[97] Langevin, 1921, 1937.

[98] Langevin, 1921. Wolfgang Pauli made the same remark in his famous article on relativity which appeared in the same year (Pauli, 1921).

[99] Shortly before Langevin's demonstration, Emile Picard, writing about Sagnac's experiment, said that "it would be interesting to explain [it] . . . through the general theory of relativity, by going as far as numerical concordances" (Picard, 1921). Thus, Langevin fulfilled this wish.

[100] On the contrary, Einstein only mentioned Michelson's experiment indirectly in his article of 1905 "On the Electrodynamics of Bodies in Motion", and favored essentially theoretical arguments.

[101] P. Langevin, E. Bauer, A. Metz in particular showed very clearly in their presentation, however, that the logic of the theory of relativity was otherwise. For example, Jean Becquerel, in his preface to André Metz' book (Metz, 1923), insisted on the fact (taught well by Langevin) that the solid base of the theory was made up of the body of knowledge in electrodynamics, as well as the discord between the concepts of space and time of classical mechanics, on one hand, and electromagnetism, on the other, and that "it is an error to believe that Michelson's experiment is the true base of the theory: it is only a verification that came before date" (J. Becquerel, preface, in Metz, 1923, p. xiv).

[102] For example, Brylinski, 1922. This eventuality was taken up by Nordmann, 1921, who was somewhat partial to relativistic ideas, but tempered them according to various criticisms (notably Painlevé's).

[103] See, for example, Le Roux, 1925.

[104] Esclangon, 1924.

[105] This is what Langevin very clearly expressed (Langevin, 1919, 1922 (b)). See also Metz, 1923. Although more reserved, Emile Picard made the same remark (Picard, 1922, p. B22).

[106] Picard, 1922, p. B 27.

[107] See, for example, Picard, 1921.

[108] Having obtained this quadratic form, Picard went on to say, "one can disregard the manner by which one was led to it" (Picard, 1921). Picard wished to ignore the principle of the theory; the latter was for him a mathematical form, without any explanatory weight.

[109] Brylinski, 1922. However, Brylinski admitted the possibility of a discontinuous ether.

[110] Cited by Picard, 1922, p. B 14. The text is Minkowski's, dating from 1908.

[111] Painlevé, 1921 (a).

[112] See the account given by Nordman, 1922 (a): after the discussion at the April 3 session, Einstein and Painlevé entrusted Langevin with presenting the solution to the problem at the beginning of the next course. The problem of Painlevé's train went beyond the framework of special relativity since there were three, not two systems of inertia (the train called for two, coming and going).

[113] Bouasse, 1923, pp. 17–19.

[114] Painlevé, 1921 (a) and (b).

[115] This letter is mentioned by Eisenstaedt, 1981, p. 174.

[116] Painlevé, 1922.

[117] Le Roux, 1922, saw in the form of ds^2 "mathematical absurdities", "singular consequences" of "the fundamental hypothesis of Einstein", which were far from being experimentally verifiable.

[118] See, for example, Einstein, 1922; Bergson, 1922. On the discussions of Einstein with French philosophers (which were an extension of discussions with physicists and mathematicians) see Paty, 1980.

[119] Picard, 1921, 1922.

[120] Le Roux, 1922.

[121] Borel, 1922.

[122] However, Einstein and Langevin, speaking to the Société Française de Philosophie, had just insisted on the fact that the theory of relativity was a physical theory (Einstein, 1922).

[123] Bouasse, 1923, pp. 8, 21–22.

[124] Berthelot, 1922.

[125] Pyenson, 1975.

[126] The only account of Hadamard's remark and Einstein's discussion was supplied by Nordmann, 1922, pp. 154–158.

[127] Brillouin, 1922 (b) and 1923 (citation in 1922 (a)); de Donder, 1922. For a description of these contributions, see Eisenstaedt, 1981, pp. 187–189.

[128] In German: *Fernparallelism*; in French: *parallélisme absolu*.

[129] See Cartan Einstein, 1979. On the work of Cartan, see in part Cartan, 1931 (a) and (c), 1930.

[130] Einstein, Letter to Cartan, May 21, 1932. At that time, Einstein was working with W. Mayer along different lines.

[131] De Broglie, 1923 (a), (b), (c) and 1924.

[132] De Broglie, 1923 (a).

[133] De Broglie, 1923, 1924, 1939 (citation in 1923 (c)). De Broglie insisted on several occasions on the importance of his "meditation" on special relativity in the formulation of his ideas.

[134] De Broglie, 1955. This link is particularly noticeable in the fact that luminous quantum has an impulse $p = h / \lambda$, which is necessarily relativistic. De Broglie wrote: "The photon could only find its place in a relativistic physics". See also de Broglie, 1939.

[135] Langevin, 1913 (b).

[136] Bauer, 1922, p. 56.

[137] Eisenstaedt, 1986.

REFERENCES

Abrantes, Paulo Cesar Coelho. *La Réception en France des théories de Maxwell concernant l'électricité et le magnétisme*. Doctoral diss., University of Paris, 1985 (mimeo).

Bauer, Edmond. 'Sur le champ électromagnétique des trajectoires stationnaires de Bohr,' *Comptes Rendus de l'Académie des Sciences* [hereafter, *CRAS*], **174** (1922), 1335–1338 (a).

Bauer, Edmond. *La théorie de la relativité*. Paris, Eyrolles, 1922 (b).

Bauer, Edmond. Pierre Auger and Francis Perrin. 'Sur la théorie de la diffusion des rayons X,' *CRAS*, **177** (1923), 1211–1212.

Becquerel, Jean. *Le Principe de relativité et la théorie de la gravitation*. Paris, Gauthier-Villars, 1922.

Bergson, Henri. *Durée et simultanéité. A propos de la théorie d'Einstein*. Paris, Alcan, 1922; 2nd ed., 1923; 3rd ed., 1925, 4th ed., 1929.

Berthelot, Daniel. *La physique et la métaphysique des théories d'Einstein*. Paris, Payot, 1922.

Biezunski, Michel. *La diffusion de la théorie de la relativité en France*. Doctoral diss., University of Paris, 1981 (mimeo).

Biezunski, Michel. 'Einstein à Paris,' *La Recherche*, 13 (April 1982), 502–510.

Biezunski, Michel. 'Einstein's Reception in Paris in 1922', in Thomas F. Glick, ed., *The Comparative Reception of Relativity*. Dordrecht, D. Reidel, 1987 (this volume).

Borel, Emile. *L'Espace et le temps*. Paris, Alcan, 1922.

Borel, Emile. 'Hypothèses physiques et hypothèses géométriques', *CRAS*, **174** (1922), 1050.

Bouasse, Henri. *La question préalable contre la théorie d'Einstein*. Paris, Blanchard, 1923.

Brillouin, Marcel. 'Champ isotrope. Sphère fluide hétérogène,' *CRAS*, **174** (1922), 1585–1589 (a).

Brillouin, Marcel. 'Gravitation einsteinienne. Statique. Points singuliers. Le point matériel. Remarques diverses,' *CRAS*, **175** (1922), 1009–1012 (b).

Brillouin, Marcel. 'Les points singuliers de l'univers d'Einstein' *Journal de Physique*, 6th ser., **4** (1923), 43–46.

Broglie, Louis de. 'Onde et quanta,' *CRAS*, **177** (1923), 507–510 (a).

Broglie, Louis de. 'Quanta de lumière, diffraction et interferences,' *CRAS*, **177** (1923), 548–550 (b).

Broglie, Louis de. 'Les quanta, la théorie cinétique des gaz et le principe de Fermat,' *CRAS*, **177** (1923), 530–532 (c).

Broglie, Louis de. *Recherches sur la théorie des quanta*. Doctoral diss., University of Paris,

1924; *Annales de Physique*, 10th ser., **3** (1925), 22–128; reed., Paris, Masson, 1963.

Broglie, Louis de. 'Les rapports entre la théorie des quanta et la relativité,' in *Les Nouvelles théories de la physique* (meeting of the Institute of Intellectual Cooperation, Warsaw, May 30–June 3, 1938), Paris, 1939, pp. 49–66.

Broglie, Louis de. 'Allocution pour le jubilé de M. Emile Borel, prononcée à la Sorbonne le 14 janvier 1940' (1940), in de Broglie, 1951a, pp. 290–297.

Broglie, Louis de. 'La vie et l'oeuvre de Paul Langevin' (1947), in de Broglie, 1951a, pp. 233–269.

Broglie, Louis de. *Savants et découvertes*. Paris, Albin Michel, 1951 (a).

Broglie, Louis de. 'Henri Poincaré et les théories de la physique,' in de Broglie, 1951 (a), pp. 45–65 (b).

Broglie, Louis de. 'Le dualisme des onde et des corpuscules dans l'oeuvres d'Albert Einstein' (discourse at the Academy of Sciences, December 5, 1955), in de Broglie, *Nouvelles perspectives en microphysique*, Paris, Albin Michel, 1956.

Brylinski, E. 'Sur l'interprétation de l'expérience de Michelson,' *CRAS*, **174** (1922), 153–154.

Cartan, Elie. 'Notice historique sur la notion de parallélisme absolu,' *Mathematische Annalen*, **102** (1930), 698–706; in Cartan, 1952–1955, vol. 3, pt. 2, 1121–1129.

Cartan, Elie. 'Le parallélisme absolu et la théorie unitaire du champ,' *Revue de Métaphysique et de Morale*, **38** (1931), 13–28; in Cartan, 1952–1955, vol. 3, pt. 2, 1167–1185 (a); also in Langevin, 1932c, fasc. V).

Cartan, Elie. 'Notice sur les travaux scientifiques' (1931), in Cartan 1952–1955, vol. 1, pt. 1, 1–98 (b).

Cartan, Elie. 'Sur la théorie des systèmes en involution et ses applications à la relativité,' *Bulletin de la Société mathématique de France*, **659** (1931), 88–118; in Cartan, 1952–1955, vol. 2, pt. 2, 1199–1229 (c).

Cartan, Elie. *Ouevres Complètes*. 3 vols. in 5. Paris, Gauthier-Villars, 1952–1955.

Cartan, Elie and Albert Einstein. *Lettres sur le parallélisme absolu, 1929–1932*. Robert Debever, ed. English trans. by Jules Leroy and Jim Ritter. Brussels, Académie Royale de Belgique and Princeton University Press, 1979.

Carvallo, E. *La Théorie d'Einstein démentie par l'expérience*. Paris, Chiron, 1934.

Chazy, Jean. 'Sur les vérifications astronomiques des théories d'Einstein,' *CRAS*, **174** (1922), 1157–1160.

De Donder, Théophile. *Théorie du champ électromagnétique de Maxwell-Lorentz et du champ gravifique d'Einstein*. Paris, Gauthier-Villars, 1920.

De Donder, Théophile. *La gravitation einsteinienne. Premiers Compléments*. Paris, Gauthier-Villars, 1922.

Dufour, A. and F. Prunier. 'Sur l'observation du phénomène de Sagnac par un observateur non entraîné,' *CRAS*, **204** (1937), 1925–1927.

Duhem, Pierre. *Le Mixte et la combinaison chimique. Essai sur l'évolution d'une idée*. Paris, Naud, 1902.

Duhem, Pierre. *La Théorie physique. Son objet, sa structure* (1906). Reed., Paris, Vrin, 1981.

Duhem, Pierre. 'Quelques réflexions sur la science allemande,' *Revue des Deux Mondes*, Feb. 1915, 657–686.

Einstein, Albert. 'Auf die Riemann-metrik und den Fernparallelismus gegründete einheitliche Feldtheorie,' *Mathematische Annalen*, **102** (1930), 685–697 (a).

Einstein, Albert. 'Théorie unitaire du champ Physique,' *Annales de l'Institut Henri Poincaré*, **1** (1930), 1–24 (b).

Einstein, Albert. 'Reply to Criticism,' in Schilpp, 1949, pp. 663–693.

Einstein, Albert. *Out of My Later Years*. Westport, Ct., Greenwood Press, 1950.

Einstein, Albert. In *Technische Rundschau* (Bern), no. 20, May 6, 1955.

Einstein, Albert. *Lettres à Maurice Solovine*. Paris, Gauthier-Villars, 1956.

Einstein, Albert *et al*. 'La Théorie de la relativité,' Séance du 6 avril 1922, *Bulletin de la Société Française de philosophie*, **17** (1922), 91–113; re-ed. in *La Pensée*, no. 210, Feb. 1980, 12–29.

Einstein, Albert and Michele Besso. *Correspondance* 1903–1955. Pierre Speziali ed. and tr. Paris, Hermann, 1972; reed., 1979.

Einstein, Albert, Edwige and Max Born. *Briefwechsel* 1916–1955 (1969); English version as *The Born-Einstein Letters*, Irene Born, tr. New York, Walker, 1971.

Einstein, Albert and W. Mayer. 'Einheitliche Theorie von Gravitation und Elektrizität,' *Sitzung berichte Preuss. Akademie Wissenschaft*, 1931, 541–557; *ibid.*, Zweite Abhandlung, 1932, 130–137.

Eisenstaedt, Jean. 'Histoire et singularités de la solution de Schwarzschild (1915–1923),' *Archive for History of Exact Sciences*, **27** (1982), 157–182.

Eisenstaedt, Jean. 'La relativité générale à l'étiage: 1925–1955,' *Archive for History of Exact Sciences*, **35** (1986), 115–185.

Esclangon, Ernest. 'Sur la déviation einsteinienne des rayons lumineux par le Soleil,' *CRAS*, **174** (1922), 1404–1007.

Ferrier, M. 'Sur les déviations des rayons lumineux passant au voisinage d'un astre,' *CRAS*, **174** (1922), 1404–1407.

Goldberg, Stanley. 'Henri Poincaré and Einstein's Theory of Relativity,' *American Journal of Physics*. **35** (1967), 934–944.

Goldberg, Stanley. 'Poincaré's Silence and Einstein's Relativity,' *The British Journal for the History of Science*, **5** (1970), 73–84.

Holton, Gerald. 'On the Thematic Analysis of Science: The Case of Poincaré and Relativity,' in *Mélanges Alexandre Koyré, L'aventure de l'esprit*, 2 vols. (Paris, Hermann, 1964), II, 257–268; reprinted in *Thematic Origins of Modern Science* (Cambridge, Harvard University Press, 1973), pp. 185–195; or, in French, see the chapter 'Aux origines de la théorie de la relativité restreinte,' in his *L'Imagination scientifique* (Paris, Gallimard, 1981), pp. 130–184.

Kahan, Theo. 'Sur les origines de la théorie de la relativité restreinte,' *Revue d'Histoire des Sciences*, **12** (1959), 159–163.

Langevin, Jean. 'Note à propos du séjour d'Einstein en France organisé par P. Langevin au printemps de 1922,' *Cahiers Fundamenta Scientiae* (Strasbourg), no. 93 (1979), 9–31.

Langevin, Luce. 'Paul Langevin et Albert Einstein d'après une correspondance et des documents inédits,' *La Pensée*, no. 161 (Feb. 1972), 3–40.

Langevin, Paul. 'L'esprit de l'enseignement scientifique,' in the multi-authored volume *L'Enseignement des sciences mathématiques et des sciences physiques* (Paris, Imprimerie Nationale, 1904); also in Langevin, 1923, pp. 424–453 [lecture at the Musée Pedagogique, February 18, 1904] (a).

Langevin, Paul. 'La physique des électrons,' *Revue Générale des Sciences*, March 15, 1905, and Langevin 1923, pp. 1–69 (b) [communication presented to the International Congress of Arts and Sciences, St. Louis, Mo., September 23, 1904].

Langevin, Paul. 'Sur l'origine des radiations et l'inertie électromagnétique,' *Journal de Physique*, **4** (1905), 165–xxx; also in Langevin 1950, pp. 313–328 (a).

Langevin, Paul. 'Sur l'impossibilité physique der mettre en évidence le mouvement de translation de la terre,' *CRAS*, **140** (1905), 1171–1173; also in Langevin 1950, p. 395–396 (b).

Langevin, Paul. 'L'évolution de l'espace et du temps,' *Scientia*, **10** (1911), 31–54; also in Langevin 1923, pp. 265–300 [lecture at the Congress of Philosophy, Bologna, 1911] (a).

Langevin, Paul. 'Le temps, l'espace et la causalité dans la physique moderne,' *Bulletin de la Societété Française de Philosophie*, **12** (1911), 1–46; also in Langevin 1923, pp. 301–344 (b).

Langevin, Paul. 'Les grains d'électricité et la dynamique électromagnétique,' in the multi-authored volume, *Les idées modernes sur la constitution de la matière* (Paris, Gauthier-Villars, 1913); also in Langevin 1923, pp. 70–170 [lecture at the Société Française de Physique, 1912] (b).

Langevin, Paul. 'L'inertie de l'énergie et ses conséquences,' *Journal de Physique Théorique et Appliquée*, new ser., 3 (1913), 553–591; also in Langevin 1923, pp. 345–405, and Langevin 1950, pp. 397–426 [lecture at the Société Française de Physique, March 26, 1913] (b).

Langevin, Paul. 'Henri Poincaré: le physicien,' *Revue du Mois*, 8th yr. (1913); also in *L'oeuvre d'Henri Poincaré*, supplement of *Revue de Métaphysique et de Morale*, **21** (1913), 675–718; and in V. Volterra, J. Hadamard, P. Langevin, P. Boutroux, *Henri Poincaré. L'oeuvre scientifique. L'oeuvre philosophique* (Paris, Alcan, 1914), pp. 115–202 (c).

Langevin, Paul. 'Le principe de relativité' [lecture at Société des Electriciens, December 1919], published as a book, *Le principe de relativité* (Paris, Chiron, 1922); also in Langevin 1950, pp. 436–466.

Langevin, Paul. 'Les aspects successifs du principe de relativité,' *Bulletin de la Société Française de Physique*, **138** (1920), 5–6; also in Langevin 1923; pp. 406–423, and Langevin 1950, pp. 427–435 [communication to the February 6, 1920, meeting of the Société Française de Physique].

Langevin, Paul. 'Sur la théorie de la relativité et l'expérience de M. Sagnac,' *CRAS*, **173** (1921), 831–834; also in Langevin 1950, pp. 467–469 (a).

Langevin, Paul. 'Préface' to Edmond Bauer, *La Théorie de la relativité* (1922), pp. i–iv (a).

Langevin, Paul. 'L'aspect général de la théorie de la relativité,' *Bulletin Scientifique des Etudiants de Paris*, April–May 1922, 2–22 [lecture on March 30, 1922, to the Association Générale des Etudiants] (b).

Langevin, Paul. *La physique depuis vingt ans*. Paris, Doin, 1923.

Langevin, Paul. 'L'oeuvre d'Einstein et l'astronomie,' *L'Astronomie*, 45th yr. (1931), 277–301.

Langevin, Paul. 'La physique au Collège de France,' in the multi-authored volume, *Le Collège de France (1530–1930). Livre jubilaire composé à l'occasion de son quatrième centenaire* (Paris, Presses Universitaires de France, 1932), pp. 61–79 (a).

Langevin, Paul. 'La relativité. Conclusion générale,' in Langevin 1932c, fasc. VI, pp. 1–17 (b).

Langevin, Paul (ed.). *La Relativité. Série d'exposés et de discussions*. 6 fasc. Paris, Hermann, 1932 (c).

Langevin, Paul. 'Sur l'expérience de M. Sagnac,' *CRAS*, **205** (1937), 304–306; also in Langevin 1950, pp. 470–472.

Langevin, Paul. *La Pensée et l'action*. Paris, Editeurs Français Réunis, 1950; new ed., Paris, Editions Sociales, 1964 (a).

Langevin, Paul. *Oeuvres Scientifiques*. Paris, C.N.R.S., 1950 (b).

Langevin, Paul and Maurice de Broglie (eds.). *La Théorie du rayonnement et les quanta*. Communications and discussions from the meeting held at Brussels from October 30 to November 3, 1911, under the auspices of M.E. Solvay. Paris, Gauthier-Villars, 1912.

Lecornu, Léon. 'Quelques remarques sur la relativité,' *CRAS*, **174** (1922), 337–342.

Le Roux, J. 'La courbure de l'espace,' *CRAS*, **174** (1922), 924–927.

Le Roux, J. 'La variation de la masse,' *CRAS*, **180** (1925), 1470–1473.

Lorentz, Hendrik Antoon. *Versuch einen Theorie der elektrischen und optiken Erscheinungen in bewegten Körpern* (Leiden, Brill, 1895); also in Lorentz 1935–1939, V, 1–137.

Lorentz, Hendrik Antoon. 'Electromagnetic Phenomena in a System Moving with Any Velocity Smaller than that of Light,' *Kon. Akademie van Wetenschappen Amsterdam. Proceedings of the Section of Science* [in English], **6** (1904), 809–831; also in Lorentz 1935–1939, V, 172–197.

Lorentz, Hendrik Antoon. *Collected Papers*. 9 vols. The Hague, Nijhoff, 1935–1939.

Mascart, Elie. 'Sur les modifications qu'éprouve la lumière par suite du mouvement de la source lumineuse et du mouvement de l'observateur,' *Annales de l'Ecole Normale Supérieure*, **1** (1872), 157–214; 3 (1874), 363.

Mascart, Elie. *Traité d'optique*. 3 vols. Paris, Masson, 1889–1894.

Metz, André. *La Relativité. Exposé élémentaire des théories d'Einstein, et réfutations des erreurs contenues dans les ouvrages les plus notoires*. Paris, Chiron, 1923.

Metz, André. 'Les problèmes relatifs à la rotation dans la théorie de la relativité,' *Journal de Physique et le Radium*, 8th ser., 13 (1952), 224–238.

Miller, Arthur I. *Albert Einstein's Special Theory of Relativity. Emergence (1905) and Early Interpretation (1905–1911)*. Reading, Ma., Addison-Wesley, 1981.

Nordmann, Charles. *Einstein et l'univers. Une lueur dans le mystère des choses*. Paris, Hachette, 1921.

Nordmann, Charles. 'Einstein expose et discute sa théorie,' *Revue des Deux Mondes*, 42nd yr., 9 (1922), 129–166.

Painlevé, Paul. 'La mécanique classique et la théorie de la relativité,' *CRAS*, **173** (1921), 677–680 (a).

Painlevé, Paul. 'La gravitation dans la mécanique de Newton et dans la mécanique d'Einstein,' *CRAS*, **173** (1921), 873–887 (b).

Painlevé, Paul. 'La théorie classique et la théorie einsteinienne de la gravitation,' *CRAS*, **174** (1922), 1137–1143 (a).

Painlevé, Paul. 'Note sur les deux communications précédentes,' *CRAS*, **174** (1922) 1161–1162 (b) [cf. Chazy 1922 and Trousset 1922].

Painlevé, Paul. *Paroles et écrits*. Paris, Rieder, 1936.

Paty, Michel. 'Sur le réalisme d'Albert Einstein,' *La Pensée*, no. 204 (April 1979), 18–37.

Paty, Michel. 'Einstein et la philosophie en France: à propos du séjour de 1922,' *La Pensée*, no. 210 (Feb. 1980), 3–12.

Paty, Michel. 'Mach et Duhem. L'épistémologie de savants-philosophes,' in *Epistémologie et matérialisme*, O. Bloch, ed. (Paris, Klinsieck, 1986), pp. 177–217.

Pauli, Wolfgang. 'Relativitätstheorie,' in *Encyclopädie des Matematischen Wissenschaften*, 5, pt. 2 (Leipzig, B.G. Teubner, 1921), pp. 539–775; reprinted in Pauli's *Collected Scientific Papers*, R. Kronig and V.F. Weisskopf, eds. (New York, Wiley, 1964), I,

1–237, plus supplementary notes (1958), pp. 238–263. Translated into English as *Theory of Relativity* (New York, Pergamon Press, 1958).

Pestre, Dominique. *Physique et physiciens en France, 1918–1940*. Paris, Archives Contemporaines, 1984.

Picard, Emile. 'Quelques remarques sur la théorie de la relativité,' *CRAS*, **173** (1921), 680–682 (a).

Picard, Emile. 'La théorie de relativité et ses applications à l'astronomie,' *Annuaire du Bureau des Longitudes*, Paris, 1922, B1–29.

Poincaré, Henri. *Cours de physique mathématique. I. Leçons sur la théorie mathématique de la lumière*. Paris, Carré, 1889.

Poincaré, Henri. 'A propos de la théorie de M. Larmor,' *L'éclairage électrique*, **3** (1895), 5–13, 285–295; 5 (1895), 5–14, 385–392; reprinted in Poincaré 1950–1965, vol. 9, 369–426.

Poincaré, Henri. 'La mesure du temps,' *Revue de Métaphysique et de Morale*, **6** (1898), 1–13; also in Poincaré 1905.

Poincaré, Henri. 'Relations entre la physique expérimentale et la physique mathématique,' in *Rapports présentés au Congrès International de Physique de 1900* (Paris, 1900); in Poincaré 1902a (chapters 9 and 10) (a).

Poincaré, Henri. 'La Théorie de Lorentz et le principe de réaction,' *Archives Neerlandaises des Sciences Exactes et Naturelles*, 2nd ser., 5 (1900), 252–278; also in Poincaré 1950–1965, vol. 9, 464–488 (b).

Poincaré, Henri. *Cours de Physique Mathématique. Electricité et optique. La lumière et les théories électrodynamiques*. 2nd ed. Paris, Carre et Naud, 1901.

Poincaré, Henri. *La science et l'hypothèse* (1902). Paris, Flammarion, 1968 (a).

Poincaré, Henri. 'L'état actuel et l'avenir de la physique mathématique,' *Bulletin des Sciences Mathématiques*, **28** (1904), 302–324 [communication to International Congress of Arts and Sciences, Saint Louis, Sept. 24, 1904]; also in Poincaré 1905.

Poincaré, Henri. *La valeur de la science* (1905). Paris, Flammarion, 1970.

Poincaré, Henri. 'Sur la dynamique de l'électron,' *Rendiconti del Circolo Matematico di Palermo*, **21** (1906) [received July 23, 1905], 129–176; also in Poincaré 1924, pp. 18–76 and Poincaré 1950–1965, vol. 9, pp. 494–550.

Poincaré, Henri. *Science et méthode* (1908).

Poincaré, Henri. 'La mécanique nouvelle,' *Revue d'Electricité*, **13** (1910), 23–28; also in Poincaré 1924, pp. 1–17 [lecture to French Association for the Advancement of the Sciences, Lille, 1909].

Poincaré, Henri. 'Lettre à Pierre Weiss' [recommending Albert Einstein, Nov. 1911], in Académie Royale des Sciences, des Lettres et Beaux-Arts de Belgique, *Einstein et la Belgique* [Exposition] (Brussels, Palais des Académies, May 16–June 19, 1979).

Poincaré, Henri. 'L'Espace et le temps' [University of London lecture, May 4, 1912], in Poincaré 1913 (a).

Poincaré, Henri. 'Les rapports de la matière et de l'éther,' *Journal de Physique*, **2** (1912), 347–xxx; in Poincaré 1913 (b).

Poincaré, Henri. *Dernières pensées*. Paris, Flammarion, 1913; reed. 1963.

Poincaré, Henri. *La Mécanique nouvelle. Conférence, mémoire et note sur la théorie de la relativité*. Paris, Gauthier-Villars, 1924.

Poincaré, Henri. *Oeuvres*. 11 vols. Paris, Gauthier-Villars, 1950–1965.

Pomey, J.B. 'Les Conférences d'Einstein au Collège de France,' *Le Producteur, revue de culture générale appliquée*, **8** (1922), 201–206.

Potier, A. 'Conséquence de la formule de Fresnel relative à l'entraînement de l'éther par les milieux transparents.' *Journal de Physique Théorique et Appliquée*, **3** (1874), 201–204.

Pyenson, Lewis. 'La réception de la relativité généralisée: disciplinarité et institutionalisation en physique,' *Revue d'Histoire des Sciences*, **28** (1975), 61–73.

Sagnac, Georges. 'Sur les interférences de deux faisceaux superposés en sens inverse le long d'un circuit optique de grandes dimensions,' *CRAS*, **141** (1910), 1302–1305.

Sagnac, Georges. 'Les systèmes optiques en mouvement et la translation de la terre,' *CRAS*, **152** (1911), 310–330.

Sagnac, Georges. 'Ether et mécanique absolue des ondulations,' *CRAS*, **169** (1919), 469–471, 529–531 (a).

Sagnac, Georges. 'Mécanique absolue des ondulations et Relativité newtonienne de l'énergie,' *CRAS*, **169** (1919), 643–646 (b).

Sagnac, Georges. 'Comparaison de l'expérience et de la théorie mécanique de l'éther ondulatoire,' *CRAS*, **169** (1919), 783–785 (c).

Sagnac, Georges. 'Le problème de la comparaison directe des deux vitesses simultanées de propagation et de la révélation de la translation de la terre,' *CRAS*, **169** (1919), 1027–1031, 1128 (d).

Sagnac, Georges. 'La relativité réelle de l'énergie des éléments de radiation et le mouvement dans l'éther des ondes,' *CRAS*, **170** (1920), 1329–1242 (a).

Sagnac, Georges. 'Les deux mécaniques simultanées et leurs liaisons réelles,' *CRAS*, **171** (1920), 99–102 (b).

Sagnac, Georges. *Notice sur les travaux scientifiques*. Paris, 1920 (c).

Sagnac, Georges. 'Les invariants newtoniens de la matière et de l'énergie radiante et l'éther mécanique des ondes variables,' *CRAS*, **174** (1922), 29–32 (a).

Sagnac, Georges. 'Les oscillations des raies spectrales des étoiles doubles expliquées par la loi nouvelle de projection de l'énergie de la lumière,' *CRAS*, **175** (1922), 89–91 (b).

Sagnac, Georges. 'Sur le spectre variable périodique des étoiles doubles: incompatibilité des phénomènes observés avec la théorie de la relativité générale,' *CRAS*, **176** (1923), 161–173.

Schlipp, Paul-Arthur (ed.). *Albert Einstein: Philosopher-Scientist*. La Salle, Il., Open Court, 1949; 3rd ed., 1970.

Tonnelat, Marie-Antoinette. *Histoire du principe de relativité*. Paris, Flammarion, 1971.

Trousset, J. 'Les lois de Kepler et les orbites relativistes,' *CRAS*, **174** (1922), 1160–1161.

Wittakker, Sir Edmund. *A History of the Theories of Aether and Electricity*. Vol. 2: *The Modern Theories, 1900–1926*. New ed., London, Nelson, 1953.

Wittakker, Sir Edmund. 'Albert Einstein,' *Biographical Memoirs of Fellows of the Royal Society* (London, Royal Society, 1955), pp. 37–67.

Zahar, Elie. 'Poincaré's Independent Discovery of the Relativity Principle,' *Fundamenta Scientiae*, **4** (1983), 146–176.

CNRS, Paris   (Translated by Wayne K. Ishikawa)

MICHEL BIEZUNSKI

EINSTEIN'S RECEPTION IN PARIS IN 1922

INTRODUCTION

Ever since their formulation, Einstein's special and general theories of Relativity have given rise to passionate confrontations. France was no exception.

The characteristics of the French reception of the theories of Relativity can be summarized as follows: They were received with a significant delay[1] (the controversy within the scientific community began in the early twenties); the physicists were not, among the scientists, the most interested; mathematicians were more motivated to study and develop Einstein's theory (for the challenge of tensorial calculus in the general theory of relativity), as well as engineers (because they were conscious of the importance of Maxwell's equations). The delay in the physicists' reception can be explained by the fact that theoretical physics did not exist as a separate discipline. But there existed instead a discipline called "mathematical physics".[2] Thus, the theory of relativity was considered, during a long period, on the border of physics. One has to wait until nuclear physics develops – i.e. in the late 30's but mostly after the Second World War – to see a change in that perspective. Elie Cartan, Emile Borel, Jacques Hadamard, Jean Becquerel were among the scientists who played a role in the reception and development of the theories of Relativity. One figure, however, is prominent, Paul Langevin, who popularized Einstein's theory of special relativity as early as 1906. His seminar was attended by only a handful of students, of very high quality: apart of the names quoted earlier, Louis de Broglie and Alfred Kastler, for example, had their first contact with Einstein's theories through Paul Langevin.

Also philosophers showed interest in Einstein's theory of relativity. Henri Bergson tried to establish links between his concept of time and Einstein's.[3] Emile Meyerson developed an important epistemological reflection[4] based on the implications of Einstein's achievements. Gaston Bachelard criticized him[5] and developed an inductivist point of view. André Metz[6] contributed to the discussion with epistemological developments of Einstein's theories. Thus, one might think that France

169

Thomas F. Glick (ed.), The Comparative Reception of Relativity, 169–188.
© 1987 *by D. Reidel Publishing Company.*

would have afforded a particularly favorable reception of the theory of relativity. But it didn't happen that way. In the first place, though among the greatest of their time, the scientists concerned with Einstein's theories made up a very small minority. One way of determining how a new theory is actually received is to examine how it is taught. From this point of view, the situation appears quite different: Langevin gave his lessons at the *Collège de France*, which is not part of the regular academic system.[7] Apart from isolated initiatives (Lémeray in 1915 in Marseille, Eugène Bloch in the late 1920s, and, occasionally, Louis de Broglie in the early thirties) it is impossible to find any traces of courses of relativity in the University before the Second World War. And it is only in the mid-1960s that the special theory of relativity was introduced in the first years of the physics curriculum.

The theory of relativity has been part of physics training at the *Ecole polytechnique* from 1920 until 1925. Then, its teaching was stopped until 1936, because it was considered not crucial enough to be kept inside very dense physics courses.

The controversial issue created by the new theory is a rare case where one can see the effects produced by a scientific revolution.

The debate was not strictly scientific, limited to learned circles, but involved a portion of the general public as well. The reactions culminated in 1922, when Einstein came to Paris to explain and discuss both the special and general theories of relativity. The analysis of this visit and of the reactions to it will be the focus of this paper.

When I started to work on this topic,[8] I thought the major point of interest was the scientific debate within the community of mathematicians and physicists. But I discovered that the controversy was alive within the lay public also and that the wealth of popular journalism devoted to his ideas during Einstein's visit helped me understand more deeply the stakes of the controversy.

Drawing somewhat on a distinction made by Yehuda Elkana,[9] the first part of my essay will analyze this controversy in terms of (i) the current state of science (from the point of view of its development), (ii) the images of science, the main epistemological trends which support contradictory conceptions of what science should be, (iii) the ideologies which play a role in science. The related issues in the present case are scientism and anti-German feelings.[10] I will thus argue that the reluctance to accept Einstein's ideas originates, not from the current state of knowledge, but from images and ideologies shared by the scientific community.

Einstein was a special case: his renown extended beyond the limited circle of scientists and educated public. He was already almost as popular in the early twenties as he is now.

In the newspapers, scientists freely exposed their ideas, prejudices and opinions on scientific and extra-scientific matters. They are particularly interesting for the window they provide on the personal reactions of scientists confronted with a major rupture in their accepted scheme of knowledge. Journalists' opinions also show the dimension of the controversy. The questions discussed were not only centered on physics; they encompassed the role of science in society, the role of scientists in international relationships, the hope for a broader access to scientific knowledge. One usually doesn't find such elements in scientific articles.

I will try to show that there is a convergence between the terms of the debate as it emerged within the scientific community and the public debate as it appeared in the popular press.

I. A SCIENTIFIC CONTROVERSY?

1. The State of Science

If one considers the question of the acceptability of Einstein's theories in the context of the current state of knowledge, the response should have been entirely positive. I will try to illustrate this assertion by an example. I don't want to refer to the scientific writings of an enthusiastic supporter of Einstein's theory, such as Paul Langevin who argued in many books, articles or lectures[11] that Einstein's developments in the special, and later, general theory of relativity were of great relevance to the central, unanswered questions in contemporary physics.

Rather I want to consider the case of scientists, well acquainted with modern physics, who were able to understand and could have accepted the special theory of relativity but did not do so. The most enigmatic example of this kind is that of Henri Poincaré. His work on the electrodynamics of the electron and on the principle of relativity would seem to place him in a particularly favorable position to accept Einstein's theory of special relativity. But he remained silent, except in a lecture given in 1909 in which he does not mention Einstein's name.[12] His silence is analyzed in previous studies.[13] Stanley Goldberg showed that Poincaré's project was so different in content than Einstein's that it was impossible for him to accept Einstein's point of view.[14] Henri

Poincaré died too early (1912) to be part of the controversy I am studying here. That is the reason why I will focus on Emile Picard, whose attitude was rather close to that of Poincaré.

One of France's most prominent scientists, he was born in 1856. He became professor of physical and Experimental Mechanics at the Sorbonne in 1881. He was elected to the Academy of Sciences in 1889 and was its permanent Secretary from 1917 until his death in 1941. In addition to his scientific work as a mathematician and mathematical physicist, he wrote a number of books and articles on the state of science aimed at a lay audience. One of them, *Modern Science, its present state*,[15] was published precisely in 1905, the year of publication of the four famous articles by Einstein,[16] including the one on special relativity. Of course, at the time he wrote his book, he could not have heard of Einstein's work, because Einstein was not yet an ordinary physicist. However, the questions that preoccupied him were not so far from Einstein's preoccupations. In this book, Picard lists a set of open questions in modern physics:

How was the seemingly arbitrary alliance of mathematical demonstration and experimental principles in classical mechanics to be made comprehensible? How could the conceptions of absolute space and time be made less abstract? Must physics be based on a deductive method? In what way were the problems associated with the hypothetical electromagnetic ether to be resolved? And how can the Michelson experiment be explained?

One sees, afterwards, that the special theory of relativity gives an answer to these questions, directly related to the "body of physics". It establishes furthermore a link between those items. In 1922, Picard knew quite well the content of Einstein's theories. In January 1922, two months before Einstein's visit to Paris and even before it was planned, a debate on the theory of relativity was held at the Académie des Sciences. A rather controversial debate must have taken place, since the headline of that day's *Echo de Paris*[17] was: "Coming back to Newton's Mechanics". The major event in that debate was a communication by Emile Picard.

In discussing the contributions made to contemporary science by Einstein's theories of special and general relativity, Picard expressed mixed feelings on the conceptions of space and time on which they are based. He qualified the "new" ideas on space and time as "concerning metaphysics rather than physics".[18] And he didn't take a clear position for or against the theories claiming that the time hadn't come for reaching a verdict. This absence of position suggests that Picard, though disturbed by its epistemological consequences, could not find any scientific argument to refute to theory or relativity.

A number of attempts were made to refute Einstein's theories. Daniel Berthelot, presenting Carvalho's works to the *Académie des Sciences*, reported that his main conclusion was that the Michelson experiment can be explained by means of the Newtonian theory.[19] This example is a case of indirect refutation: not a straightforward attack, but an effort to avoid a change of theory. It is, in the strict sense, a conservative position.

A third example of refutation is the analysis of the results of an experiment concerning the gravitational redshift of spectral rays of solar metals conducted by Pérot in *Observatoire de Meudon*. The *compte rendu* at the *Académie* took place during Einstein's visit to Paris, on April 3, 1922.[20] The results of this experiment giving the value of the redshift were of the same order of magnitude as the experimental error. For some scientists, including Deslandres, director of the Meudon Observatory, this was an opportunity to reaffirm that prudence counseled waiting for further results before adopting the theory – while he was reported to say that the bases of the theory of relativity were disputable and fragile.[21] Very different comments were made on the same experiment by others: Léon Brillouin affirmed that the measured value "fits, taking into account the errors resulting from experiment, with the value announced by Einstein",[22] while Paul Langevin stated that "the redshift predicted by the new theory conforms exactly to the results of the experiment".[23]

The points of disagreement didn't yield a scientific alternative resulting in a coherent frame that could lead to a consensus and could definitely answer whether the theory should be accepted or rejected. This doesn't mean that the theory would be unfalsifiable in the Popperian sense, which might constitute a reason for denying it scientific status. An attempt at such a falsification was presented to Einstein by Jacques Hadamard during the scientific colloquium organized at the *Collège de France*.[24] He imagined what might happen to the gravitation equation in a case where the mass became infinite. In that case, the equation would cease to be valid. Einstein responded the next day by showing Hadamard a calculation (of which, unfortunately, no trace remains) showing that this situation is physically impossible.[25] However,

"Hadamard's catastrophe", as Einstein liked to call it, was a prefigura-
tion of the hypothesis of the black holes.

This attempt at a "falsification" is very different from the other critics
of the theory of relativity. Hadamard was not an opponent to the theory
of relativity. On the contrary, he wanted to know to what extent the
theory could be used. His attempt at a refutation was located inside the
theory. In all other cases, attacks on the theory of relativity didn't reach
its theoretical core, but only its epistemological implications. Thus, the
debate was not limited to purely technical issues, in order to compare
theories which were placed on the same level, but raised the question of
the frontier of physics: most physicists thought they had a clear idea of
what physics was and what it should be. The theory of relativity was not
to be included in that frame.

Given the state of science in France, scientists were in a position to
accept Einstein's theory of relativity. This theory could have helped
them to answer Picard's open questions and to further the development
of modern physics. The sources of the reluctance to accept new ideas
must therefore lie somewhere else.

2. The Prevailing Images of Knowledge

When scientists express their feelings about the way science should be
developed, they refer – implicitly or explicitly – to existing traditions.

In France of the 1920s, the images of knowledge, which, as Elkana
argues, are socially determined, culture-dependent and time-dependent, consisted essentially of a mixture of Cartesian mechanism and
Comte's positivism. They had constituted the "mental set" of generations of French scientists, imparting to French science a distinctive style.
The development of Celestial Mechanics was considered the acme of
scientific achievement. As we shall see, the mechanistic tradition was
advanced both in behalf of and against the reception of Einstein's
theories of relativity, while the positivist tradition was urged mostly
against the reception of the new ideas.

The importance of Cartesian philosophy in France is well-known. It
gave rise to the mechanistic point of view in science: every phenomenon
can be reduced to a description in terms of elementary movements, and
this is true not only for physics, but for chemistry and biology as well.
The supremacy of what was called in France "rational mechanics" – as a
reference to Descartes' philosophy – had been made explicit by Auguste

Comte in his classification of the sciences. In Comte's scheme mechanics was considered totally different from and superior to physics. It was a *"science rationnelle"* in the sense that it established a connection between an imperfect reality and an ideal, mathematical, model. Mechanics is also notably flexible: new experimental results can be integrated by means of *ad hoc* rectifications of the mathematical model. This flexibility of classical mechanics accounts for much of its appeal.

For this reason, many scientists remained attached to this tradition. Paul Painlevé, for example, tried to present Einstein's special relativity in the frame of the freedoms allowed by classical mechanics in his lessons at the *Ecole polytechnique*: "To define a means of measurement of durations and of lengths, Einstein has to adopt a new principle, and he chooses, from experiments, the principle of conservation of the speed of light".[26] Thus, he denies the theoretical change. Einstein's first motivation was to solve the problem of the non-invariance of Maxwell's equations with the principle of relativity, which is a quite different point of view that the one expounded by Painlevé.

Paul Langevin, who was a defender of the theory of relativity was also trained in the mechanistic tradition. But, for him, the most important issue included in that tradition was to try to unify various physical theories. Einstein was for him in this continuity because, after Maxwell's achievements on the unification of electricity and magnetism, the unificationist trend shifted from a pure mechanistic origin to an electromagnetic and mechanical base.

Secondly, there is the positivist tradition. In his book *De l'Explication dans les sciences*,[27] Emile Meyerson characterized the positivist stance as motivated by the intention to exclude metaphysics from science. According to positivism, physics is based on "facts", the relationships between facts being the physical laws. Now, if facts are "given" – i.e. exist independently of us – they only have to be observed and there is no room for physical concepts. Meyerson, of course, rejected this view and argued that, as a matter of fact, science does not conform to such precepts. Meyerson's influence, however, was limited to philosophical circles and the positivist conception of science prevailed among the major part of the French scientific community.[28]

Now, what one considers as an appropriate legitimation of scientific knowledge depends upon the view of science one holds. Thus, if science is limited to establishing relationships between facts, there is no need to be in possession of a unique theory for a given domain. The choice of a

theory can be made, for example, on the grounds of utility. This empirical position is a liberal one. It recognizes the possibility that several
different theories can describe the same physical phenomena. On utilitarian grounds, then, the introduction of special relativity into physics
can be regarded as not well motivated; because one had to overthrow
the well-established concepts of space and time. Speed of light didn't
appear as a significant scale for experiments to most physicists, except
the few ones working on radioactivity, and later, on nuclear physics.
Furthermore, if one regards physics as a phenomenological construction
derived only from experimental results, it is obvious that a theory such
as relativity can be regarded as a metaphysical dream, without any
connection to what was considered to be physics at the time: it dealt
only with very high velocities, which were beyond experimental reach
until the mid 1930s.

Henri Bouasse wrote, in a pamphlet entitled "*La question préalable
contre la théorie d'Einstein*":[29] "Finally, it is us, laboratory physicists,
who will have the last word: we accept theories which are commonplace
to us; we refuse those which we can't understand and are, for that
reason, useless to us (. . .). We, i.e. all physicists, since physics has
existed."

Some holders of this "utilitarian" point of view referred to the heritage
of Henri Poincaré, defining their point of view as the conventionalist
one. In fact, Poincaré was much more subtle. He did not pretend to
possess the uniquely valid point of view on science. As Langevin said,
he was "eclectic". Poincaré's position can be described as a lack of
confidence in theories in general. He abandoned the opinion that one
theory could be considered as "true", definitely, and then admitted that
several theories could follow, to explain the same phenomenon.

Among his epigones was Edouard Guillaume who openly fought
against Einstein's theory. In correspondence with Paul Langevin in
1917,[30] he advanced the idea of a modification of the ether due to the
movement of a body, which could explain the Lorentz contraction. In
his theory, the speed of light becomes relative but time remains absolute. He invokes Poincaré's authority to justify his position, saying that
this idea had been expressed by Poincaré at the St. Louis Exhibition in
1904, and secondly that Poincaré would certainly have accepted the
process of replacing an idea by another.

But Poincaré's position was not as clear as Guillaume says: Poincaré
admitted that the "old mechanics" could be replaced by a "new one",

which could itself be replaced by another one, and so on. Then, he accepted the possibility of a temporal succession of theories, which implies that one could replace another one. It seems to me that this is a nuance to introduce to a narrow interpretation of Poincaré's conventionalism. Nevertheless, Poincaré never quoted Einstein's theory, even if in that lecture he was speaking of it. The argument of generalized equivalence between all theories in physics is thus used only to show that Einstein's theory is likely to be replaced by any "simpler" theory, i.e. one closer to familiar concepts.

These images of science appear to be much more relevant to the attitudes towards the new theory than the different ways of putting the questions unsolved by contemporary physics. They concern the way scientists think science should be.

3. The Scientistic Ideology

The scientistic ideology still represented a strong influence at the time of the debate on Einstein's theories. It was inherited from the 19th century. Scientism developed as an anti-religious tool, and constituted a basis of agreement for the majority of scientists who strongly believed that progress of mankind was inseparable from the progress of science. Science was considered the best means for man to dominate nature. The so-called "experimental method" was defended as the only way of considering science.[31]

The change in the conception of physics brought by Einstein's theory of relativity was incompatible with the scientistic point of view because it entailed a reconsideration of the most fundamental concepts in physics – space and time. Therefore it appeared as an attack against the religious ones, and science was not allowed, in the scientistic point of view, to generate any doubt, especially in the non-scientists' minds. I use the word "ideology" to characterize scientism because this philosophical point of view was a weapon employed against other influences, particularly religious ones, not only on science, but on society as well.

The form it took during Einstein's visit in Paris is indicated in a speech given at the time by Henri Le Châtelier, a distinguished chemist, whose remark is quoted in *Le Journal* of March 22:[32]

I never had the ambition of overturning any existing law. On the contrary, any defect, even only partial, in a law, gives me the same feeling as a stain on a new garment. I have directed all my research toward confirming and enlarging my predecessors' work. I have

done so from inclination and without any selfish motives; through practice, I have
discovered that this approach is infinitely preferable. It is insane to think that everyone of
us can create new sciences by overturning the structure painfully built by generations of
scientists. Any progress needs the involvement of a great number of researchers; science
is, as Berthelot said, a collective work.[33]

His remark is followed by a comment of the journalist using these
arguments against Einstein's theories.

There are other reasons that have played a role in forming opinions
towards Einstein's achievements. These are related to the historical and
political context of France in the first decades of this century.

4. *The Historical and Political Context of Einstein's Visit to Paris*

Paul Langevin, who had been teaching the special theory of relativity in
his seminar at the *Collège de France* since 1906, first invited Einstein to
come to Paris as early as 1914. At that time, there were strong relationships between scientists throughout Europe, epitomized in the First
Solvay Physics Congress which met in Brussels in 1911. Einstein's visit,
however, had to be cancelled when the First World war broke out.
During the war, international scientific relations declined: 93 major
German scientists signed the famous "Manifesto to the Civilized
World" in defense of Germany; similarly, French scientists were writing
texts opposing "French genius" to the "German spirit" in science.
Pierre Duhem,[34] for example, argued that the extreme rigor of the
so-called "geometrical spirit" of German science made it a slave to the
deductive method. According to Duhem, because they were not based
on facts, German theories amounted to little more than metaphysical
speculation. French science, on the contrary, was characterized by its
"esprit de finesse", which leads to a clarification of the relationship
between observed facts and theories.

The postwar period was still a period of international tension. France
had suffered 1 300 000 casualties during the war, the greatest disaster in
its history. And there was the vexing issue of reparations: as long as they
remained unpaid, Germany had to be banned from the community of
nations. Even after the "victory", anti-German feelings were widespread in France. Very influential since the war against Germany in
1870, they were further exacerbated after the First World war. The
scientific community was not untouched by this current of opinion.
Furthermore, French scientists were among the most virulent propo-

nents of a boycott of German scientists, which ended only in June 1926[35], to start again – on new grounds – during the Nazi period (from 1933 on).

It was in this context that Einstein was invited again by Paul Langevin in 1922. The main argument Langevin used to convince the reluctant opinions was that science had to play a role in the progress of mankind and peace. For the sake of safety, however, Langevin wanted the visit to be strictly "private", reserved to a few members of the scientific élite. But Einstein was no longer an ordinary physicist. After the 1919 Royal Society Session and his trip to the United States in 1921, he had become a legend. And the press made of his visit an event.

When Einstein's visit was announced publicly, several newspapers presented him as a "Swiss scientist" living in Germany.[36] The camouflage provoked a controversy. Other newspapers replied by arguing that France should not be ashamed of receiving the greatest German scientist, because science was universal and profitable to mankind as a whole.[37] The argument reveals the esteem in which science was held even after the criticism it had sustained as a result of having been involved in the war.[38] With the exception of the extreme right wing and the most virulent anti-Semitic press, the major newspapers were well disposed toward Einstein from the moment of his arrival. Everybody was "surprised" to see that he didn't fit their preconceptions of a German university professor. Only one incident marred his visit. Einstein was supposed to be received at the *Académie des Sciences* on April 3rd, 1922, where 30 members were planning to leave as soon as he entered. When Einstein learned what awaited him, he decided not to go. The session, nevertheless, took place, and the physicists present discussed the results of the experiments conducted in Paris by Pérot on the gravitational redshift of solar metals. The analysis of the *compte rendu* of the session is very revealing: the discussion among those present including Deslandres shows that they felt General Relativity still had to be "demonstrated" and that the demonstration had to be "conclusive". Indeed, their image of science implied that, in order to be accepted, the theory of General Relativity had to conform to the scheme of the experimental method as it had been expressed by Claude Bernard. Perot's experiments were deemed inconclusive; the same verdict had been passed on the results of the British solar eclipse expedition of 1919.

There was consensus on a necessary carefulness before adopting Einstein's theories. One must remember the conditions in which this session was held to maintain that, all members of the Academy who were, for political reasons, opposed to Einstein, were not among his supporters from a scientific point of view. This point illustrates a characteristic division of political opinions in French academic society which begins with the French Revolution and which plays a crucial role in the intellectual debates.

Boycott of German scientists was a reality. Einstein's visit to Paris was a symbol for a future reconciliation. Not everyone in France, nor in Germany for that matter was convinced that this was the right thing to do. The very fact that Einstein could be received at all in Paris was most surprising. Indeed, the situation was worsening. Walther Rathenau, who convinced Einstein to go to Paris, was killed in June 1922. Less than one year after Einstein's visit to Paris, the French army occupied the German industrial area in the Ruhr valley. The anti-German feelings were not reserved to marginal nostalgics; it was a strong and influential current.

In short, one sees that the obstacles to the reception of Einstein's theory of relativity came both from prevailing images of science – mechanism and positivism – and from contemporary ideologies – anti-German feelings and the scientistic ideology.

The content of the articles published in the French press reveals somewhat different aspects. Einstein's visit constituted, for the week he was in Paris, the principal topic in the 40-odd Paris daily newspapers. One of the most controversial issues discussed concerned the understanding of the theory of relativity, and furthermore, of science. Through the study of the impact of a scientific revolution among the lay public, it is possible to establish a link between the scientific debate and the public reception.

II. THE POSSIBILITY OF POPULARIZATION

The theory of relativity had the reputation for being understood by an only handful of people – estimates ranging from one to twenty. This belief was an important element in the Einstein legend as early as 1921 when he went for the first time to the United States and may already have existed in Germany.

1. The Irrelevance of the Usual Popularization Pattern

Einstein's visit created a new realization: certain kinds of science could not be popularized. Several articles argue that theory of relativity defied all efforts to be made evident to laymen. Two reasons were usually advanced: it was argued that the mathematics of both the special and general theories of Relativity are indissociable from the theoretical content so that no merely verbal presentation of the ideas is possible; and the so-called "paradoxes" were believed to conflict too violently with common sense. The point I want to make here is that there is something more behind this reputation. Commonly, popularization is thought to be a process of translation from the language of science into the vernacular. This model works as long as science itself is an established, completed product; but it breaks down during "scientific revolutions" and during lesser controversies within science due to the fact that there is no agreement over what scientific language should be. Such was the case for the theory of relativity.

Thus, the famous "misunderstandings" – everything is relative, etc. – were not due to a bad popularization. I will show that they are simply not efforts at popularization at all, but something directly related to images of science and ideologies which, as I mentioned before, had influenced reception within the scientific community itself. I want to show that the "distortions" as they appear in the press resulted from the specificity of the context of the early twenties. Three factors are especially noteworthy.

The first factor is the revolutionary side of Einstein's theory. It was perceived as an obstacle to popularizing because its "destructive character" was anathema to the traditional scientistic conception of popularization, which seeks to show the achievements of triumphant science.

The second factor is the "novelty" of the theories. In 1922 they were still new theories for the average physicist. Physicists' resulting and frequent admissions that they didn't understand the theory contributed to the difficulty that popularizers felt themselves to be operating under.

The third factor has to do with the fundamental character of the theories. It had far-reaching implications for concepts which were taken for granted, such as space, time, observer-dependence, simultaneity. The corresponding classical concepts were part of ordinary language, while the new, revolutionary ones could not easily be expressed in this way. The fact that Einstein rejected the usual concepts of classical science forced him to express himself. Scientists, and especially physicists, are used to specific forms of communications including a high level of formalism. Everybody is supposed to agree on the basic concepts and there is no need to speak about them. This is the field of the philosophers. Indeed, Einstein was accused of having rejected the use of the

traditional language of science. In *L'Oeuvre* of April 2nd, a journalist
quotes a scientist saying that Einstein, who "should have spoken in the
international scientific language, i.e. formulas written on a blackboard,
preferred to express himself in a language which is not his; he wanted to
speak to the 'gens du monde' (educated public)" and therefore could
not be understood by his peers". Another example is the following: in
the April 4 issue of *L'Oeuvre*, there is a long report of Einstein's
lecture. It includes this quotation: "You see this man? He is falling!
Well, for him, gravitation doesn't exist any more, because he is falling.
Isn't it funny?" The journalist adds this comment: "but we understand
at once that it is sufficient that the body brings with it its axes of
coordinates to suppress gravitation." This reflection shows that clarity
was of lesser importance than the display of technical jargon, egre-
giously employed in this case. What Einstein actually said was perfectly
clear, but the form he used to express it didn't fit the scientific norm and
was not understood for that reason.

In order to confirm the hypothesis that one can speak of context-
dependent images of science, I will present the form taken by these
three characteristics I have emphasized – revolutionary, novel and
fundamental aspects – outside the scientific community.

2. Effects Produced by the Scientific Controversy Outside Science

1. The revolutionary aspect of Einstein's theory was used by those who
considered themselves as political revolutionaries. One can read, for
example, in *l'Humanité*, one of the Communist daily newspapers[39] that
Einstein had given a mathematical form to the philosophy of "absolute
relativism", a cornerstone of Marx's dialectical materialism and a major
weapon in his refutation of absolutes. The author writes that he will
leave it to others to give a detailed analysis of Einstein's ideas. His text,
then, should not be understood as a popularization one but rather as a
free use of Einstein's concepts, interpreted for his own revolutionary
purposes. He uses a semantic trick, consisting in a literal interpretation
of the words. The freedom of interpretation of the scientific concepts is
perhaps due to the fact that they were not yet accepted within the
scientific community. Many scientists claimed they didn't know what
relativity was about, and didn't want to know. Journalists could feel
free to present openly his way of understanding the theory. The absence
of consensus within the scientific community authorized the use of
"semantic slidings".

2. The novelty of Einstein's theories was particularly noted by those interested in new ideas. The post-war period was a time of cultural renewal, the surging forth of modernity: dadaism, psychoanalysis, expressionism, and Nietzschean philosophy were being introduced to France simultaneously. The theory of relativity was compared[40] to the recently issued movie, the *Cabinet of Dr. Caligari*. Both are noticed for being up to date and for having been "produced" in Germany! In the literary chronicles of the time, translations and expositions of Einstein's work as well as Freud's, crowd the review columns. At the same time, the people declaring their interest in Einstein's theories are qualified as "snobbish" both by the "regular" scientists and also by the journalists for whom this way of thinking is well outside their mental set.[41]

3. The belief in the impossibility of expressing in a simpler way the basic concepts of the theory led to widespread plays with words. For example, in almost every newspaper, one finds this sentence: "Time doesn't exist, Einstein says", usually associated with cartoons. In one of them, published in *L'Intransigeant*,[42] an old lady applying make-up in front of her mirror quips: "How smart Einstein is, he said that time doesn't exist". Another turns on the fact that the French language uses the same word (*temps*) for time and for weather:[43] A man holding an umbrella waits until the rains stops then says: "Einstein is right: time doesn't exist". Here again, these statements evidently are not attempts at scientific popularization, for Einstein is not quoted as saying: Absolute time doesn't exist. The omission of the word "absolute" can be explained by a confrontation of the debate among the scientific community. For most physicists, there was no difficulty with the concept of time. A frequent variable in their equations and used daily in their discussions, it had acquired an entrenched and unquestioned character. To raise the issue of its validity seemed unnecessary and capricious. The numerous newspaper references to the sentence "Time doesn't exist, Einstein says", don't pretend to explain anything about Einstein's theories. Their ironical form is not oriented toward Einstein, but rather indicates symptomatically what terms were at stake in the representations of science, both for scientists and for the lay public. There was no place to discuss the concept of time, accounting to the organization of the physical sciences: physicists had forgotten that physics once dealt with time. Identifying absolute time with experienced time means that the Newtonian concept has been incorporated into common sense. These "jokes" about time do not indicate that people did not under-

stand relativity but rather that Einstein's preoccupation with time remained a puzzle. This question could not have been raised by a true
physicist!

3. Outside and Inside Science

These three points – revolution, novelty, and fundamental character –
relate to the influence of prevailing images of science and of ideologies
as obstacles to the reception of Einstein's ideas within the scientific
community.

Scientific revolution was not appealing to those who believed, in the
positivist manner, that science was nearly completed. Classical mechanics, the paradigm of perfected knowledge, seemed an undamaged structure and one without need or room for the theory of relativity.

Furthermore, the theory had its origin in Germany. This fact enlivened prevailing anti-German feelings. Arguments in defense of tradition had a distinct isolationist flavor nourishing the insularity that had
contributed so much to France's general scientific decline from the
mid-19th century.

Finally, the absence of theoretical physics in France proved crucial.
Mathematical physics thrived, but theoretical physics could not be fit
into the positivist hierarchy, where physics came below mechanics. This
hierarchy was institutionalized in France's university system, which had
chairs in General Mathematics, Rational Mechanics, Mathematical
Physics, and, at the top, Mathematical Astronomy and Celestial Mechanics. This structure served as an obstacle to Einstein's theories,
which forced a new perspective on the internal organization of the
different branches of physics.

4. On March 31, 1922, Einstein Was Understandable

There is a curious twist connected to Einstein's visit. The press releases
of his public lecture of March 31 at the *Collège de France* were almost
unanimous:[44] Einstein was understandable! Almost everybody admired
the way he developed his ideas, his clear mind, the slow but good French
he spoke, etc.; and almost everybody had the feeling of having
understood something of his scientific conceptions. The interest he
aroused transformed, for a while, the usual conceptions of the popularizers. Coupled with the fact that a German scientist was being honored
in Paris less than four years after the end of The First World war, the

future of the theory seemed promising in France. Unfortunately, the potential was not actualized, neither in science nor in politics. It remained an extraordinary event.

NOTES

[1] Stanley Goldberg: *The Early Response to Einstein's Special Relativity 1905–1911. A Case Study in National Differences (Germany, United Kingdom, France, USA)*. Ph.D Dissertation, Harvard University, 1969 (HU 90.9562 Widener Library, Harvard University).

[2] Lewis Pyenson shows that this fact acted as a brake in the reception of the theory of relativity in 'La réception de la relativité généralisée: disciplinarité et institutionalisation en physique', *Revue d'histoire des sciences*, 1975, **XXVIII/I**, 61–73, p. 70.

[3] Henri Bergson: *Durée et Simultanéité*, Félix Alcan, Paris, 1922. See also P.A.Y. Gunther (ed.): *Bergson and the Evolution of Physics*, The University of Tennessee Press, Knoxville, Tennessee, 1969.

[4] Emile Meyerson: *La déduction relativiste*, Payot, Paris, 1925.

[5] Gaston Bachelard: *La Valeur inductive de la Théorie de la Relativité*, Vrin, Paris, 1928.

[6] André Metz: *Les nouvelles théories scientifiques et leurs adversaires. La Relativité*, preface J. Becquerel, E. Chiron, Paris 1926.

[7] The *Collège de France* was created precisely in the 16th century by the king François 1er to make possible the introduction of new ideas that couldn't find a place in the university system: it was still the case with Einstein's theories.

[8] Michel Biezunski, Ph.D. dissertation: *La diffusion de la théorie de la Relativité en France*, Université Paris 7, 1981.

[9] Yehuda Elkana: 'A Programmatic Attempt at an Anthropology of Knowledge', in Y. Elkana and E. Mendelsohn (Eds): *Sciences and Cultures, Sociology of the Sciences*, vol. 5, Dordrecht, Reidel, 1981, pp. 1–76.

Yehuda Elkana analyzes science as a cultural system in distinguishing three factors (i) the body of knowledge, (ii) the socially determined images of knowledge and (iii) values and norms included in ideologies which do not directly depend on the images of knowledge. "At any given moment there is a state of knowledge with its methods, solutions open problems, nets of theories and, at its core, scientific metaphysics (. . .) The two or more different theoretical networks are engaged in a critical dialogue. Depending on the stage of the science, on the time, the place and the culture, there will probably be several dominant research programmes and there will be a consensus which are in a critical dialogue with other groups.

"Beliefs held about the task of science (understanding, prediction, etc.) about the nature of truth (certain, probable, attainable, etc.) about sources of knowledge (. . .) are all part of the time-dependent, culture-dependent images of science. It is the image of science which decides what problems to choose out of the infinity of open problems suggested by the body of knowledge (. . .) Ideologies, political considerations, social pressures, values and norms strongly influence the emergence of the dominant images of knowledge."

[10] Although of different origin, scientism and anti-German feelings were very influent in the early twenties. Scientism was on the decline, but still represented a major trend.

[11] The first article on the theory of relativity was published by Paul Langevin in 1911. It is

entitled: 'L'Evolution de l'Espace et du Temps'. *Scientia*, **X**, 1911, 31–54. Other texts include *Le Principe de Relativité* (lecture in 1919 at the Société française des électriciens), Chiron 1922. Paul Langevin gave a lecture during Einstein's visit in Paris at which Einstein was present: 'L'aspect général de la théorie de la relativité', *Bulletin scientifique des étudiants de Paris*, **2**, avril-mai 1922, 2–22.

[12] Henri Poincaré: *La Mécanique nouvelle. Conférence, mémoire et note sur la théorie de la relativité*, Intr. de Edouard Guillaume, Gauthier-Villars, Paris, 1924.

[13] Gerald Holton: 'On the Thematic Analysis of Science: The Case of Poincaré and Relativity', in *Mélanges Alexandre Koyré*, Hermann, 1964, II, 257–268.

[14] Stanley Goldberg: 'Henri Poincaré and Einstein's Theory of Relativity', *American Journal of Physics*, **35** (10), 934–944, Oct. 1967.

Stanley Goldberg 'Poincaré's silence and Einstein's Relativity', *The British Journal for the History of Science*, **5** (17), (1970).

[15] Emile Picard: *La Science moderne, son état actuel*, Paris, Flammarion, 1905.

[16] Albert Einstein: *Eine neue Bestimmung der Moleküldimensionen*, Bern: Wyss (inaugural dissertation, Zürich Universität.)

Albert Einstein 'Über einen die Erzeugung und Verwandlung des Lichtes betreffenden heuristischen Gesichtspunkt', *Annalen der Physik*, ser. 4, **XVII**, 132–148.

Albert Einstein 'Über die von der molekularkinetischen Theorie des Wärme geforderte Bewegung von in ruhenden Flüssigkeiten suspendierte Teilchen', *Annalen der Physik*, ser. 4, **XVII**, 549–560.

Albert Einstein 'Elektrodynamik bewegter Körper', *Annalen der Physik*, ser. 4, **XVII**, 891–921 (followed by the short article: 'Ist die Trägheit eines Körpers von seinem Energieeinhalt abhängig?', *Annalen der Physik*, ser. 4, **XVIII**, 639–641.)

[17] *L'Echo de Paris*, January 7, 1922. In the following days, *L'Echo de Paris* published two articles about Emile Picard's attitude towards the theory of Relativity: 'M. Emile Picard et la Relativité', January 14, 1922 and 'La théorie de la relativité – D'Einstein à Emile Picard', January 28, 1922.

[18] In *Le Matin*, January 10, 1922: "M. Emile Picard, Secrétaire Perpétuel de l'Académie des Sciences, donne son opinion sur les théories d'Einstein."

[19] *L'Echo de Paris*, January 7, 1922, *op. cit.*

[20] This session was held in a very particular atmosphere: Einstein, who had been invited after much hesitation, refused to go: a boycott had been planned by about 30 academicians because Einstein was a German scientist. Einstein's friends did not attend that session.

[21] 'L'effet Einstein n'a pas encore été vérifié', *Le Temps*, April 5, 1922.

[22] Léon Brillouin: 'Les théories d'Einstein et leur vérification expérimentale', *La Science et la Vie*, 63, **XXII**, June–July 1922, 19–29.

[23] Paul Langevin: 'L'aspect géneral de la théorie de la relativité', *op. cit.*

[24] One can read a detailed *compte-rendu* by Charles Nordman: 'Einstein expose et discute sa théorie', in *Revue des deux mondes*, **IX**, 1922, 129–166.

[25] Charles Nordmann, *ibid.*

[26] Paul Painlevé et Charles Plâtrier: *Cours de Mécanique*, Gauthier-Villars, 1929.

[27] Emile Meyerson: *De l'Explication dans les Sciences*, Paris, Payot, 1921.

[28] I want to emphasize that there is a difference between "positivism" in the French and in

the logical positivist's sense. French positivism is Comte's: knowledge is restricted to relationships among observed facts and is arranged in a natural hierarchy beginning with mathematics and ending with sociology.

[29] Henri Bouasse: *La Question préalable contre la théorie d'Einstein*, A. Blanchard, 1923.

[30] The Langevin-Guillaume correspondence is part of the Langevin Archive, still in possession of his family. I thank Mrs. Luce Langevin for having authorized me to access this material.

[31] The *Introduction à la Médecine expérimentale*, by Claude Bernard, written in 1885, "defines the fundamental principles of any scientific research", referring to the Académie Française (Petit Larousse illustré, 1977).

[32] Le Châtelier's remark is quoted in *Le Journal*, 22 March 1922, in an article by Lucien Chassaigne, entitled *'Belle modestie d'un savant'*: "Je n'ai jamais eu l'intention de renverser aucune loi existante. Au contraire, la mise en défaut, seulement partielle, d'une loi me fait l'effet d'une tache sur un vêtement neuf. J'ai dirigé toutes mes recherches dans le but de confirmer et d'étendre le travail de mes devanciers. Je l'ai fait par goût et sans aucun calcul, mais à l'usage j'ai reconnu que cette méthode était infiniment avantageuse. C'est folie de croire que chacun de nous peut renverser l'édifice péniblement établi par des générations de savants. Tout progrès nécessite la collaboration d'un grand nombre de chercheurs; la science, comme le disait Berthelot, est une oeuvre collective."

[33] Marcelin Berthelot (1827–1907) was a chemist and a politician. He is considered with the physiologist Claude Bernard (1813–1878) as the one of the initiators of the scientist point of view in science.

[34] Pierre Duhem: 'Quelques réflexions sur la science allemande,' *Revue des deux mondes*, 25 février 1915.

[35] Brigitte Schroeder-Gudehus: *Les Scientifiques et la Paix*, Presses Universitaires de Montreal, 1978.

[36] This is the case of *Le Temps*, *L'Ere nouvellle*, and *La France*.

[37] *Le Populaire*, *L'Eclair*, *L'Internationale*, *Le Matin*, etc.

[38] *La Victoire*, *Le Journal du Peuple*.

[39] The quotation comes from the April 1, 1922 issue.

[40] *Le Journal*, March 28, 1922.

[41] The number of references is too big to be quoted exhaustively. The following articles are some of the ones refering to the question of snobism:

– Marcel Coulaud: 'La première d'Einstein au Collège de France', *Bonsoir*, April 3, 1922.

– Julien Benda: 'Einstein et les salons', *Le Gaulois*, October 28, 1921.

– 'N'entre pas qui veut aux conférences d'Einstein', *Le Journal du Peuple*, April 3, 1922.

– Emile Borel: 'Einstein et les gens du monde', *L'Oeuvre*, April 4, 1922.

– 'Einstein et le snobisme', *Paris-Midi*, April 2, 1922.

– 'Les chaussettes d'azur en joie', *Le Peuple*, April 6, 1922 ("Chaussettes d'azur" is an ironical turn for "bas bleus", itself an ironical way of qualifying the women involved in the Feminist movement.)

– Victor Snell: 'Mode et Relativité', *Le Populaire*, March 31, 1922.

– J.B.: 'Snobisme nouveau', *Le Temps*, April 7, 1922.

[42] *L'Intransigeant*, in the April 6 issue.

[43] In the April 5, 1922 issue of *L'Oeuvre*.

[44] The only notable exception is an article published in *L'Oeuvre* of April 1, 1922. But it is remarkable that on April 4, 1922, 4 articles (a whole page) are devoted to Einstein's theories, correcting this opinion.

BARBARA J. REEVES

EINSTEIN POLITICIZED: THE EARLY RECEPTION OF RELATIVITY IN ITALY

In Italy as elsewhere amid the turbulence following World War I, the reception of Einstein's theories of relativity took on distinctive political colorings in certain settings. This paper analyzes two such episodes. The first focuses on the language used in scientists' writings, for audiences wider than the community of practitioners, about what had earlier been termed a "revolutionary" theory in the context of the new climate of revolution in postwar society and politics. Stalking the metaphors they used, such as revolution and evolution, destruction and construction, facilitates understanding of the variety of impacts of relativity within the scientific community, especially on scientists' views of the relation of theory to experiment and the nature of scientific growth and change. Scientists' publicly expressed opinions of scientific development were transformed by the postwar political situation from an easy acceptance of revolutionary change in science to a rejection of it and an adherence to an evolutionary, progressive view. Furthermore, this new gradualist view stimulated the doing of science: the mathematical physicist Tullio Levi-Civita wrote what became an influential paper in all Latin countries, entitled, "How a conservative could reach the threshold of the new mechanics."

The second episode follows the association of Einstein, considered as a "relativizer" of traditional conceptual absolutes and claims of objectivity in science, with contemporary "relativizing" movements in philosophy, cultural analysis, literature, the arts, and politics. The interplay of the language and the categories of absolute and relative in nonscientific evaluations of Einstein and in cultural and political analyses reveals the writers' general philosophical, cultural, or political predilections as well as their individual momentary assessments of the rapidly changing political and cultural climate in Italy in the years immediately following the end of the war. The culture critic Adriano Tilgher linked Einstein's theory of relativity to the philosophical relativism of Hans Vaihinger and the cultural relativism of Oswald Spengler, and Mussolini was pleased to associate his actions as *Duce* of Fascism with these "great philosophies." This communality of language suggests a new way of

189

Thomas F. Glick (ed.), The Comparative Reception of Relativity, 189–229.
© *1987 by D. Reidel Publishing Company.*

thinking about the interaction of scientific and nonscientific movements.

The two episodes taken together offer an unusual way to survey the boundaries between scientific fields and to watch the ground being prepared for the establishment in Italy of the new field of theoretical physics in the disciplinary no-man's land between mathematical and experimental physics. The social and cognitive protection of science, and of the theory of relativity in particular, sought by relativity's supporters, and the positive association of the theory with the Fascist movement by Mussolini in 1921 and confirmed in 1924, were complementary and equally essential aspects of the successful establishment in 1926 of the first chair in theoretical physics in Italy.

A significant facet of the impact of relativity in Italy was Einstein's visit there in 1921, at the invitation of a University of Bologna committee headed by Federigo Enriques, to give lectures on special and general relativity and cosmology. Speaking in Italian learned when he had lived in Pavia as a teenager, Einstein presented the public lectures on Saturday, Monday, and Wednesday, October 22, 24, and 26. On Sunday he participated in a special private session of the Academy of Sciences, to which he had been elected a corresponding member the previous spring in anticipation of his visit. On October 27, by invitation of the University of Padua and other academic institutions there, he presented a single lecture on relativity in the great hall of the university, the same room in which Galileo had lectured on his new mechanics over three centuries earlier.

THE THEORY OF RELATIVITY: RELATIVE OR ABSOLUTE?

Einstein's theory of relativity was a theory that, in its restricted or "special" form, challenged the fundamental Newtonian concepts of absolute space, absolute time, and simultaneity and made central the role of the observer in the measurement process. The theory in its generalized form, general relativity, assimilated physics to geometry by way of what was seen to be great mathematical complexity for the sake of experimental results just barely within the limits attainable in measurements. General relativity was a theory of great mathematical beauty and elegance and profound physical consequences to the initiated, and of enormous and undue mathematical complexity and physical nonsense to most Italian physicists and astronomers. It suggested that space was curved or "warped" and not our familiar flat Euclidean space or even the pseudo-Euclidean flat space-time of Minkowski, that our familiar force of gravity, or gravitational field, was nothing but a change in the metric or curvature of space-time and not a real "force," or a real "field," at all, and that the path of light was bent in moving through the curved space-time that had been represented as a strong gravitational field. Both forms attacked the two-centuries-old edifice of Newtonian physics at its foundations, transforming concepts which had become intuitive or simply facts for many physicists, such concepts as absolute flat space like a container or stage, absolute time flowing equably, the absolute simultaneity of spatially separated events, the concept of force as action at a distance, and the inverse square law of universal gravitation.

Relativity could thus hardly escape being noticed outside the confines of the very restricted circle of its practitioners, especially given the dramatic presentation of the results of the British eclipse expeditions at the Royal Society meeting on 6 November 1919,[1] characterized by Abraham Pais as a proper if metaphorical canonization of Einstein,[2] and the news coverage following. Likewise, because of the notoriety the theory gained as a challenger of traditional conceptual absolutes and of claims of objectivity in science, it was linked by some to contemporary relativizing intellectual movements in other fields as well as in politics.

However, since Einstein himself was no relativist, it is necessary briefly to consider how the theories of relativity might just as well be deemed theories about absolutes as about anything "relative."

The term "theory of relativity" or *Relativitätstheorie* was not originally Einstein's but seems to have been used first in 1906, at a meeting of the German Physical Society. There Max Planck, in a paper comparing experimental measurements on the masses of high-speed electrons with various theoretical predictions, called the theories of Lorentz and Einstein (which made the same mathematical predictions) *Relativtheorien*. In the discussion, the experimentalist A.H. Bucherer called the Lorentz-Einstein theory *Relativitätstheorie*, and Einstein picked up the term for his own theory.[3] The title of Einstein's 1905 paper on what we now call special relativity was "On the electrodynamics of moving bodies." There he claimed that the difficulties that had arisen in reconciling classical Newtonian mechanics with late-nineteenth-century electrodynamics were due to an incorrect understanding of the relative motion of bodies, in particular, of bodies moving with constant velocities with respect to each other. In the paper he discussed systems in relative motion, and the idea that the measured values of lengths, time intervals, and simultaneity in a given system were relative to the state of motion or rest of the observer with respect to that system. His presentation was based, in a way not common at the time, on two postulates, the "principle of relativity," a term also used by Lorentz and Poincaré in their electron theories, and the principle of the constancy of the velocity of light, independent of the motion of its source. For Einstein, the principle of relativity meant that the laws of physics, the mathematical formulas, retain the same form which they have in one coordinate system or frame of reference when referred to any other coordinate system moving uniformly, that is, with constant velocity, with respect to the first. An alternative expression is that the laws of physics are independent of the state of rest or uniform motion of the observer. A more technical way to make the point is to say that the laws of physics are invariant in form under a Lorentz transformation, the equations which allow the change of frames of reference in uniform motion with respect to each other. Thus, the "principle of relativity" is really an invariance principle. Likewise, the principle of the constancy of the velocity of light independent of the motion of its source is an invariance principle. And so the special theory of relativity is about invariances, which can be understood as absolutes.

Similarly, general relativity expands the possible relative motions of reference frames to any accelerated motion; it thus covers the case of gravitation, which is simply accelerated motion. The point is the same as before: that the laws of physics meet the mathematical conditions of generalized covariance is an invariance principle, an absolute. Meaning is in the mind of the observer, and whether the theory of relativity was viewed as relativizing or absolutistic, especially by nonscientific commentators, can be interpreted as revealing their more general philosophical, cultural, and political inclinations.

RELATIVITY: THE REVOLUTIONARY THEORY

In Italy as elsewhere before World War I, special relativity (often as the so-called Lorentz-Einstein electron theory) was presented to audiences of nonspecialists (including physicists) as revolutionary, as a theory in the process of destroying and rebuilding the edifice of science. In 1911 the mathematician Tullio Levi-Civita cautiously called the efforts of Ernst Mach and his colleagues Federigo Enriques, Gian Antonio Maggi, and Giovanni Giorgi to put classical mechanics on a more rigorous foundation and to eliminate the absolute reference frame "evident reformism," whereas Lorentz's and Einstein's solutions were "revolutionary."[4] Lorentz's "ingenious explanation, introducing the notion of local time and that famous contraction of his," nevertheless had

the air of a trick: a trick of genius, but always a *coup de pouce*, as Poincaré was to characterize it.

But . . . Einstein somewhat displaced the interpretation to draw from the artifice a profound philosophical conception: the principle of relativity, intrinsically unassailable, however much it throws our habitual intuitions into confusion.[5]

Levi-Civita understood that Lorentz's and Einstein's theories were inverses of each other: Lorentz's theory had as its goal what was for Einstein the first postulate, the principle of relativity, that absolute motion, especially absolute motion of the earth with respect to the electromagnetic ether assumed at rest, could not be observed.

The young experimental physicist Michele La Rosa of Palermo spoke favorably in print up to 1912 of the "magnificent revolution" brewing in theoretical physics,[6] asserting that relativity theorists seemed to have enough building blocks left over from their sweeping demolition of classical physics, enough "ultimate and unchanging elements" which had "absolute value" and "universal significance" to rebuild "a new more beautiful and majestic edifice."[7] Echoing Planck's 1910 address to the Königsberg meeting of the Deutsche Naturforscherversammlung, La Rosa saw relativity theory "no longer . . . as a lamentable scourge of destruction but as an instrument of order and creation."[8]

This language of revolution was used apparently without prejudice, in order to reflect the profound nature of the scientific innovations being considered. Perhaps such characterizations even made the theory fashionable, since equally revolutionary innovations were being made in such other intellectual and cultural fields as philosophy, literature, art, and music. Planck was certainly a political and cultural conservative, though he had made his own scientific revolution, and La Rosa was too, at least later. Levi-Civita, more cautious scientifically, had strong socialist sympathies, as did a few others, such as Roberto Marcolongo, who had already worked in electron theory and was a strong supporter of relativity after the war. Yet the relativity revolution seemed at this time to threaten only science, to borrow an observation from I. Bernard Cohen in his recent book.[9]

In prewar Italy, there were two sorts of vocal reactions against relativity. The first resembled the 1911 presidential address of W.F. Magie of Princeton to the American Physical Society and Section B of the American Association for the Advancement of Science. Magie found the abandonment of the ether "a great and serious retrograde step in the development of speculative physics." The relativists were asking physicists "to abandon what has furnished a sound basis for the interpretation of phenomena and for constructive work in order to preserve the universality of a metaphysical postulate."[10] The young Italian experimental physicist Orso Mario Corbino might have agreed in 1907 with this part of Magie's complaint: he, too, found the principle of relativity metaphysical. For Corbino, as for the mathematician and philosopher Federigo Enriques in his 1906 book *Problemi della scienza*, the concept of the absolute ether "for its simplicity remains in any case today the only conceivable basis for a concretely developed electromagnetic theory."[11] Corbino in 1907 was sympathetic to the goals of the electromagnetic world picture, which as he understood it permitted the eventual detection of absolute motion and excluded the principle of relativity; by 1912, however, he had changed his mind, apparently convinced by relativity's Italian supporters and by experimental measurements of the masses of high-velocity beta rays published between 1908 and 1910.[12]

Max Abraham, German-born theoretical physicist, student of Max Planck's at Berlin, and professor of mechanics at the Milan Polytechnic from 1909 to 1915, exemplified the second kind of response, which went beyond the simple rejection of relativity to the creation of alternative theories. Abraham never abandoned his commitment to the ether and the electromagnetic world picture, as Corbino and Enriques eventually did. Abraham had developed a theory of the motion of electrons, incorporating an electromagnetic ether, in 1902, which was still regarded as a viable rival to the "Lorentz-Einstein theory" when he arrived in Milan to teach.[13] Like many others before the war, Abraham found special relativity revolutionary, overturning "fundamental conceptions of kinematics and dynamics," and "the proven foundations of all physical measurement." He noted, however, that "the apparent generality of the solution to the problem of space and time met halfway the philosophical impulse of the time for the unification and summarizing of knowledge."[14] In Italy Abraham consolidated his position as a public opponent of special relativity with the development, beginning in 1911, of an electromagnetic theory of gravitation, based on Einstein's

1911 derivation of the bending of light in a gravitational field, a consequence of the identity of inertial and gravitational mass and the mass-
energy relation $E = mc^2$. This Abraham saw as taking "the ax to one of the
roots" of special relativity, thus signalling its downfall.[15] He claimed his
own theory

would be the realization of the 'body Alpha' postulated by C[arl] Neumann to fix the
system of absolute reference. . . . And thus the relative becomes absolute. Or better: the
concepts of absolute and relative are fused in a higher unity.[16]

In the pages of *Scientia* in 1914 Abraham attacked the still uncompleted
general theory of relativity for its failure to live up to its promise, but he
predicted an "honorable burial" for special relativity because of its
service in the critique of the concepts of space and time.[17]

Abraham may have been seen by his experimentalist colleagues
unwilling to give up the ether, such as the dean of Italian experimentalists Augusto Righi, as "able to give theoretical dignity to their position."[18]
When La Rosa changed his mind on special relativity in late 1911, he
began to send Abraham drafts and reprints of his antirelativity papers and
received explanations, corrections, and support in return. In thanking La
Rosa for his little book on the ether, Abraham concluded:

You have made yourself defender of this poor slandered and still scorned ether, which has
had to bear the defects of all past theories. I hope that your book succeeds in restoring its
reputation.[19]

RELATIVITY: THE EVOLUTIONARY THEORY

After World War I, as with so much else, the dimensions of the
presentation of relativity were transformed. The newspapers, first of all,
seemed to have had few qualms about the language they used. The
London *Times* headlines for the article on the 6 November 1919 Royal
Society "canonization" of Einstein read, "Revolution in science/New
theory of the universe/Newtonian ideas overthrown," and a subhead
read "space 'warped.'"[20] Although the *New York Times* avoided any
form of the word "revolution," the headline of its second story certainly
implied an upsetting theory:

Lights all askew in the heavens/Men of science more or less agog over results of eclipse
observation/Einstein theory triumphs/Stars not where they seem or were calculated to be,
but nobody need worry/A book for 12 wise men/No more in all the world could comprehend it, said Einstein when his daring publishers accepted it.[21]

The conservative Milan daily *Corriere della sera* followed the lead of the *Times* of London in its first report on 11 November in its "London Courier" column, under the headline, "Scientific revolution/Light is matter and not motion/Prediction of a scientist/The cosmic ether passes into history."[22]

Yet the equanimity with which relativity was termed revolutionary by its sympathizers among scientists was dissolved by the war. Levi-Civita had been in correspondence with Einstein in the spring of 1915,[23] and he followed the final stages of Einstein's development of the general theory later that year, even though the war divided them. Levi-Civita had then made relativity central to his own research, publishing fourteen papers on it by the end of 1918, just at the time when he was called to the University of Rome.[24] The Rome geometer Guido Castelnuovo, who had also interested himself in relativity before the war, promoted a series of lectures by Levi-Civita and Marcolongo, mathematical physicist at Naples, for the Mathematical Seminar of the University of Rome with the support of Vito Volterra, its director and preeminent mathematical physicist. The lectures took place in the spring of 1919,[25] as if to prepare the Italians for the announcement of the positive results of the eclipse expeditions at the Royal Society the following November.

Levi-Civita presented on 8 March what became a very influential paper, also translated into French and Spanish, entitled "How a conservative could reach the threshold of the new mechanics." This semipopular paper aimed at convincing those familiar with variational methods in classical mechanics that a legitimate desire to seek conceptual synthesis and to generalize from well-established classical formulas of Hamiltonian mechanics could lead one through rather simple mathematics to Einstein's theory of gravitation. The mathematics may have been relatively simple, but the paper was scientifically and philosophically profound, because it constructed a bridge between Einstein's gravitational theory and classical mechanics rather than focusing on the cleft between them. Levi-Civita prefaced the technical part of the paper with the remark that no scientist could be fearful of the new, but that researchers had to be conservatives: they had to protect established scientific patrimony with close analysis and severe criticism of any effort to tear down previously successful theories.[26]

This article set the tone for all favorable scientific presentations of relativity in subsequent years. Whatever the connotations of the use of the term "revolutionary" for Einstein's theories had been in Italy before the war – intellectual, political, modernist, or even just fashionable[27] – the political and cultural climate in Italy after the war would not permit even a socialist such as this one to use the language of revolution when advocating relativity. If relativity's supporters used political metaphors at all, they chose the opposite: "conservative." Levi-Civita and his student Attilio Palatini,[28] Marcolongo,[29] Castelnuovo,[30] Luigi Donati,[31] and the geometers Guido Fubini[32] and Gino Fano[33] – all practitioners and promoters of relativity in their own work, in their teaching, and in popular articles for journals such as *Scientia* and *Elettrotecnica* and articles and interviews for newspapers – all stressed continuity with the past, improvement, generalization, progress: in effect, *evolution* rather than revolution.

The mathematical physicist and electrical engineering professor Luigi Donati, in lectures delivered to the Bologna section of the Associazione Elettrotecnica Italiana in December 1921 and January and February 1922, presented relativity as "a successful attempt at compromise" between classical physics and the facts observed in opposition to it. For Donati, since the development of science, and of physics in particular, was asymptotic toward objective truth,

the new theories were stripped of that revolutionary character with which they appeared to be clothed, and, in their rigidly logical structure, they assume the appearance of a necessary adaptation to reality.[34]

Marcolongo, in an article printed in the newspaper *La provincia di Padova* the day after Einstein lectured at the University of Padua in October 1921, asserted that relativity was "not a revolution, but a slow, constrained evolution," to which physical theories, as all things human, were subject. In spite of the call of "a certain astronomer" to "save the law of Newton," for Marcolongo

the great law of Newton is not in any danger, and . . . there is nothing to be saved because all is secure. On the contrary, one of the most beautiful characteristics of the new theories is that they conserve the glorious edifice constructed by Newton, while improving it with modifications which are qualitatively very slight and conceptually grandiose.[35]

In articles on special relativity in *Scientia* in 1924, he conceded that relativity might seem revolutionary, but it was also "judiciously conservative and therefore, in a certain sense, less subversive than the Galilean revolution" of the seventeenth century had been of Aristotelian physics. For Marcolongo, the theory of relativity was part of the ongoing process in science of approaching truth, of "being integrated into a

larger, more complete, more comprehensive vision of the entire experimental world."[36] Thus, "scientific truth has to be understood not as an
agreement between thought and things, but as a continuous process of
evolution to richer and more harmonious forms."[37]

Consider finally the case of Giuseppe Armellini, the principal astronomer who supported relativity. He was called to Padua as Levi-Civita's
successor as professor of mechanics, went to Pisa in 1920 as professor of
celestial mechanics, and in 1922 was named professor of theoretical
astronomy at Rome and director of the observatories there. At Levi-
Civita's suggestion to the publisher Ulrico Hoepli,[38] Armellini wrote the
preface to the Italian translation of the text on relativity by the German
astronomer August Kopff. There he affirmed his belief that general
relativity represented real progress in science; by implication it was
therefore not revolutionary. It had given a solid "experimental" foundation to classical mechanics by eliminating "that encumbering parasite,"
the absolute frame of reference; it thus had solved the great problem of
the relations between classical mechanics and electrodynamics by unifying them. Armellini also asserted that general relativity clarified the
nature of gravity, "that mysterious force" called by Leibniz and many
since 'occult,' by showing that gravity and inertia were the same
phenomenon.[39] He stressed the importance of relativity as lying more in
the "logical" realm than in the experimental; many astronomers, himself included, still had some doubts about the so-called experimental
tests of general relativity.[40] However, far from overturning Newton's
gravitational theory, "the great discovery of Einstein thus puts the seal
on the brilliant work of Newton and constitutes, so to speak, the
ultimate completion of a marvelous and imperishable edifice!" As
Armellini put it, just as Newton showed that the planets move, according to Kepler's laws, *as if* they were attracted by an inverse square law,
so "relativity shows exactly that, *omitting consideration of negligible
quantities*, the motion of the celestial bodies occurs 'as if' they were
attracted by the law named above."[41] "Thus . . . the gravitational
theory of Einstein has taught us that the law of Newton is not rigorously
true, but it is valid only to a very high approximation." Again, on the
logical plane, the unity that relativity brought about was impressive: not
only did it unify classical mechanics with electrodynamics, but it unified
physics with geometry.[42] Armellini was convinced that relativity was
becoming the foundation of theoretical physics, and was making scientists see the universe from a completely new point of view.[43]

Among scientists only Federigo Enriques seems to have used the term "revolution" in a positive sense when referring to relativity after the war. In his introduction of Einstein on 22 October 1921, during the latter's visit to the University of Bologna to give three lectures, Enriques was willing to represent relativity as a revolution against Kant, for whom absolute space and time were essential as forms of perception; but even here Enriques had already spent the bulk of his remarks stressing continuities with the past of science – relativity was a generalization of Newtonian mechanics, a better approximation to truth, more true than classical mechanics, the unifier of physics and geometry, and so on. And he further softened the force of the revolutionary metaphor by placing Einstein's work in the context of philosophical discussions of relative motion going back to Parmenides, thus making the philosophical revolution against Kant appear to be the inevitable result of the evolution of philosophical thought.[44]

Even Einstein himself, in his first lecture in Bologna, was quoted to the same effect. He was reported as saying that he did "not want to dethrone Newton . . . but only to continue him; he [did] not want war but peace."[45] Indeed, Einstein denied at every opportunity that relativity was revolutionary. In his 1917 popular book, which appeared in Italian in 1921 as *Sulla teoria speciale e generale della relatività*, with a preface by Levi-Civita, he wrote that "the most beautiful fate of a physical theory is to point the way to the establishment of a more inclusive theory, in which it lives on as a limiting case."[46] In his article on relativity commissioned by the *Times* of London just after the announcement at the Royal Society of the results of the eclipse expeditions, Einstein concluded by stressing that

no one must think that Newton's great creation can be overthrown in any real sense by this or any other theory. His clear and wide ideas will forever retain their significance as the foundation on which our modern conceptions of physics have been built.[47]

On his arrival in the United States in April 1921, Einstein was quoted as saying that relativity "was a step in the further development of the Newtonian theory."[48] The next day he was quoted at greater length on the matter:

There has been a false opinion widely spread among the general public that the theory of relativity is to be taken as differing radically from the previous developments in physics from the time of Galileo and Newton, that it is violently opposed to their deductions. The contrary is true. Without the discoveries of every one of the great men of physics, those who laid down preceding laws, relativity would have been impossible to conceive, and there would have been no basis for it. Psychologically, it is impossible to come to such a theory at once, without the work which must be done before.[49]

At Columbia University, Michael Pupin introduced him "as the discoverer of a theory which is 'an evolution, not a revolution of the science of dynamics.'"[50] In London in June of that year, in spite of the fact that Lord Haldane insisted that relativity was "more revolutionary than [the conceptions] of Galileo, Copernicus, or Newton,"[51] Einstein countered with the opposite view. According to the report in *Nature*, he

seemed, too, with earnestness and obvious sincerity, to disclaim for himself any originality, and he deprecated the idea that the new principle was revolutionary. It was, he told his audience, the direct outcome and, in a sense, the natural completion of the work of Faraday, Maxwell, and Lorentz.[52]

Thus Einstein was entirely consistent in these years in his presentation of relativity as an evolutionary development in physics, to colleagues, general audiences, and the press, and to Germans, Britons, Americans, and Italians.

This mode of presentation functioned in the first instance to protect relativity, and science in general, from political turmoil. Pitting one scientific theory against another in analogy to dethroning a king, subverting a government, or overturning a social order made science seem too much like contemporary political affairs. Einstein himself, in his Bologna lectures, pointed to the benefits of the evolutionary view of scientific change. For him, relativity was "something that distances the human spirit from life, this life of sorrows and passions, of small and great struggles, of grudges and implacable hatreds. . . ."[53] Marcolongo explicitly tried to quell the revolt raised by the passionate call to arms "in defense of the law of Newton" issued by "a certain astronomer" and others in the Italian scientific community with his assurances that there was nothing to be saved because all was secure. Neither theoretical science nor the scientific community, therefore, was threatened with revolution, according to these advocates of relativity. Thus, both inside the scientific community itself and in the image of it projected to the wider public, they were attempting to maintain, or even impose, what Michel Biezunski has called in another context the "social-epistemological order."[54] As Biezunski has emphasized, the process of popularization of a theory still being debated is "part of the struggle to make the new ideas accepted,"[55] and it is entirely understandable that relativity's supporters after the war would have wanted the theory not be or seem revolutionary.

The reasons for the strategy adopted by relativity's Italian proponents after the war may have been the following. They may have judged that the scientific enterprise would best be served by being presented as stable enough to permit gradual change and thus to be explicitly not revolutionary; this approach would protect science from the postwar European connotations of "revolution" as the overthrow of the entire bourgeois order. Or they may have actually come to believe that science did change gradually, that progress was possible without upheaval, that relativity was in fact not so destructive of the old scientific order as it had appeared to be before the war. By 1921 both experimental and theoretical evidence could be, and was, invoked in its support. The experimental evidence was well known. The theoretical evidence included Levi-Civita's 1919 paper showing that general relativity could be understood as a generalization of classical Hamiltonian mechanics. These two motives were probably present in varying proportions in each of relativity's Italian supporters. In any case, their strategy asserted the existence of order and stability in science, and they committed themselves to maintaining that order and the possibility of progress in science through gradual change in a world full of upheavals in other spheres. Furthermore, the evolutionary metaphor not only shaped the views of scientific change put forward by these men, but it was also capable of generating significant scientific work, such as Levi-Civita's 1919 paper.

The opponents of relativity likewise used the same political metaphors of revolutionary and conservative to describe the theory and their positions on it, with meanings suitably transformed. For them, relativity *was* revolutionary and they, as conservatives, did not want revolution. Carlo Somigliana, mathematical physicist at Turin and a contemporary of Marcolongo's, referred to himself as an "unconvinced conservative" on postcards to Levi-Civita thanking him for reprints of his pro-relativity articles. Somigliana noted that Woldemar Voigt had used in 1887 what were later called the Lorentz transformations to describe the propagation of waves in incompressible media when the source is in motion, a fully classical problem. He therefore insisted that both Galilean and Lorentz transformations had peacefully coexisted in classical physics and that Lorentz transformations could hardly be used to destroy classical mechanics.[56] Somigliana found his role as critic, pointing out slips in logic or physical reasoning. He was baffled by what he saw as the "unlimited faith" and "almost mystical enthusiasm" of ordinarily sensible men trained in the rigors of the exact sciences[57] – such ordinarily sensible men as his colleagues at Turin Fano and Fubini, or Levi-Civita, or his contemporaries Marcolongo and Volterra, both about age 60 in 1920 and sharing with Somigliana the same training in the Italian "school" of elasticity in the 1880s.

Somigliana's rather gentle political rhetoric became stronger among some experimental physicists and observational astronomers who called relativity "revolutionary" to condemn it. It was a revolution they had not made and did not want. Two prominent university professors undertook experimental work or began preparations for experiments, and a third adopted Walter Ritz's ballistic theory of light, all in order to attempt to show that Einstein's relativity was unnecessary. The experimentalists were not naive measurers, working without hypotheses, as one historian has recently claimed.[58] They were, however, usually committed to a world view which included an ether at rest, absolute space and time, recourse to experiments as both the source and arbiter of physical theories, and expectations of visualizable, easily understandable models of physical processes. They were realists, believing scientists discovered or sought to discover elements of a reality outside themselves. They believed that their concepts, whether of mechanics or of electrodynamics, were concrete, intuitive, and commonsensical. They accused Einstein of being metaphysical, too abstract, too mathematical. Occasionally they agreed with the mathematicians and mathematical physicists that theory construction was a more or less creative process of the human mind. But for the experimentalists theories required decisive experimental proofs for validation, and theoretical beauty or mathematical elegance or even grand synthesis were not adequate arguments for supporting relativity, as they were for the mathematicians. Relativity, for the physicists and astronomers, meant revolutions in their intuitive concepts, in their methodologies, and in their world views. Few scientists can adapt easily to such situations, and the Italian experimentalists were no exception.[59]

Augusto Righi, dean of Italian experimentalists, himself never indulged in polemics against relativity, but the theory, particularly special relativity, was clearly incompatible with his belief that the foundations of natural philosophy had to remain accessible to all, and that they had to be based on "fundamental intuitive ideas," "the logical consequences unconsciously derived from the intellectual baggage accumulated by the race over centuries of observations and the rational use of human intelligence."[60] Righi devoted the last two years of his life to a critique of the Michelson experiment, the experimental foundation of relativity for those who required experimental foundations. He believed he had found an error in Michelson's theory, so that the null result should have been expected, and he was planning an *experimentum crucis* to test his interpretation at the time of his death in June of 1920. Righi's authority was frequently invoked by antirelativists in the next several years, and it was not until 1925 that an Italian physicist claimed in print to be able to show that Righi had made a mistake and that his reanalysis of the interferometer experiment in fact supported Michelson fully.[61]

The experimentalists' polemic against relativity issued rather from Michele La Rosa, professor at Palermo and thirty years younger than Righi.[62] La Rosa had given up his cautious support for the revolution in theoretical physics by late 1911, apparently upon finding the way out offered by the ballistic theory of light advanced by Walter Ritz, as described in articles by Daniel F. Comstock and Richard C. Tolman in the *Physical Review*.[63] The ballistic theory was relativistic in classical terms, that is, it employed Galilean transformations in describing bodies in motion, and therefore it maintained classical mechanics fully secure. After the war and the appearance of general relativity, La Rosa used very sarcastic language in writing about relativity for popular audiences. In a brief essay contribution to the compilation of the opinions about relativity held by seventeen Italian scientists and philosophers, published as an appendix to the translation of Kopff's text on special and general relativity, La Rosa called the concepts of general relativity not true ideas but "pseudoideas or phantasms, repugnant to honest good sense." He continued,

No, my good companions of misadventure, don't worry yourselves; the image of the physical world that relativity has given us is not made for us poor mortals. It belongs to a world which is not ours, to the fourth dimension.[64]

Among astronomers the rejection of relativity was almost universal, save for Giuseppe Armellini, discussed above. The priest Giovanni Boccardi, director of the observatory and professor of astronomy at the University of Turin, indulged publicly in a sarcasm similar to La Rosa's: the theory was one of innumerable cosmological fancies presented to the public, developed *a priori*, justified by nothing. It described how things might be in a hypothetical world different from ours; it might be well founded, but it was not true.[65]

Both the astronomers and the experimental physicists seemed to believe that scientific theories and laws emerged somehow from experi-

mental "facts" and observational information. Thus the *a priori*, hypothetical, postulational nature of special and general relativity was
unacceptable to them. So also was the unfamiliar non-Euclidean
geometry which Einstein claimed represented physical space. Boccardi
published an article in the Turin newspaper *La stampa* three weeks
before Einstein's first lecture on 22 October 1921 focusing on these
questions, entitled "In defense of the law of Newton (à propos the
theory of Einstein)," which led to a pair of exchanges with the geometer
Guido Fubini.[66] Boccardi referred to the "revolution" said to have been
brought to the physical sciences by Einstein's theory, and to the dubbing
of Einstein as "the Newton of our days," and he found the state of mind
of the public in the face of these claims "harmful to science itself,
because it is destined to give license to so many other sensational
discoveries." He objected to the four dimensions of space-time, on the
ground that "another Einstein could then come forward with a new
hypothesis, and so we would have space of 5 dimensions, or even of *n*
dimensions." He found the arguments for general relativity, such as the
advance of the perihelion of Mercury, "the least of our worries,"
because there were better explanations, and the experimental tests "too
small to be true."[67]

The president of the Italian Astronomical Society Vincenzo Cerulli
wrote a scathing article for the Kopff volume in which he decried the
"ultrasensible and unrepresentable" interpretation of the action at a
distance force of Newtonian gravitation. Cerulli considered Einstein's
general theory of relativity not as revolutionary or even progressive but
as a "degenerative crisis" that would lead the muse of astronomy Urania
back into those bonds from which Newton had freed her by admonish-
ing *"Hypotheses seu metaphysicae seu qualitatum occultarum in philosophia experimentali locum non habent."*[68] As the experimental physicists
did not recognize the ether as an hypothesis, so the astronomers did not
recognize the action at a distance force of gravity as an hypothesis. Both
were considered inductive generalizations from experience, from
"facts," or even simply *as* facts *of* experience.

Thus, among the scientists and mathematicians directly concerned
with relativity, revolution had become a dirty word. Relativity's sympathizers, including Einstein himself in his Bologna lectures, stressed
progress and improvement, continuity and evolution, rather than subversion, upset, discontinuity, and revolution. Relativity's opponents
instead emphasized precisely its subversive, discontinuous, revolution-

ary aspects and made it clear that they believed such developments were "degenerative," "repugnant," and unnecessary.

RELATIVITY: THE CONSTRUCTIVE THEORY

The engineering/architectural metaphor of construction, destruction, and rebuilding, has both revolutionary and conservative overtones, depending on the emphasis. The mathematicians were wont to use the idea of theory construction in referring to relativity, and to emphasize the greater synthesis and harmony of the new structure in comparison to the old. A few spoke of Einstein's new construction of the universe. A particularly vivid use of the metaphor was elaborated by the electrical engineering professor Ferdinando Lori in an article that appeared in the newspaper *Il Veneto* the week that Einstein spoke at Padua in October 1921.[69] Lori characterized relativity as binding together in union or wedlock disparate experimental results, the Michelson-Morley experiment which failed to reveal the motion of the earth through the ether, the anomalous advance of the perihelion of Mercury, and the nonrectilinear light propagation along the sun's limb, with the mathematical theories of spaces of more than three dimensions, non-Euclidean geometries, and the tensor calculus of Levi-Civita and his teacher Gregorio Ricci of Padua. The mathematical theories were "instruments improved by the work of refined artificers in separate and distinct workshops," "like the stones which compose and adorn an architectural monument."

The artist drags them from their rest in the bowels of the mountain . . . : he squares them, cuts and incises them, connects them, and the work which results, multiplying their beauty, creates its own beauty. Thus the Giant of thought, carrying his own structure among the greatest edifices of science that up to the moment of his pilgrimage were like temples of separate deities, draws from them the elements that he needs to erect in his greater temple the throne of a more powerful divinity.

Lori here shaded off into religious language, picking up the ever-present incomprehensibility issue: "One who is not initiated can only worship in silence." But he emphasized that there *were* initiates, and those who had faith in the daring of human intellect would be able to boast of the height and light of this most singular summit conquered by human genius.

In a later lecture on relativity, Lori put the idea of theory construction in sharper relief by asking about the truth and reality of mathematical constructions and their applicability to other fields.[70] He started from the premise that the cultivator of mathematics did not know or want to find out or care whether what he asserted was true, while denying that this was a nihilistic position. On relativity, he claimed that pure mathematicians were content that Einstein had demonstrated how one could reach a broader, more comprehensive picture of natural phenomena from a theory of four-dimensional space, but that physicists and others wanted more: to know about the real nature of space and time, or space-time. Returning often to the problem of mathematical language and its translatability into ordinary language, Lori phrased the entire lecture in terms of questions. He wanted to try to enlarge the circle of those who understood relativity beyond those who knew the tensor calculus; if this were not possible, then the theory was of purely mathematical interest and had no physical or philosophical content. His questions made it clear he believed that was not the case.

RELATIVITY: THE PHILOSOPHICAL THEORY

All the rhetoric, the political metaphor, the polemic, and the real conceptual innovations of special and general relativity could hardly have failed to have had an impact in Italy outside science itself, especially in philosophy, religion, and politics.

Of course some philosophers had been interested early in the "revolution against Kant," the ontological claims about the nature of space and time, as well as in epistemological attacks on objectivity and on the separation between the observer and his observations, between subject and object.

The philosophical debate in Italy[71] must be situated in the context of the neoidealist near-hegemony in philosophy and the longstanding neoidealist devaluation of scientific and mathematical knowledge and research. Epitomized in the *Logic* of Benedetto Croce, published as a book in 1908 but based on a paper first published in 1905, this devaluation denied that science or mathematics were creative activities of the human spirit, that they had philosophical interest or value as knowledge, and that they could lead to truth. Scientific and mathematical research resulted not in genuine knowledge but only in classification schemes or techniques useful for practice.[72] Therefore, most Italian neoidealists just dismissed the question of the philosophical consequences of relativity, unlike the situation in England, for example.

There such idealists as Wildon Carr, Collingwood, Eddington, and Whitehead made relativity considerations central to their writings in this period. Ugo Spirito, young neoidealist follower of Giovanni Gentile, found the efforts of Wildon Carr and his countrymen Alessandro Bonucci and Antonio Aliotta to advocate or at least discuss an idealist interpretation of relativity "arbitrary" and the idea of an idealist science "a contradiction in terms," since an idealist science would no longer be science but philosophy. Furthermore, he denied any connection between the physical principle of relativity and philosophical relativity, finding the physical theory of relativity to be "only pure realism."[73]

The obvious idealist position, that the theory of relativity exemplified scientific activity as a creative process of the human spirit, was championed principally by the mathematicians. One physicist, Sebastiano Timpanaro, a student of Righi's, held this view of scientific activity in general although he had difficulties with the theory of relativity, maintaining an "ethero-ballistic" theory of light, a variant of the ballistic theory of Ritz supported by La Rosa. Timpanaro also founded the journal L'Arduo, which aimed at promoting this view of science and carried many articles on relativity in its brief life (1914, 1921–1923).[74] Relativity was also used by a few philosophical pluralists, such as the anti-neoidealist Antonio Aliotta, to bolster their own positions.[75]

Those in quest of a philosophy of science which left room for traditional religion could respond to relativity in at least two ways. The obvious way was to reject it and seek a way to maintain classical mechanics with its attendant absolutes, as La Rosa did by adopting the ballistic theory of light. After all, for Newton, absolute space was the sensorium of God. An alternative course was taken by Giuseppe Gianfranceschi, a University of Rome trained Jesuit physicist who later became director of Vatican Radio, professor of physics and rector of the Vatican's Gregorian University, and president of the Pontifical Academy of the New Lincei. Gianfranceschi accepted relativity the mathematical theory as a useful guide to research but asserted that it had no value as a physical theory, since it was composed of purely subjective postulates and did not reproduce the real world. As a philosophical theory he found it a *teoria demolitrice*, a destroyer of all knowledge of the external world. He insisted on an interpretation based in the first instance on human failings: since we are human, we cannot expect to discover or measure absolute motions. However – and here his interpretation led him toward a concept similar to Carl Neumann's body

Alpha, an imagined absolute reference frame attached to the fixed stars – an entity or a being (or a Being) with nonhuman, that is to say superhuman, powers of perception, which/Whom Gianfranceschi called "the Observer," could see all motions in the world, both relative to other bodies and absolutely, with respect to it/Himself.[76]

RELATIVITY: THE CULTURAL THEORY

Unlike the mathematicians, physicists, astronomers, and engineers, neither the philosophical nor the religious writers seem to have made much of the political metaphors of revolution and reaction, revolutionary and conservative. But the appropriation of the metaphors provided by relativity to buttress a discussion of relativism as a broad cultural movement was explicitly set out in an essay published in July 1921 and then incorporated into a little book by the theater and culture critic Adriano Tilgher, entitled *Relativisti contemporanei*, which appeared a few days after Einstein's late October visit. Tilgher introduced Einstein as the third *duce* of the formidable relativist assault on tradition coming from Germany and working to renew the foundations of knowledge: he linked Einstein's theories of relativity in science to the philosophical relativism of Hans Vaihinger's *als ob* or "as if" and to the cultural relativism of Oswald Spengler's *Decline of the West*, adding Giovanni Gentile's actual idealism to the list of contemporary relativisms.

Tilgher declared relativism to be the major cultural event of the time, emphasizing the revolutionary character of its results for the usual representation of all aspects of the world. He inverted the way the political metaphor was used by the scientists: at this time he called the theory of relativity revolutionary to praise it. For Tilgher, the essence of Einstein's theory was the introduction of subjectivism into the core of the sciences of nature, the last stronghold of objectivity, thus making nature no longer independent of mind. He correlated the movement to relativity in science both chronologically and intellectually with pragmatism in philosophy, titanism or the revolt against tradition in the arts, the capitalism of the trusts, imperialistic and nationalistic politics, and World War I. All these movements shared one worldview – the refusal to admit a single truth, a single justice, a single good, in short, any absolute theoretical or practical principle or order. Since everything became a matter of will and action, Gentile's activist philosophy actual idealism likewise fit very nicely.

Tilgher's essay on Einstein was published in the conservative Turin newspaper *La stampa* in early July 1921,[77] nearly three months before Boccardi's article in the same paper, "In defense of the law of Newton," and three and a half months before Einstein's visit to Italy. The essay was reprinted in the Bologna newspaper *Il resto del carlino* the day of Einstein's first lecture there.[78] The first edition of *Relativisti contemporanei*, containing the essays on Spengler, Vaihinger, Einstein, and Gentile, appeared the next week, while Einstein was still in Italy.[79] The book provided a wealth of scientific and philosophical metaphors that might be employed in any context. It had immediate political repercussions.

RELATIVITY: THE POLITICAL THEORY

Days after the book's appearance, the Fascist movement met in Rome for its third annual congress, at which the delegates voted to transform the movement into a political party. This congress represented a personal victory and a return to power for Benito Mussolini. Three months earlier he had resigned from the movement's executive committee over opposition to his "pact of pacification" with the socialists, although he had continued to publish his newspaper *Il popolo d'Italia*. The newspaper had been carrying discussions about the principles and program of fascism since long before the parliamentary elections in May 1921, in which Mussolini and other Fascists were first elected to the Chamber of Deputies. It can be used as a convenient gauge of his political opportunism from the time he founded it in November 1914 as a renegade interventionist socialist, and especially from the establishment of the Fascist movement in March 1919.[80] It will be useful here to review briefly the early development of Fascism in Italy.[81]

Fascism in 1919 was a movement of the disaffected, of marginal men, appealing especially to upwardly mobile ex-revolutionary syndicalists, the elite among the war veterans – the officers and the *arditi* or shock troops – and the intellectual proletariat, often Futurists. It was an "antiparty," disdainful of creeds and dogmas, rebelling against the constraints of rules, organizations, and bureaucracies. Yet its thrust, its "state of mind," was distinctly antimonarchical, antiparliamentary, anticlerical, antidemocratic, antibourgeois, and antisocialist. Antidemocratic, it stressed natural hierarchies in politics and society. Antisocialist, it countered internationalist and neutralist sentiments with interventionist and nationalist rhetoric, yet pressed for such syndicalist goals as the eight-hour workday, a minimum wage, and representation of workers in the running of industrial firms. Antibourgeois, it called for a progressive tax on capital and government confiscation of 85 percent of war profits. Anticlerical, it called for government takeover of the property of religious congregations. Antiparliamentary, it proposed the abolition of Parliament and the granting of legislative power in specific areas to national technical councils.

One by one, all these attitudes save the nationalist and the hierarchical were transformed as Fascism became a mass movement after late 1920 and voted itself into a party in November 1921 in accord with Mussolini's strategy of the *via parlamentare*. (They changed even more by October 1922, when Mussolini received the reins of power "legitimately" from the hands of the king.) The December 1921 party program was drawn up precisely "to characterize its 'credo.'"[82] There were distinct elements of continuity with the Fascism of 1919. According to the program,

the nation is not the simple sum of living individuals, nor the instrument of parties for their own ends, but an organism comprehending the indefinite series of generations of which individuals are fleeting elements.

The program called for the establishment of professional and economic guilds or corporations having the right of election to the national technical councils, which were to be concerned with all the productive aspects of the nation. It contained proposals for social legislation for the benefit of workers, and syndicalist proposals for the involvement of workers in the planning and functioning of industries. It asserted that Fascism did "not believe in the viability of or the principles that inspire the so-called League of Nations," nor in those "of the red, white, or other colors of Internationals."[83] It called for administrative decentralization in order to reduce the bureaucracy. It sought the restoration of the prestige of the national state.

Yet the program of December 1921 showed pronounced discontinuities with the Fascism of 1919 as well. The program sought the restoration of the authority of government, to the breakdown of which the Fascist movement had contributed much. It recognized the freedom of the Church in the exercise of its spiritual ministry. It validated the social function of private property and opposed collectivization, endorsing the denationalization of the railroads and telephones and private competition for the state monopolies in the mail and the telegraph, yet called for state protection of certain industries and agriculture from dangerous foreign competition. It called for balancing national and local budgets. In its statute, the party defined itself as "a volunteer militia at the service of the nation," whose activities were "support[ed] on three pivotal points: order, discipline, hierarchy."[84] The 1921 program thus combined laissez-faire and liberal elements with the more familiar nationalist and syndicalist thrust of the 1919 mind set.

According to historian Adrian Lyttelton, in 1919 Mussolini was content to have Fascism appear "as a movement which stood for the pragmatic pursuit of immediate reforms, without reference to abstract ideological goals,"[85] that is, for action rather than theory. In a speech to the first Fascist national congress in October 1919, Michele Bianchi, ex-syndicalist labor organizer, called the *fasci* "an organization without ready-made doctrines," so that "problems are faced by them not in series but according to whether the moment is ripe."[86] This tactical flexibility allowed Mussolini and other Fascist leaders of his persuasion not only to play on the widespread opposition to entrenched party oligarchies, but also to manipulate characterizations of Fascism to suit the audience. In a March 1921 editorial, Mussolini could still claim with great effect that

we permit ourselves the luxury of being aristocrats and democrats, conservatives and progressives, reactionaries and revolutionaries. . . . Fascism is not a church. It is rather a training ground. It is not a party. It is a movement. . . .[87]

By late 1921 many of the ideological goals and even elements of the mind set of early Fascism had been radically transformed, as Fascism became a mass movement and then a party. Yet Mussolini's tactics remained unchanged. In the fall of 1921 he had been attempting to conciliate various factions in the burgeoning movement and to orchestrate the transformation of the movement into a party, while binding himself as little as possible to specific programs and principles, in order to maintain his "freedom of political action."[88] The overtures to the socialists embodied in the "pact of pacification" had failed by the end of the summer, and it appeared opportune to turn to the right wing of the Popolari (the Catholic people's party), and to the Nationalists, industrialists, and the bourgeoisie in general.

On 22 November, ten days after the Rome congress, Mussolini sought to justify these changes of principles and objectives in a front-page

editorial in *Il popolo d'Italia* entitled "On the track of the great philosophies: relativism and Fascism." Tilgher had called attention to the rise
of the Fascist movement in his chapter on Gentile's actual idealism,
writing that Fascism was only absolute activism transplanted onto political terrain. Even though the editorial did not mention Einstein by name,
perhaps because the paper had published a disapproving account of
Einstein's Bologna lectures only ten days before,[89] Mussolini was evidently pleased to be associated with the philosophy of relativity/
relativism; he considered himself, as he put it, if not a theoretical
relativist, at least a practical one. In Germany, he wrote,

relativity is a most daring and destructive theoretical construction, . . . in Italy it is just a
fact. Fascism has been a super-relativist movement because it has never sought to give
definite 'programmatic' dress to its complex and powerful states of mind, but has
proceeded by intuitions and fragments. . . . Everything that I have said and done recently
is relativity by 'intuition.'

Mussolini then used the philosophy of relativism to attack the claims of
socialism to be "scientific." Referring to a speech he had given some
months before, in which he had asserted that there was "nothing in the
world more grotesque than to call socialism scientific," he continued,

If in fact one understands by relativity the end of scientism [*scientificismo*], the decline of
the myth 'science,' understood as the discoverer of absolute truths, I can boast of having
applied this criterion to the examination of the socialist phenomenon.

He pointed to Fascism's eclecticism and its disdain for fixed categories
as well as its disdain for men who thought themselves the bearers of
absolute truths as examples of this very relativist mentality. He closed
the editorial by picking up Vaihinger's linkage of relativism to Nietzsche's will to power, writing that "Italian Fascism has been and is
the most formidable creation of an individual and national 'will to
power.'"[90]

It may now seem bizarre that eleven months before the March on
Rome, at a time when the quest for power was clear, Mussolini was
boasting that Fascism was a movement without a program and was
priding himself on acting by intuition. Yet this editorial was not a
rhetorical aberration devised to make cultural capital of an intellectual
fashion, as I have already indicated by highlighting his continuing
insistence on "freedom of political action." Indeed, in two articles on
the December 1921 program published in *Il popolo d'Italia* just before
and just after the program itself, he made many of the same points.

First, Fascism wanted to affirm itself as force and capacity for life (to live, to know how and be able to live, is already a very great program!); then, on the bases of the fundamental principles that inspired its action, Fascism constructed little by little the edifice of its theoretical and practical program.

He then argued that Fascism was not the usual sort of political phenomenon:

The words right or left, reaction or revolution, are not terms applicable to the Fascist program, which is reactionary in comparison with the propositions of socialism and profoundly innovative in comparison with other platforms.[91]

Six days later, in the "Preface to the Program," Mussolini emphasized that the program was "a collective work,"

a successful attempt, despite always [being] difficult, to conciliate and balance theory with practice, the ideal with the contingent, the necessary absolute of principles with the inevitable 'relative' of life.

It was "a program, not a masterpiece"; it did not make claims to originality, since "today it is especially impossible to be 'original' in politics." There was "nothing to be said," he continued,

about the fundamental principles of the Fascist program it is a matter of theoretical positions that Fascism takes with respect to the state, the nation, the government, the guilds [*corporazioni*]. We do not tarry over theoretical disquisitions on the concept of the state or of property. We find them satisfactory and that is enough for us.[92]

Having just asserted that it was not necessary to dally over theoretical discussions, Mussolini reinforced the argument for flexibility:

It is hardly necessary to explain that the Fascist program is not a theory of dogmas of which discussion is no longer tolerated. Our program is in continuous formulation and transformation; it is subjected to a labor of unceasing revision, the only way to make of it a living thing, not a dead ruin.

For Mussolini, the program provided Fascists with not only

a political principle for living, but also a moral one. It is not sufficient to have a program: it is necessary to exert the will to prepare the means to realize it in the briefest possible time.

Thus the will to action and power – absolute activism – took precedence for Mussolini over any single programmatic statement, which was always open to review according to the needs of the moment.

Mussolini apparently did not abandon these views of the pragmatic and provisional nature of programs or of the nontraditional character of Fascism as neither left nor right in the first years after the March on

Rome of October 1922. Indeed, he permitted the reprinting of the "relativism and Fascism" editorial in a 1924 collection of his writings and speeches from the period 1914 to 1922.[93] A relativist philosophy which lent support to pragmatic political moves and emphasized the primacy of action over theory seemed eminently suited to the goals of the Fascist movement and party as Mussolini conceived them in these early years.[94]

An apparent attack on Mussolini's position in the "relativism and Fascism" editorial was contained in an article entitled "Relativism and politics," published in January 1922, in the first issue of *Gerarchia*, the new monthly magazine supplement to *Il popolo d'Italia*. The author, painter and writer Ardengo Soffici, claimed that "the doctrine of relativism" had been "founded by a group of Germans and Jews, or German Jews, with Einstein at the head." He considered it to be the central element of "an esthetic and intellectual offensive" mounted by the Germans after their defeat, a "premeditated German Jewish plot against our intelligence." For Soffici, a philosophy of the relative was a contradiction in terms; rejecting the possibility that any scientific discovery or theory could modify our mental conception of the world, which was a creation of the human spirit, he asserted that Einstein himself did not support the relativist philosophical interpretation drawn from the theory of relativity.[95]

As for politics, Soffici claimed that no doctrine which denied the concept of truth could serve as a foundation for political action: it could lead only to anarchy, bolshevism, and chaos – which was the reason that the Germans and Germanophiles (read Tilgher), and the neutralists and defeatists, were propagating it in Italy. A political movement that aimed at reconstituting the law and order of a state, and not simply subverting it, had to have a "well specified truth" as the foundation for political action. It could be a truth of any kind, on the condition that it was not subordinated to the vicissitudes of scientific discoveries. Those might modify our ideas about the mechanics of the world, but they could never alter or destroy the essentially spiritual internal conceptions on which human faiths, civil religions, and great collective passions rested. For Soffici, nothing ought to influence politics less, or did influence the positive political conceptions of his time less, than a theory such as Einstein's or its insidious even if suggestive derivatives, because they were essentially disintegrating and anarchical. Besides asserting that the relativity of space and time was nonsensical to the Italian psyche, intuitive as it was, Soffici emphasized several reasons why a political doctrine of relativity was entirely incompatible with Fascism. Whereas Fascism was a party essentially Italian, national, in favor of order, and realistic, the theory of relativity denied the spiritual foundation of Fascism by being a theory of the uncertain (and therefore non-Italian), of the indefinite (and therefore non-national), of the variable (and therefore non-orderable), and of the fluid (and therefore non-concrete and non-realistic). "For Fascism, in contrast," Soffici wrote, "the absolute exists and cannot not exist. . . . Upon it rest the principle of the *patria*, the principle of order, the principle of hierarchy, the principle of authority, and the civil superiority of our people and our nation." Even though he recognized how useful a good political relativism could appear to some people, who remained unnamed, Soffici asserted that there had been enough of "the old pragmatism;" what was now needed was a politics of "simple, plain, immutable principles," in which the end overshadowed the means – the politics of Machiavelli. All the rest, especially relativism, was empty talk and a trap.[96]

The chauvinist, anti-German, and antisemitic thrust of Soffici's article, together with its emphasis on the need for clear and distinct absolute principles, strongly echoes the Action Française. Soffici was undoubtedly familiar with this elitist movement of the extreme right, based on integral nationalism and antisemitism and established in the wake of the Dreyfus affair, from his association with the Parisian literary and artistic avant-garde dating from his seven years in Paris (1900–1907) and continuing in extended stays and correspondence in later years. He was especially close over fifteen years to the poet and critic Guillaume Apollinaire, who was an admirer of Charles Maurras, the movement's founder.[97]

Soffici's own search for values had led him from the cosmopolitanism of the international avant-garde in Paris, to the innovations of Parisian cubism and the dynamism of Italian Futurism, to the chauvinism of interventionism, and finally to the certainties of nation, order, hierarchy, and authority that he wanted to find in Fascism.[98] Soffici had been writing political columns for *Il popolo d'Italia* nearly monthly since May 1921; he had specifically addressed the questions of order and authority, the *patria*, property, and hierarchy in essays that appeared on 30 August, 24 September, and 13 October. On 30 August the last section of his column, bearing the subhead "Agreement," read:

The spirit that is beginning again to predominate in the arts, in letters, in philosophy, in science is a spirit of traditional order, of classical discipline (not to say academic): it aims from all sides at the restoration of the fundamental realistic principles of authority and of hierarchies. It seems then natural and necessary that politics too ought to set out on this spiritual course. There are in fact some signs of this historic movement. What is to be said is that there will be ever less room for the revolutionary and anarchical romanticism of German-Jewish origin which by now has had its time.

But the real fruit out of season is Bolshevik Futurism.[99]

The long paragraph contains the essence of the January 1922 *Gerarchia* article, including even the chauvinist anti-Germanism and antisemitism. Soffici's later *Popolo d'Italia* articles explored the themes of the *Gerarchia* article in more depth. Thus, his "Relativism and politics" article cannot be viewed simply as an attack on the positions taken in Mussolini's "Relativism and Fascism" editorial. After all, Soffici's article appeared in the first issue of the much-touted monthly supplement to Mussolini's newspaper, and he could scarcely have disapproved totally of the contents. Second, in *Gerarchia* it was paired with an essay on "Relativism and decadence" by Silvio Pagani, under the overall rubric "On the margin of relativism." Pagani dealt in broad strokes with the rise and decline of peoples, asserting that the philosophies of relativity in Germany and elsewhere would lead not to disorder and decadence, but to a new greatness of humanity, as a result of understanding the dialectic between the search for unity and the affirmation of individuality.[100] Pagani's positive assessment of the outcome meant of course that he treated Tilgher and Einstein sympathetically, in contrast to Soffici's attacks. An unsigned editorial note at the end of the pair pointed out that the two articles,

in spite of the apparent contradiction, in a certain sense complement each other. The one examines very closely the concrete problems of human action and psychology, the other the essential problems of thought in itself.[101]

The editor recognized that both of the analyses redounded to the credit of Fascism, even though Soffici and Pagani valued what they took to be relativism in opposing ways.

Analogously, Soffici and Mussolini emphasized contrasting aspects of what each took to be relativism and reached different conclusions on its value. While Soffici focused on the lack of certain, absolute values in relativism in general, which to him meant the lack of secure principles on which to base political action, Mussolini, ever the political pragmatist, stressed action itself – tactics – understanding with Tilgher that programs could not be absolutes because they were human creations and therefore transient, and that they contained no criteria for support save by the force of their imposition by those who possessed the will to do so. Mussolini had set his sights on gaining power and then acting to maintain it by whatever means were necessary from moment to moment. Soffici was taking a longer view, seeking firm principles on which to base a new social order. And he remained faithful to his version of Fascism and to Mussolini to the end. Both Soffici's and Mussolini's views seem to have been necessary for the success of Fascism.

Perhaps taking a cue from Soffici, Tilgher wrote an essay in late 1921 or early 1922 entitled "Relativism and revolution," which was included in the third 1922 edition of *Relativisti contemporanei*. Exhibiting greater sensitivity to immediate political currents than he had earlier, Tilgher now claimed that the modern philosophies of relativism that he had been discussing were essentially elements of a revolutionary philosophy, because they justified an ethic of individual action in attempting to impose one's own views on others. Action became an end in itself, as in Gentile's actual idealism. But Tilgher now claimed that the content of such a revolutionary philosophy could only be negative: it could aim solely at the overturn of the existing order and could provide no principles on which to establish a new one.[102]

Thus neither for Soffici nor for Tilgher could relativism offer a philosophical foundation for a new political order. The 22 November editorial on relativism and Fascism notwithstanding, for Soffici and others Fascism had to become a party which at least appeared to be based on absolutes, spiritual principles which were not subject to the vagaries of fashion or scientific discoveries, and which were most certainly not German or Jewish.

It appears then that Fascism's encounter with Einstein, with the theory of relativity and the philosophies of relativism drawing on Einstein for support, played a role in the articulation of Fascist ideology in this critical period less than a year before the March on Rome. We might consider Soffici and Tilgher as traditionalists, convinced that a political movement needed to be governed by clear and distinct ideas, from which programs for action and actions themselves would flow. They do represent one strain of Fascist thought. And nobody chose to notice that the nationalistic political "absolutes" which served as its foundation were a relativization of the universals of the Enlightenment and liberal democracy.

Mussolini apparently did not take explicit cognizance of those two manifestos, and he continued to play pragmatic politics, of course. We might consider that he knew better what it meant to seize power in the twentieth century, that he was more the heir of Nietzsche than of Descartes. The relativist philosophical base maintained its ground among the competing versions of Fascist ideology at least into the middle twenties.[103] It was thus available for exploitation back in the sciences, whence it had come.

RELATIVISTIC POLITICS AND THEORETICAL PHYSICS

And exploited it was, at the moment when it could foster what was to prove to be one of the most significant institutional innovations in Italian physics in the twentieth century, the establishment of the chair in theoretical physics at the University of Rome.

In early twentieth century Italy, theoretical physics, the expected disciplinary home of relativity, did not exist in an institutional sense; it was a sort of no-man's land between the traditional university disciplines, reified in chairs and institutes, of experimental physics on the one hand and mathematical physics and mechanics on the other. Levi-Civita, Marcolongo, Palatini, and Max Abraham, the German-born theoretical physicist teaching at Milan, for example, were professors of mechanics, Fubini of analysis, Castelnuovo, Enriques, and Fano of geometry. Volterra, professor of mathematical physics, taught a course on relativity in 1919–1920 and 1920–21 but turned to other subjects; Somigliana, also professor of mathematical physics, never taught it. Luigi Donati, professor of mathematical physics and electrical engineering at Bologna, seems to have been the earliest to incorporate relativity in his courses, in 1910–11, but he was not replaced on his retirement in 1921.

Calls for the establishment of chairs in theoretical physics began in Italy soon after 1905, as it happened, but they came to nothing at the time. It seemed almost as though there had to be a candidate or at least another revolution in theoretical physics before it would be possible to have the chair established in the extremely rigid and centralized Italian university system. The candidate was Enrico Fermi, who finished university only in 1922 and did his earliest theoretical work in relativity. The fields were atomic physics (not yet quantum mechanics) and relativity, but the political context as set out above was just as essential.

The long-awaited full-scale university reform that empowered individual faculties to write their own statutes and develop their own instructional programs was finally begun in 1923 by the neoidealist philosopher Giovanni Gentile, Mussolini's first Minister of Public Instruction. The provisional statute of the University of Rome was approved by ministerial decree on 22 November 1924, having been submitted by the university several months earlier.[104] It did not include theoretical physics among the course offerings of the faculty of science.

The possibility of creating a chair in theoretical physics at Rome seems to have presented itself first in the fall of 1924. The faculty of science found itself in the unusual position of having three regular positions available. That one should go to physics was "intuitive," according to the physiologist Giulio Fano, the dean.[105] A decision on the field was postponed, apparently to marshall support for theoretical physics. Two months later the internal political maneuverings became clear. Orso Mario Corbino, professor of experimental physics, director of the physics institute, senator, and former Minister of Public Instruction, spoke in favor of theoretical physics for this chair as well as for the need of another (which would be the third) chair of experimental physics, to be established later. Castelnuovo presented a motion in the names of himself, Levi-Civita, Armellini, Enriques, Severi, and another mathematician, to the effect that the faculty wanted to provide "in a special way" for instruction in that field of physics which was "throwing so much light on the problem of the constitution of matter and energy," emphasizing the "fertile collaboration of theory with experiment" in the new field. The motion was passed with only one negative vote, that of the second professor of experimental physics, Antonino Lo Surdo, who wanted the chair for experimental physics.[106]

Bureaucratic catch-22 situations between the faculty and the ministry delayed the opening of the national competition for the chair past the end of 1925. Meanwhile, Fermi had worked in Göttingen with Max Born and in Leiden with Paul Ehrenfest and had placed second in the competition for the chair of mathematical physics at the University of Cagliari, on the island of Sardinia. The committee divided 3–2 in awarding first place to Giovanni Giorgi, Marcolongo's student, and inventor of the MKS system, contributor to relativity, and thirty years Fermi's senior. The two, reputedly Levi-Civita and Volterra, preferred Fermi "because of the importance and originality of his investigations." The majority, reputedly Somigliana, Marcolongo, and Giovanni Guglielmo, professor of experimental physics at Cagliari, preferred Giorgi for "his greater maturity, . . . his scientific production, [and] the

speculative and philosophical character of his mind," while highly
appreciating the scientific output of Fermi and holding "the most
optimistic expectations for [his] future."[107]

The commission's report, atypically, was undated, but it had to have
been ready before, probably weeks before, the meeting of the Consiglio
Superiore della Pubblica Istruzione of 17 February 1926, at which it was
approved for correctness of procedures. It would appear, however, that
the report or its contents leaked out some two weeks earlier, since it
seems to have been referred to in an article entitled "La 'relatività' e
l'azione," which was published in the Rome Nationalist newspaper La
tribuna on 5 February 1926. [108] The report was situated there as part of a
debate over the relative merits of supporting pure versus applied fields
in science and engineering, theory versus practice. The author, engineer
and Nationalist theoretician Giovanni Ottone, used the difficulties of
the theory of relativity to argue for the importance of scientific studies of
a theoretical nature, in order to benefit

> that practical field to which many people give exclusive value, ever forgetting how in our
> days the multiplication of applications has been the effect of the progress of theoretical
> ideas, and that now the possibility of discoveries due to chance – while leaving to this
> capricious divinity its part – is ever more being confined to the fantasies of lazy people.

"The intellectual material," Ottone continued, "the sum total of theoretical principles which must be at the disposal of anyone who wants to
attempt something new, original" was so large as to worry those concerned with educational programs in the exact sciences and engineering.
Yet the abstruse language of advanced mathematics was "the only one
with which the relations, or the ultimate reasons for things, can be
expressed while – exactly as is the case with the doctrines of Einstein –
ordinary [language] shows itself insufficient." It had become essential
not only for research in physics, mechanics, and electrical engineering,
but also for "progress in industries, transport, instruments of war, in
whatever is needed for our national and individual existence." The
chauvinism of Nationalist and Fascist politics was useful to Ottone as he
noted that it was the Italians Ricci and Levi-Civita who had invented the
tensor calculus that Einstein employed in general relativity, and that
Castelnuovo had recently explored the theory's significance. The practical lesson was that a vast theoretical training was needed to follow,
understand, or evaluate the importance of relativity. And Ottone
blamed "decades of democratic indulgence, of socialist aversion to high

culture, of the exaltation of manual labor to the detriment of intellectual [labor]," "the revolt against intelligence," for what he interpreted, in a "recent ministerial report on the outcome of competitions, especially for chairs in physics," as the lack of many such people. In the competition for the chair at Cagliari, only Fermi and Giorgi had been considered outstanding, and only Fermi and Giorgi, and Fermi's contemporary and friend Enrico Persico, had worked on relativity.

The Nationalist Association had merged with the Fascists in 1923 and since then had been providing a significant part of the ideology for the party and the government. But Ottone's article was not simply a very esoteric argument for the support of theoretical studies as beneficial to practice, or a defense of intellectual work and ideas against the anti-intellectuals in the Fascist movement. The nation demanded progress in the development of industry, transport, and the war machine, but, Ottone reminded his readers, Spengler had shown that progress was not inevitable, as the socialist claimed. Progress required an effort of will and devotion to theoretical studies, for the sake of the nation. This position was endorsed by none other than Mussolini himself, in the 1921 editorial "Relativismo e fascismo," reprinted in his *Diuturna* in 1924. There, Ottone noted, Mussolini had affirmed that Fascism was relativism *par excellence*, Life, and Action, the only reference point being the Nation.

Of course, opponents could undoubtedly cite chapter and verse of Mussolini's voluminous writings to make the opposite point. Yet the winter and spring of 1926 seem to have been important for the Fascist support of high culture in general. The Academy of Italy was established by a decree-law of 7 January 1926, converted into law 25 March 1926, though it was inaugurated only on 28 October 1929. A Circular to universities was issued by the Minister of Public Instruction on 1 May 1926 declaring that "the Government intends to spur scientific activity in university laboratories" and asking for immediate reports on ongoing and proposed research.[109] And the opening of the competition for a new chair in theoretical physics at the University of Rome was announced in a ministerial decree dated 23 June 1926.[110]

It thus appears that Mussolini's highly rhetorical association of himself and the Fascist movement with the relativizing philosophies of Einstein, Spengler and others may have been reappropriated for science by interpreting Mussolini's support for relativism as a declaration of support for mathematical and theoretical studies in the service of practice. This reappropriation gave them a more secure position in the spectrum of Fascist culture than they might otherwise have had, and Italian physics may inadvertently owe much more to Mussolini than has ever before been acknowledged.

ACKNOWLEDGMENT

I should like to acknowledge the generous assistance in the preparation of this paper given by Michelangelo De Maria, Arturo Russo, Giorgio Israel, Giovanni Battimelli, Carlo Tarsitani, Gigliola Fioravanti, Giovanni Paolini, Marina Tesauro, Sebastiano Timpanaro Jr., the Fondazione Einaudi of Turin and its former director Mario Einaudi, John Heilbron, Martin Harwit, L. Pearce Williams, Sander Gilman, Michel Biezunski, Judith Goodstein, Thomas F. Glick, David Vampola, Stanley Goldberg, Alan Beyerchen, the Interlibrary Loan Staff of the Ohio State University Library, and the History Department of the Ohio State University and the Physics Department of the University of Rome for their hospitality.

I am grateful to Lucia Lo Casto La Rosa of Palermo, and to the Accademia Nazionale dei Lincei in Rome and the Rector of the University of Rome for permission to make use of documents in their possession.

This work was supported in part by a University Postdoctoral Fellowship at the Ohio State University and by National Science Foundation Grant SES-8410156.

NOTES

[1] See John Earman and Clark Glymour, 'Relativity and eclipses: the British eclipse expeditions of 1919 and their predecessors,' *Historical studies in the physical sciences* **11**, 1980, 49–85.

[2] Abraham Pais, *"Subtle is the Lord . . .": the science and the life of Albert Einstein* (New York: Oxford University Press, 1982), 305.

[3] Arthur I. Miller, *Albert Einstein's special theory of relativity: emergence (1905) and early interpretation (1905–1911)* (Reading, MA: Addison-Wesley, 1981), 88.

[4] Tullio Levi-Civita, 'Estensione ed evoluzione della fisica matematica (nell'ultino cinquantennio, con speciale riguardo al contributo italiano,' *Atti Società Italiana per il Progresso delle Scienze*, 5^0 riunione, Rome 1911, 237–254; *Opere matematiche: note e memorie* (Bologna: Zanichelli, 1956–74), **3**, 275–291, at 285–286.

[5] *Ibid.*, 288.

[6] Michele La Rosa, 'Le moderne vedute della fisica e di loro rapporti con le altre scienze,' *Calendario astronomico* 1912 (Palermo: Virzì, 1912), p. 3 of offprint.

[7] Michele La Rosa, 'Storia di un ipotesi: l'etere,' *Annali della Biblioteca Filosofica di Palermo* **1**, 1912, 209–223, at 222–223, lectures delivered 27 February and 14 March 1911.

[8] *Ibid.*, 223. Both the characterization of relativity as revolutionary and the idea of relativity's no longer demolishing but rather ordering and rebuilding echo expressions used by Planck in his address to the 1910 Königsberg meeting of the Deutsche Naturforscherversammlung, "Die Stellung der neueren Physik zur mechanischen Naturanschauung," published in the *Physikalische Zeitschrift* **11**, 1910, 922–932; *Physikalische Abhandlungen und Vorträge* (Braunschweig: Vieweg, 1958), **3**, 30–46, at 31 and 44; translated by R. Jones and D. H. Williams as 'The place of modern physics in the mechanical view of nature,' *A survey of physical theory* (New York: Dover, 1960, first edition 1923), 27–44, at 28 and 43.

[9] I. Bernard Cohen, *Revolution in science* (Cambridge: Harvard University Press, 1985), 14–15.

[10] See W.F. Magie, 'The primary concepts of physics,' *Science* (ns) **35**, 1912, 281–293, at 290.

[11] Orso Mario Corbino, 'Le recenti teorie elettromagnetiche e il moto assoluto,' *Rivista di scienza* [later *Scientia*] **1**, 1907, 162–169, at 166–167; Federigo Enriques, *Problems of science*, tr. Katharine Royce from 1906 Italian ed. (Chicago: Open Court, 1914), ch. 6.

[12] See, e.g., letter Corbino to Vito Volterra, 28 dicembre 1912 and the undated three-page ms in Corbino's hand, item 52, beginning "Col proposito di verificare la legge di dipendenza tra la massa e la velocità . . .," which analyzes experiments through 1910 and concludes, "It seems therefore that experimentally the question has been resolved in favor of the principle of relativity or at least against the formula of Abraham – ." Vito Volterra Collection, Accademia Nazionale dei Lincei, Rome. See also letter from Corbino to Michele La Rosa, 6 aprile 1912, which shows Corbino to be a supporter of relativity in the course of criticizing a draft of what was to be La Rosa's May 1912 *Nuovo cimento* paper, "Fondamenti sperimentali del secondo principio della teoria della relatività." La Rosa Correspondence, Palermo.

[13] See Levi-Civita's report to the Società Italiana di Fisica meeting at Padua in September 1909, 'Sulla costituzione delle radiazioni elettriche,' *Nuovo cimento* (5) **18**, 1909, 163–169, at 168–169; *Opere matematiche* **3**, 129–134, at 133.

[14] Max Abraham, 'Die neue Mechanik,' *Scientia* **15**, 1914, 8–27, at 16; French translation, *ibid.*, supplément, 10–29, at 18.

[15] *Ibid.*, 22; French tr., 24.

[16] Max Abraham, 'Sulle onde luminose e gravitazionali,' *Nuovo cimento* (6) **3**, 1912, 211–219, at 219.

[17] Abraham, 'Die neue Mechanik,' 26; French tr., 29.

[18] See Giovanni Battimelli and Michelangelo De Maria, 'Max Abraham in Italia,' *Atti del III Congresso Nazionale di Storia della Fisica*, Palermo, 11–16 ottobre 1982, 186–192, at 188, and their more extensive forthcoming study.

[19] Letter from Abraham to La Rosa, 13 maggio 1912, and earlier letters 27 novembre 1911 and 18 marzo 1912. La Rosa Correspondence, Palermo.

[20] *Times* (London), 7 November 1919, reproduced in Pais, *"Subtle is the Lord . . ."*, 307.

[21] *New York Times*, 10 November 1919, 17, reproduced in Stanley Goldberg,

Understanding relativity: origin and impact of a scientific revolution (Boston: Birkhäuser, 1984), 313.

[22] *Corriere della sera* (Milan), 11 November 1919, 2.

[23] See Judith Goodstein, 'The Italian mathematicians of relativity,' *Centaurus* **26**, 1983, 241–261.

[24] See letters Guido Castelnuovo to Tullio Levi-Civita, 5 novembre 1918 and 29 dicembre 1918. Levi-Civita Collection, Accademia Nazionale dei Lincei, Rome.

[25] See letter Roberto Marcolongo to Tullio Levi-Civita, 29 gennaio 1919. Levi-Civita Collection, Accademia Nazionale dei Lincei, Rome.

[26] Tullio Levi-Civita, 'Come potrebbe un conservatore giungere alla soglia della nuova meccanica,' *Rendiconti del Seminario Matematico della Università di Roma* (1) **5**, 1918–19, 10–28, *Opere matematiche*, **4**, 197–216.

[27] For a "cultural revolution" in another area in the prewar period, see Edmund E. Jacobitti, *Revolutionary humanism and historicism in modern Italy* (New Haven: Yale University Press, 1981).

[28] Attilio Palatini, 'La teoria della relatività nel suo sviluppo storico. Parte prima: la relatività della prima materia,' *Scientia* **26**, 1919, 195–207; 'Parte seconda: la relatività generale,' *ibid.*, 277–289.

[29] Roberto Marcolongo, 'La teoria della relatività in senso stretto,' *Rendiconti del Seminario Matematico della Università di Roma* (1) **5**, 1918–1919, 55–76; 'I fondamenti analitici della teoria generale della relatività e le equazioni del campo gravitazionale,' *ibid.*, 77–94. *Relatività*, 1st ed. (Messina: Principato, 1921); 2d ed., 1923; as appendix II the latter reprints, with some additions, 'Uno sguardo sintetico alla teoria speciale e generale della relatività,' *Esercitazioni matematiche* **2**, 1922, 127–148, lectures delivered in April 1922 to the Circolo Matematico di Catania. 'La relatività ristretta. Parte prima: suo punto di partenza sperimentale,' *Scientia* **35**, 1924, 249–258; 'Parte seconda: modificazione che essa apporta nei concetti di spazio e di tempo,' *ibid*, 321–330.

[30] Guido Castelnuovo, 'Sulla teoria della relatività,' sunto di due conferenze tenute alla sezione di Roma [of the Associazione Elettrotecnica Italiana], il 26 e il 28 aprile 1922, *Elettrotecnica* **9**, 1922, 417–422. 'L'espace-temps des relativistes a-t-il un contenu réel?' *Scientia* **33**, 1923, 169–180. *Spazio e tempo secondo le vedute di Alberto Einstein* (Bologna: Zanichelli, 1923).

[31] Luigi Donati, 'Introduzione alla teoria della relatività,' sunto della comunicazione tenuta il 18 dicembre 1921, *Elettrotecnica* **9**, 1922, 147–149; 'La relatività speciale,' sunto di due conferenze tenute . . . alla sezione di Bologna il 15 e il 29 gennaio 1922, *ibid.*, 286–289; 'Relatività generale,' sunto della conferenza tenuta . . . alla sezione di Bologna il 12 febbraio 1922, *ibid.*, 401–402.

[32] Guido Fubini, 'Sul valore della teoria di Einstein,' *Scientia* **35**, 1924, 85–92. See also the polemical exchange with the astronomer Giovanni Boccardi, noted below.

[33] Gino Fano, 'Vedute matematiche su fenomeni e leggi naturali,' discorso letto per l'inaugurazione dell'anno accademico 1922–23 il 6 novembre 1922, Turin, Università, *Annuario* 1922–23, 15–45.

[34] Donati, 'La relatività speciale,' 286.

[35] Roberto Marcolongo, 'La conferenza Einstein,' *La provincia di Padova*, 27–28 ottobre 1921, 3.

[36] Marcolongo, 'La relatività ristretta. Parte prima,' 250.

[37] Marcolongo, 'La relatività ristretta. Parte seconda,' 330.

[38] See Ulrico Hoepli to Tullio Levi-Civita, 8 marzo 1922, and undated draft of Levi-Civita's reply. Levi-Civita Collection, Accademia Nazionale dei Lincei, Rome. In a letter of 25 March 1922, Hoepli told Levi-Civita that he was planning to include in the Italian edition a section containing "the thought of the greatest Italian scientists and philosophers" on relativity. Levi-Civita suggested, in his draft reply dated 27 March 1922, the names of twelve Italian scientists "who have worked on or meditated on the theory." The list included Castelnuovo, Fubini, La Rosa, Maggi, Marcolongo, Palatini, Volterra, and the mathematicians Pietro Burgatti of Bologna, Francesco Paolo Cantelli of Rome, and Umberto Cisotti of Milan, the experimental physicist Raffaele Augusto Occhialini, and the astronomer Vicenzo Cerulli. Levi-Civita Collection, Accademia Nazionale dei Lincei. Of these, only La Rosa's, Burgatti, and Cerulli had essays published. La Rosa's and Cerulli's essays for the volume will be discussed below. Other contributors included the 22-year-old theoretical physicist Enrico Fermi, the astronomer-priest Giovanni Boccardi, the Vatican astronomer Pio Emanuelli, the culture critic Adriano Tilgher, and the philosophers Antonio Aliotta, Alessandro Bonucci, Ugo Spirito, Sebastiano Timpanaro, and Erminio Troilo.

[39] Giuseppe Armellini, 'Prefazione,' in August Kopff. *I fondamenti della relatività einsteiniana*, Italian ed. tr. and ed. Rafaele Contu and Tomaso Bembo (Milano: Hoepli, 1923), xvi.

[40] *Ibid.*, xv, xviii.

[41] *Ibid.*, xviii.

[42] *Ibid.*, xx–xxi.

[43] *Ibid.*, xxii.

[44] [Federico Enriques,] 'Einstein e l'interpretazione subiettiva della scienza,' *Periodico di matematiche* (4) **2**, 1922, 77–80. See also the nearly identical 'Le conferenze di Alberto Einstein a Bologna: parole di presentazione di Federigo Enriques,' *Rivista di filosofia* **13**, 1921, 271–274.

[45] Dante Manetti, 'Le conferenze di Alberto Einstein sulla relatività (Nostra intervista col prof. Tullio Levi-Civita),' *Il Messaggero* (Rome), 30 October 1921, 3. The phrase does not appear in the two very similar but not identical published summaries of Einstein's lectures prepared by the physicist Giorgio Todesco, 'Tre conferenze di A. Einstein sulla relatività,' *Annuario scientificio ed industriale* **58**, 1921, **1**, 395–412; and 'Sulla teoria della relatività (dalle conferenze di Einstein a Bologna). Prima conferenza: il principio speciale di relatività,' *Periodico di matematiche* (4) **2**, 1922, 125–135; 'Seconda conferenza: la relatività generale,' *ibid.*, 221–231; 'Terza conferenza: la concezione relativistica dell'universo,' *ibid.*, 231–236.

[46] Quoted in Gerald Holton, 'Einstein's search for the *Weltbild*,' *Proceedings of the American Philosophical Society* **125**, 1981, 1–15, at 14, from Einstein, *Ueber die spezielle und die allgemeine Relativitätstheorie* (Braunschweig: Vieweg, 1917), 52. A slightly different translation appears in *Relativity: the special and general theory, a popular exposition*, tr. Robert Lawson (New York: Crown, 1961), 77.

[47] 'Einstein on his theory/Time, space, and gravitation/The Newtonian system,' by Dr. Arthur Eddington, *Times* (London), 28 November 1919, 13–14, at 14.

[48] 'Prof. Einstein here, explains relativity/"Poet in science" says it is a theory of space and time, but it baffles reporters/Seeks aid for Palestine/Thousands wait for hours to welcome theorist and his party to America,' *New York Times*, 3 April 1921, 1, 13, at 13.

[49] 'Einstein sees end of time and space/Destruction of material universe would be

followed by nothing, says creator of relativity/Theory "logically simple"/Science burdened hitherto by complicated assumptions, he asserts – entertains many visitors,' *New York Times*, 4 April 1921, 5.

[50] 'Einstein in lecture explains his theory/Professor demonstrates with chalk as audience in Horace Mann School applauds/It's a theory of method/Experience supplies the postulates and reasoning draws the conclusions, he asserts,' *New York Times*, 16 April 1921, 11.

[51] 'Einstein and Newton/The new theory of space/Lecture at King's College,' *Times* (London), 14 June 1921, 8.

[52] 'Prof. Einstein's lectures at King's College, London, and the University of Manchester,' *Nature* **107**, 16 June 1921, 504.

[53] Dante Manetti, 'Le conferenze di Alberto Einstein sulla relatività,' *Il messaggero* (Rome), 30 ottobre 1921, 3. Einstein already knew at first hand the grudges and passions of opponents to his scientific theories and his pacifist political views. He had already had his life threatened. See 'Urged murder of Einstein, pays $16 fine in Berlin court,' *New York Times*, 8 April 1921, 17, which reported the fine levied on Rudolph Liebus,"an anti-Semitic leader," for having offered a reward for the murders, "as a patriotic duty," of Einstein, Wilhelm Foerster, the pacifist, and Maximilian Harden, the pro-British editor of *Die Zukunft*. See also Alan Beyerchen, *Scientists under Hitler: politics and the physics community in the Third Reich* (New Haven: Yale University Press, 1977).

[54] Michel Biezunski, 'Popularization and scientific controversy: the case of the theory of relativity in France,' *Expository science: forms and functions of popularization*, ed. Terry Shinn and Richard Whitley (Dordrecht: Reidel, 1985), 183–194, at 183. Sociology of the sciences, **9**.

[55] *Ibid.*, 183.

[56] Carlo Somigliana, 'Sulla trasformazione di Lorentz,' *Rendiconti Accademia dei Lincei, cl. sci. fis.* (5) **31**, 1922, sem. 1, 409–414. Somigliana's argument in favor of the Newtonian interpretation was contested by Gian Antonio Maggi, 'Sulle varie interpretazioni della trasformazione di Lorentz,' *ibid.*, **32**, 1923, sem. 1, 196–197.

[57] Carlo Somigliana, 'I fondamenti della relatività,' *Scientia* **34**, 1923, 1–10, at 2.

[58] Roberto Maiocchi, 'Il ruolo delle scienze nello sviluppo industriale italiano,' *Storia d'Italia*, Annali 3, *Scienza e tecnica nella cultura e nella società del Rinascimento a oggi*, ed. Gianni Micheli (Turin: Einaudi, 1980), 863–999, at 937.

[59] The responses of experimental physicists, especially Augusto Righi and Quirino Majorana, and astronomers to relativity have been surveyed by Cataldo Godano, 'La reazione, in Italia, alla relatività: 1917–25, i fisici sperimentali,' tesi di laurea in fisica, Università di Roma, 1984.

[60] Augusto Righi, 'Sulle basi sperimentali della teoria della relatività', *Nuovo cimento* (6) **19**, 1920, 141–162, at 142.

[61] Giorgio Valle, 'Complementi alla teoria di Righi sull'esperienza di Michelson,' *Nuovo cimento* (6) **26**, 1925, 39–73.

[62] The scientific work conducted by La Rosa in opposition to relativity, on the light emitted by binary stars, warrants an extended study, making use of his unpublished correspondence, to situate it in the context of the international scientific debate on the postulate of the constancy of the velocity of light, as well as in the context of the polemic in the Italian scientific community.

[63] Michele La Rosa, 'Fundamenti sperimentali del secondo principio della teoria della relatività, *Nuovo cimento* (6) **3**, 1912, 345–365, fascicle of May 1912.

[64] Michele La Rosa, in Kopff, *Fondamenti*, 351–354, at 352.

[65] Giovanni Boccardi, in Kopff, *Fondamenti*, 336–337, at 336.

[66] Giovanni Boccardi, 'In difesa della legge di Newton,' *La stampa*, 28 settembre 1921, 3; Guido Fubini, 'Astonomia agli astronomi: la teoria della relatività da Galileo ad Einstein,' *ibid.*, 9 ottobre 1921, 3; Boccardi, 'Il procedimento logico della teoria di Einstein e l'allarme degli astronomi,' *ibid.*, 13 ottobre 1921, 3; Fubini, 'La relatività einsteiniana e gli spazii a quattro dimensioni,' *ibid.*, 20 ottobre 1921, 3; Boccardi, 'La relatività einsteiniana: per concludere,' *ibid.*, 22 ottobre 1921, 3.

[67] Boccardi, 'Difesa,' 3.

[68] Vincenzo Cerulli, in Kopff, *Fondamenti*, 340.

[69] Ferdinando Lori, 'Alberto Einstein e la teoria della relatività,' *Il Veneto*, 26–27 October 1921, 1; reprinted in *Il nuovo patto*, 1922, 249–254.

[70] See Ferdinando Lori, speech at Accademia di Padova, 14 May 1922, *Il nuovo patto*, 1922, lxxxvii–xci. See now Roberto Maiocchi, *Einstein in Italia: la scienza e la filosofia italiana di fronte alla teoria della relatività* (Milan: Franco Angeli, 1985), which became available to me only after the completion of this paper.

[71] The philosophical debate on and around the theory of relativity requires extended investigation.

[72] See, for example, Benedetto Croce, *Logic as the science of the pure concept*, tr. Douglas Ainslie (London: Macmillan, 1917), 33–37, 341, 351–352.

[73] Ugo Spirito, 'Le interpetrazioni [*sic*] idealistiche delle teorie di Einstein,' *Giornale critico della filosofia italiana* 2, 1921, no. 2, 63–75; 'Le "integrazioni" idealistiche delle teorie di Einstein,' *ibid.*, no. 4, 99–101. In 1929 Spirito changed his mind and unified science and philosophy; see 'Scienza e filosofia,' *ibid.* 10, 1929, 430–444, reprinted with other articles defending the position in *Scienza e filosofia*, 2d ed. (Florence: Sansoni, 1950; 1st ed. 1933). Gentile reached this point explicitly in 1931; see 'Filosofia e scienza,' *Giornale critico della filosofia italiana* 12, 1931, 81–92.

[74] See Sebastiano Timpanaro, 'Il valore della teoria di Einstein,' *Atti del V congresso internazionale di filosofia*, Naples, 1924, 536–541, reprinted in *Scritti di storia e critica della scienza* (Florence: Sansoni, 1952), 211–216. See also Sebastiano Timpanaro jr., 'In margine alla "Cronache di filosofia italiana",' *Società* 11, 1955, 1067–1075; *ibid.* 12, 1956, 155–166; and an article in press. Timpanaro and his efforts to propagandize for an idealist conception of science warrant an extended study.

[75] See, for example, Antonio Aliotta, *Relativismo e idealismo* (Naples: Perrella, 1922); *La teoria di Einstein e le mutevoli prospettive del mondo* (Palermo: Sandron, 1922).

[76] Giuseppe Gianfranceschi, *La teoria della relatività: volgarizzazione e critica* (Milan: 'Vita e pensiero,' [1922]), 60–64. See also Gianfranceschi, 'Sulla relatività generale di Einstein,' *Atti Pontificia Accademia dei Nuovi Lincei* 73, 1919–1920, 177–184, the unsigned 'La teoria della relatività nell'ordine meccanico,' *Civiltà cattolica*, 1921, 3, 413–422, and 'Alberto Einstein a Bologna e la sua nuova teoria della relatività,' under rubric Rassegna di cultura, *Osservatore romano*, 2 novembre 1921, 2, all of which make many of the same points.

[77] Adriano Tilgher, 'Alberto Einstein,' *La stampa*, 7 July 1921, 2.

[78] Adriano Tilgher, 'La nuova teoria,' *Il resto del carlino* (Bologna), 22 October 1921, 3, under the headline 'Einstein parlerà per la prima volta al pubblico italiano/La teoria della relatività esposta dall'autore all'Archiginnasio di Bologna.'

[79] It was announced as a new book in *La stampa* (Turin) on 26 October 1921, 3, with the

headline 'The theory of Einstein and contemporary relativism,' and in *Il resto del carlino* (Bologna) on 27 October 1921, 3.

[80] For an account of Mussolini's political shifts in these years, see in English Adrian Lyttelton, *The seizure of power: Fascism in Italy 1919–1929* (New York: Scribner's 1973), 42–82. In Italian Renzo De Felice, *Mussolini il rivoluzionario 1883–1920* and *Mussolini il fascista I. La conquista del potere 1921–1925* (Turin: Einaudi, 1965–66), remain indispensable for this period.

[81] See Lyttelton, *The seizure of power*, 44–47.

[82] 'Programma e statuti del Partito Nazionale Fascista,' *Il popolo d'Italia*, 27 dicembre 1921, 1–2. Now reprinted in *Opera omnia di Benito Mussolini*, ed. Edoardo and Duilio Susmel (Florence: La Fenice, 1951–1963), **17**, 334–350, at 334.

[83] *Ibid.*, 336.

[84] *Ibid.*, 340.

[85] Lyttelton, *Seizure of power*, 45, citing *Opera omnia* **13**, 61–63, 18 aprile 1919.

[86] Lyttelton, *Seizure of power*, 45, citing *Dai fasci al PNF* (Rome, 1942), 133ff.

[87] Mussolini, 'Dopo due anni,' *Opera omnia* **16**, 210–212, 23 marzo 1921, at 212.

[88] De Felice, *Mussolini il fascista I*, 187.

[89] See Carlo Bolognesi, 'Einstein e la sua relatività,' *Il popolo d'Italia*, 12 novembre 1921, 3. The column subjected the theory of relativity to ridicule for being "based on" a contraction of five parts in one billion, "something entirely incalculable and inadmissible." The possibility of choice of reference frames represented another 'Einsteinian error,' since there was no reason to select any but the earth. The relativity of sizes violated common sense, and the conception of the universe as finite but unlimited was "the greatest absurdity." The Einsteinian conception of science was "imbued with that Humean and Berkeleian skepticism that, while it does not hold any measure stable, ends by putting every existence in doubt." For Bolognesi, "science must have as its distinct and integrative character stability, which means absolutivity and not at all relativity." Bolognesi credited Einstein, however, with "the first observation of the so-called phenomenon of the deflection of light in a gravitational field." But that and the work related to the perihelion of Mercury were not enough to declare the law of Newton inadequate, because that law was based on what had to be the foundation for the disposition of all things: number, weight, and measure. These arguments and others reappear in Emilio Ungania, *Einstein e la sua relatività: esame critico; l'errore copernicano* (Bologna: Cappelli, 1922). Bolognesi had written an article appreciative of Ungania's peculiar theories a month before: 'Tutto è l'energia: la materia non esiste,' *Il popolo d'Italia*, 2 ottobre 1921, 3. Ungania was undoubtedly one of the circle-squarers and angle-trisectors that relativity seemed to bring out of the woodwork; see Roberto Marcolongo, 'Fra relativisti ed antirelativisti (parole pronunciate per l'inaugurazione della sezione della relatività),' *Atti del V congresso internazionale di filosofia*, Naples, 1924, 419–427, at 421.

[90] Benito Mussolini, 'Nel solco delle grandi filosofie relativismo e fascismo:,' *Il popolo d'Italia*, 22 novembre 1921, 1; reprinted in *Opera omnia*, **17**, 267–269.

[91] Mussolini, 'Programma,' *Il popolo d'Italia*, 22 dicembre 1921, 1; reprinted in *Opera omnia* **17**, 321–322, at 322.

[92] Mussolini, 'Prefazione al programma,' *Il popolo d'Italia*, 28 dicembre 1921, 1. The text which appears in *Opera omnia* **17**, 351–353, omits a phrase in this passage and adds the adjective *tecniche* to modify *corporazioni*.

[93] See Benito Mussolini, *Diuturna: scritti politici raccolti e ordinati da Arnaldo Mussolini e Dino Grandi, prefazione di Vincenzo Morello* (Milan: Imperia, 1924).

[94] This point is argued at length in Emilio Gentile, *Le origini dell'ideologia fascista (1918-1925)* (Rome-Bari: Laterza, 1975).

[95] Ardengo Soffici, 'Relativismo e politica,' *Gerarchia* 1, 1922, 29–32, at 29–30.

[96] *Ibid.*, 30–32.

[97] Eugen Weber, *Action Française: royalism and reaction in twentieth century France* (Stanford: Stanford University Press, 1962), 111, and ch. 1–4 in general. Willard Bohn, 'Free-word poetry and painting in 1914: Ardengo Soffici and Guillaume Apollinaire,' *Ardengo Soffici: l'artista e lo scrittore nella cultura del 900* (Florence: Centro Di, 1976), 209–226.

[98] See especially Ruggero Jacobbi, 'Ardengo Soffici fra tradizione e rinnovamento,' *Ardengo Soffici: l'artista*, 15–28.

[99] Ardengo Soffici, 'Olla podrida,' *Il popolo d'Italia*, 30 agosto 1921, 3.

[100] Silvio Pagani, 'Relativismo e decadenza,' *Gerarchia* 1, 1922, 26–28.

[101] (Editorial note), *Gerarchia* 1, 1922, 32.

[102] Adriano Tilgher, 'Relativismo e rivoluzione,' *Relativisti contemporanei*. 3d ed. revised and enlarged (Rome: Libreria di Scienze e Lettere, 1922), 73–79.

[103] See Gentile, *Origini, passim.*

[104] See the draft *statuto* for the University of Rome, Archivio Centrale dello Stato, Ministero della Pubblica Istruzione, Direzione Generale dell'Istruzione Superiore, Div. II, pos. 2, b.406, f. Roma, Statuto vecchio, 1924–1927.

[105] Verbali delle sedute della Facoltà di Scienze, Università di Roma, vol. 10, 4 febbraio 1924–9 novembre 1927, seduta del 27 ottobre 1924. Archive of the University of Rome.

[106] *Ibid.*, seduta del 22 dicembre 1924.

[107] 'Relazione della commissione giudicatrice del concorso per professore non stabile alla cattedra di fisica matematica della R. Università di Cagliari,' Italy, Ministero della Pubblica Istruzione, *Bollettino ufficiale* 53, 1926, parte II, Atti di amministrazione, vol. 1, 793–799, at 798. Undated but before the 17 February 1926 meeting of the Consiglio Superiore della Pubblica Istruzione, at which it was approved. Emilio Segrè, *Enrico Fermi physicist* (Chicago: University of Chicago Press, 1970), 40–42.

[108] Giuseppe Ottone, 'La "relatività" e l'azione,' *La tribuna*, 5 February 1926, 3.

[109] See Verbali della Facoltà di Scienze, Università di Roma, vol. 10, seduta del 10 maggio 1926. Archive of the University of Rome.

[110] See 'Avvisi di concorsi a posti di professore non stabile presso università ed istituti superiori,' Italy, Ministero della Pubblica Istruzione, *Bollettino ufficiale* 53, parte II, Atti di amministrazione, vol. 2, 2058–2060, fascicle dated 1 July 1926.

Ohio State University

THOMAS F. GLICK

RELATIVITY IN SPAIN

1. INTRODUCTION

The reception of relativity in Spain was a phenomenon dominated by mathematicians with extensive linkages to Italian mathematics; as a result, general relativity dominated scientific discussions. Many of the leading figures were conservative Catholics and, indeed, one could make the case that, in ideological terms, relativity became a conservative cause in Spain. In order to determine why this should have been so we must investigate the sources of demand for information about relativity. In general, the most conspicuous "consumers" of Einstein's theories in Spain were engineers, also a Catholic, socially conservative group. Engineers, we will see, differed greatly in their ability to understand relativity, although enthusiasm for Einstein was not affected by intellectual disability. Inasmuch as the scientific reception was significantly tailored to the demands of the engineering community, who, in turn, produced most of the popularizations of relativity for the general public, we treat the reception of relativity as a seamless web, a culture-wide phenomenon engaging different levels of society and domains of discourse simultaneously, with multiple feedbacks among them. To view the reception of relativity thus – as a phenomenon which ripples through an entire culture – we must disabuse ourselves of the view that the distinction between "scientific" and "popular" receptions is a meaningful one. The ideas clearly lose physical meaning the farther they stray from mathematical language. But a contextual approach must focus on what was perceived or found useful by different groups appropriating relativity, without attempting to assess the physical validity of the discussion at each point. To do so would only belabor the obvious.[1]

2. THE SCIENTIFIC RECEPTION OF RELATIVITY IN SPAIN

The earliest allusions to special relativity appeared in papers presented by Esteve (Esteban) Terradas and Blas Cabrera in 1908 at the first meeting of the Spanish Association for the Progress of Science. In

231

Thomas F. Glick (ed.), The Comparative Reception of Relativity, 231–263.
© 1987 *by D. Reidel Publishing Company.*

papers on black box radiation and theories of the emission of light, Terradas (1883–1950), at the time professor of Acoustics and Optics at the University of Barcelona, alluded to the special theory only as a new deduction of the "principle discovered by Lorentz," adding that Einstein and Laub " have recently applied it to establish more general laws of electrodynamics." Cabrera (1878–1945), professor of electricity and magnetism at the University of Madrid, presented a paper on the theory of electrons, explicating Maxwellian and Hertzian electromagnetic concepts of light. In his lecture, Einstein's theory is again introduced only as a refinement of Lorentz's electron theory and Cabrera still presupposed at this time the existence of an ether.[2]

By 1912 at the latest, however, both men were clearer on the theory's meaning and significance. In that year Cabrera stated, with reference to special relativity, that there was no experiment to detect the ether, and Terradas published a long review of Max von Laue's book on special relativity in which he noted that "The principle of relativity is accepted today by almost everyone. In chairs of physics its language has been generally adopted." Still, according to Antoni Roca, Terradas' conception of relativity at this time was wholly consistent with the theory of electromagnetism, a new language for talking about electricity. He did not yet see it as the basis for a new mechanics. He did not abandon the Lorentzian framework until the middle of the decade, as he followed developments in general relativity.[3]

As for Cabrera, he too had recognized the revolutionary nature of Einstein's ideas by the mid-teens. In a series of lectures on electricity delivered in Madrid in January 1917 he noted that the failure of attempts to determine absolute motion had created the need to "reorganize science" in order to rid it of "such evident contradictions". The confusion which the theory, with its seemingly paradoxical reinterpretation of simultaneity, incites in us is simply the product of "a mental habit, of the supposed independence of space and time."[4]

When, in the wake of the 1919 eclipse results, relativity was in vogue, Terradas and Cabrera were responsible for much of the diffusion of Einstein's ideas among the scientific community. In Barcelona during the winter of 1920–21, Terradas gave a 30-session course on "Relativity and the New Theories of Knowledge" under the auspices of the Catalan regional authority, the Mancomunitat. The course was a survey of Einstein's results, with emphasis on the special theory. In contrast with many Spanish comentators of this period, Terradas was unequivocal in his acceptance of special relativity: "Ether does not exist, and neither does absolute space, inasmuch as we cannot in any way demonstrate its existence by physical means."[5] This kind of monographic course on relativity was not unusual in Spanish scientific centers of the time and demonstrates that the scientific community assimilated the new ideas rather quickly. This explains why, in an important address on the present state of physics delivered to the Academy of Exact Sciences in November 1921, Cabrera focused on the social, rather than the cognitive, concomitants of relativity's reception. Citing resistance to relativity in the name of "immutable principles", he insisted on the urgent need to create a favorable environment "to give a greater impulse to the advancement of national science," and noted the difficulty of doing this "in the midst of an absolutely indifferent society, without receiving the heat given off by the favorable or adverse criticism of those who directly surround us."[6]

Cabrera's comment raises a number of issues. First, the locus of the resistance mentioned by Cabrera would not seem to have been within the scientific community. Among the leaders of the three recipient disciplines, physics, mathematics and astronomy, only the astronomer Josep (José) Comas Solà was outspokenly critical of relativity.[7] Physics was a thinly-staffed field, with no theoreticians (Cabrera was a experimentalist). As Antonio Lafuente has observed, there was no defense of ether by any respectable physicist nor was any to be expected in a country where Maxwellian physics had scarcely been diffused.[8] Mathematical physics, as well as mechanics, were taught in mathematics departments. In thematic terms, at any case, relativity was more congruent with the interests of mathematicians and it was principally among them that the scientific discussion of relativity took place. The Mathematical Society of Madrid was the leading center of relativistic discussion. As we will see, the close connection of the mathematical leadership with Italian relativists inclined them in relativity's favor. The same could be said of astronomers and British astronomy; Pedro Carrasco (1883–1966), director of the Madrid observatory and the most authoritative Spanish interpretor of the eclipse results, had worked with Frank Dyson in England and was inclined to accept his interpretations. The defenders of immutable principles, with the exception of Comas, had no voice in the higher councils of Spanish science. Cabrera most likely had politicians, not scientists, in mind. The second issue raised by Cabrera refers to both the thinness of the Spanish scientific community and of its public support. No science policy could be formulated in an environment where science was inadequately discussed by the political

community at large; nor could the public be expected to make any
accurate judgment of scientific ideas, such as relativity, when there were
so few scientists capable of discussing them. In such an environment one
person's word was as good as the next and Newtonians, or those
invoking Newton to promote traditionalist social and political views,
could appeal directly to the public for support. Public support for pure
science was a key issue in the reception of relativity in Spain and one of
the keynotes of Einstein's personal encounter with Spanish society. In
that sense, not only is there no way to disentangle social from cognitive
threads of this story, but to do so would be to distort its reality.

3. SPANISH MATHEMATICS AND ITALY

The resurgence of science in Spain in the first two decades of the
twentieth century, although real, was limited to a number of particularly
successful disciplines: mathematics, experimental physics, neurology,
physiology. All of these benefitted from the creation of a governmental
organism designed specifically to modernize scientific disciplines by
backing research at home and by sending young scholars abroad to
master new techniques and acquire a modern scientific ethos. This was
the Junta para Ampliación de Estudios, founded in 1907 under the
stimulus of the award of the Nobel Prize in medicine to Santiago Ramón
y Cajal the previous year. Mathematics was the first discipline effectively to modernize itself under the Junta's aegis. In 1915 Julio Rey
Pastor founded the Mathematical Laboratory and Seminar with Junta
support and over the next few years was able to send a large number of
his graduate students to study abroad, mainly in Italy.
    There were a number of reasons why Italy was the harbour of
preference for Spanish mathematicians. First, as Rey Pastor explained,
the Italians' style of teaching, the nurturing relationship they built with
students, and general cultural congruence contrasted with the hierarchical organization of German academe (which Rey Pastor knew well, as
an early beneficiary of the Junta's foreign scholarship program), the lack
of strong bonding between student and professor, which taking the
language gap into account, presented formidable obstacles to Spanish
students whom he wanted to return home victorious, rather than
defeated.[9] Moreover, as Giorgio Israel has noted, there were broad
areas of common interest uniting the two schools of mathematicians, in
particular the rejection of "purism" in geometry and a concern (appar-

ent in both Rey Pastor and Federigo Enriques, for example) for unifying the objectives of geometry and analysis.[10] Rey Pastor and his school had good relations with Enriques, Francesco Severi, Vito Volterra and, in particular, Tullio Levi-Civita. The latter three visited Spain on a number of occasions and Enriques was Rey Pastor's guest in Buenos Aires in 1927. Rey Pastor selected his disciples' research and thesis topics personally and always chose problems of current interest. In accordance with this program, he sent Fernando Lorente de Nó to Italy on a Junta grant in 1919 to study relativity with Levi-Civita. (In Rome, Lorente produced a paper, apparently unpublished, on the movement of a point in an Einsteinian field.) The two other mathematicians among Rey Pastor's students who worked on relativity, Pere (Pedro) Puig Adam (whose dissertation of 1922 was on four problems in the mechanics of special relativity) and Fernando Peña corresponded with Levi-Civita, at the behest of their advisor, Josep (José) M. Plans, Rey Pastor's second-in-command at the Mathematical Laboratory.[11]

Plans (1878–1934) was a mathematical physicist who held the chair of rational mechanics in Madrid (the chair of mathematical physics, in the physics department, was held by the astronomer Carrasco) and was Einstein's third great Spanish paladin. According to one of his disciples he was the only person in Madrid in the 1920s (Terradas was in Barcelona) capable of teaching relativity at an advanced level and was able to introduce original interpretations and formulations into his discussions.[12] Plans won a prize offered in 1919 by the Academy of Sciences for a work explaining "the new concepts of space and time". His manuscript, later published as *Nociones fundamentales de la Mecánica relativista* (Madrid, 1921), was mainly devoted to the special theory. The prize was offered by the exact, not the physical science, section of the Academy, which accounts for the mathematical slant of Plans' manuscript which omits detail regarding experimental facts. Still, the ninth chapter is a discussion of the equations of movement in a gravitational field and their application to the deformation of the perihelion of Mercury and the deflection of light rays. The point is interesting because both the prize and Plans' response to it were conceptualized before the eclipse observations had directed public and scientific attention to experimental results.[13]

But Plans' major role was yet to come and was the result of the intersection of his interests with the stimulation given to Spanish science by Levi-Civita's lectures in Barcelona and Madrid in February 1921 which, besides drawing attention to relativity's scientific importance, was a further occasion for strengthening ties between Italian and Spanish mathematics. The intimate nature of that bond was exhibited by the extraordinary interest that students showed in his course, titled "Questions of Classical and Relativistic Mechanics", particularly the portions dealing with relativity. In Barcelona, a fourth session, not included in the original course of three lectures, was added, "owing to the insistence of the audience. . . . It was, for this reason, an intimate session in which there was established that contact between professor and students which creates dialogue."[14]

Levi-Civita's visit was the occasion for the emergence of a distinct circle of scientists who would diffuse relativity in Spain. Of sixteen invitees to a dinner tendered the Italian on February 1 in Madrid, nine were mathematicians including virtually the entire leadership of Spanish mathematics, all of whom were favorable to relativity. (Besides the mathematicians, led by Rey Pastor and Plans, Cabrera and Julio Palacios represented physics, Carrasco, astronomy; four engineers, three of whom were interested in aeronautics, completed the guest list.)[15] The large representation of mathematicians at the dinner illustrates a number of the idiosyncracies of the reception of relativity in Spain. First, mathematics was by this time the strongest and largest discipline of the exact and physical sciences. Second, its members already had established close connections with Italian mathematics. Third, mathematicians were needed by physicists, engineers and others interested in understanding the new ideas to explain to them the distinctive language – absolute differential calculus – in which these ideas were expressed. Spanish mathematicians constantly emphasized, in their public utterances, the services lent by their discipline to Einstein's theory. Fourth, and consequently, scientific interest in relativity after 1921 was preponderantly in the general theory, partly because of the period in which relativity's major impact was felt (1921–1924), and partly because the predominance of mathematicians in its reception ensured a focus on what was agreed to be the mathematically most interesting aspect of the theory.

Plans stressed just this point later in the same year at the Oporto congress of the Spanish Association, in his inaugural lecture of the mathematics section entitled "The Historical Development of Absolute Differential Calculus and its Current Importance." In the history of the interpenetration of mathematics and the physical sciences, Plans began, there have been cases of two different kinds. In the first, a physical

problem stimulates a useful abstraction: vibrating chords and the properties of heat gave rise to the Fourier series. In the second, the opposite is
true. Here an abstraction lends itself, sometime after its initial conceptualization, to the explication of some physical problem. Plans' examples are the non-Euclidean geometry of Lobatchevsky and Bolyai in
special relativity and that of Riemann, and absolute differential calculus, in general relativity. Relativity, Plans asserted, was "the scientific
event of greatest consequence at the present time" and stressed "the
great services lent, in the hands of the Italian school of Ricci and
Levi-Civita, to Einstein's theory of relativity and gravitation, by absolute
differential calculus, which, as our colleague Sr. Terradas has so wisely
said, has become the language appropriate to the study of Reimannian
space-time of four dimensions, just as ordinary vectorial calculus serves
for the Euclidean space of three dimensions."[16]

It is instructive to contrast this programmatic statement by Plans with
Cabrera's statement of the same year, cited above. For the experimental
physicist, relativity was both epistemologically important and also crystalized a number of social issues regarding the place of science in
society, but for physics the structure of the atom was an issue of greater
current significance. For the mathematician, general relativity was the
central scientific issue of the day because of the signal role of mathematics in its conceptualization. In 1924, Plans wrote a manual of
absolute differential calculus, obviously designed to make general relativity accessible to physical scientists and engineers. Like his earlier
volume on relativistic mechanics, this book had its origin in a prize
offered by the Academy of Sciences, "desiring a book wherein one
might acquire the primordial ideas of this powerful mathematical re-
source developed some years ago by the mathematical school of the
University of Padua."[17]

Mathematicians, reviewing this book, knew exactly what its significance (to them) was. Lorente de Nó observed that the rapid development of Spanish interest in relativity was the rationale for both the
competition and the book. That interest encountered not only conceptual difficulties but others of a formal nature derived from the use of
absolute differential calculus. Lorente's review served to further emphasize the point that mathematics had provided the means for resolving the conceptual problems of relativity. Another Italy veteran, José
María Orts, stressed that once popular relativity fever had died down,
"the new doctrines condensed around a smaller circle of those few who

were able to handle the elements necessary to penetrate their depths."
Plans himself was one of this small nucleus of "conscientious relativists"
– that is to say, mathematicians who were able to understand Einstein
and Levi-Civita.[18] There can be no doubt about the role subsequently
played by Plans' volume in Spanish scientific education. According to
one of his students, Plans' manual had, by 1934, placed absolute
differential calculus "within the reach of persons of middling scientific
education. . . . Even the most recent graduating classes learned from
this book how to handle this potent instrument of calculation, which two
years ago I included in my syllabus of Mechanics as an auxilliary
discipline."[19]

When speaking of relativity, Spanish mathematicians continually
stressed their discipline's service to it and to science in general: to
demystify relativity, to build a positive public image of the maturity of
Spanish mathematics, and to present mathematics as a practical instrument for the development of practical knowledge, including the theory
of relativity.[20] As Jeffrey Crelinsten observed with reference to American astronomers, relativity "focused the attention of disciplinary leaders
upon the strengths and weaknesses of their professional community."[21]
The ability of Spanish mathematicians not only to master what was
popularly perceived as incomprehensible but also to have, as mathematicians, contributed to the foundations of the theory was the great
symbol of the maturity of their discipline.

4. CIVIL DISCOURSE, CATHOLICISM AND SCIENCE

The public discussion of relativity in Spain took place in a climate of
intellectual openness unprecedented in modern times. From the Revolution of 1868 to the end of the century the elite had been divided into
two warring factions, conservative and liberal, and all ideas, scientific
ones included, became ideologized and were pressed into service as
weapons of ideological warfare. The exemplar of an ideologized idea
during this period was, of course, Darwinism, attitudes towards which
obeyed a strict ideological cleavage with Catholic conservatives opposing and anticlerical liberals favoring it as a matter of conscience. In such
a context, ideas were rarely considered on their merits, not even by
scientists. By the end of century, however, a consensus arose among
Conservative and Liberal educational policy-makers to remove science
from the ideological arena in the interest of national modernization. In

this the two hegemonic parties were joined by members of the parliamentary far left and right. Since 1868, let us note, the Republican left had been the normal locus of support for new scientific ideas and the traditionalist right the center of opposition to them. By century's end however, these implacable ideological enemies had formed a tactical parliamentary alliance with respect to academic freedom. Both Catholics and Republicans wanted to preserve room in academia for expression of their philosophical positions and were wary of "Jacobin" state control of education on the part of the two hegemonic parties. The practical effect of this consensus in favor of academic freedom and civil discourse in science can be appreciated in the policies of the Junta para Ampliación de Estudios [Commission for the Broadening of Studies, literally], founded in 1906 to fund scholarly research and to provide scholarships so that scientists and others could improve their conceptual and research skills in foreign centers. The Junta was governed by a self-perpetuating board of twenty trustees, named for life and drawn from the entire political spectrum from the Catholic right to the Republican left. All decisions were made by unanimity and there was no political test for scientific applicants.[22] The Junta's impact was swift and dramatic. The leaders of our three receiving disciplines, Rey Pastor, Cabrera and Carrasco, for example, all held travel grants from the Junta before World War I.

The positive effect of this process on the reception of scientific ideas is clear and can be appreciated from the discussion of relativity. Although Catholic revanchists of the far right, mainly priests, opposed relativity in the name of a traditional cosmology, politically conservative orthodox Catholics were conspicuous among the supporters of relativity. Plans was well-known for his exaggerated religiosity. Educated by Jesuits before entering the University of Barcelona, he acquired (according to one of his students) a "love for everything Jesuitical".[23] To Terradas, who was scarcely less fervent in his piety, Plans, "armed with the divine excellences of the elect paladins of the Catholic faith," assumed the demeanor of an apostle.[24] Yet, except for one notable lapse, Plans steered clear of controversial issues involving the conflict of religion and science. The exception had to do with his translation of Eddington's volume, *Space, Time and Gravitation*, each chapter of which was preceded by a literary quotation. The quote heading chapter I was from Descartes and stated: "In order to reach the Truth, it is necessary, once in one's life, to put everything in doubt – so far as possible." Plans, who found this statement contrary to his religious convictions, consulted the Jesuit mathematician and relativist Enric (Enrique) de Rafael, who assured him that Descartes "excepted religious ideas" although readers may well be unaware of this. Not satisfied, Plans added a footnote registering his disagreement with Descartes: "It is unnecessary to say that the translator does not associate himself with this principle." Plans was duly attacked in the liberal press for having added the disclaimer, but the minor nature of this episode confirms the view that nearly a quarter century of civil discourse had rendered such ultra-orthodox strictures gratuitous. Plans, we might further speculate, chose Eddington's volume for translation because it bore the comforting message that absolute values and standards were still intact, a message of reassurance that the Catholic public wanted to hear.[25]

Ideological opposition to relativity was limited to a small band of clerical ultras and their secular epigones who were still unable to disentangle religion, politics and science and who opposed any scientific innovation that appeared to put the traditional value system in doubt. In the Darwinian polemics of the nineteenth century, the line between those who participated in civil discourse in science and those who didn't was sharply drawn between believers and non-believers. We can appreciate the new line or frontier between partisans of religious intransigence and of academic freedom by looking at Catholic or clerical publications in the 1920s. *Razón y fe* [Reason and Faith], the leading Jesuit journal, published an article by Enric de Rafael favorable to relativity, another by the engineer and educator J.A. Pérez del Pulgar who accepted the mechanics of Einstein-Minkowski as true but who still had problems of a philosophical nature with the concept of the limit of velocities, and a third by Eustaquio Ugarte de Ercilla, who had opposed both Darwin and Freud in previous articles in this journal but who here criticized *exalted* anti-relativists for their intolerance and lack of understanding.[26] The *Revista Calasancia*, the general interest journal of the Piarist Fathers, also offered their readers three distinct positions: Benjamín Navarro, priest and chemist, was openly favorable to Einstein; Father Ataúlfo Huertas was ambivalent towards relativity (although he too criticized its extremist clerical opponents), and the chemist José María Goicoechea opposed the new theory because it contradicted his own odd cosmology, in which matter was irreducible. (Spanish chemists tended to associate relativity with energeticism and to oppose it on that basis.)[27] The Madrid daily *El Debate*, associated with Catholic Action, had a number of columnists favorable to relativity and others opposed. This newspaper published, thanks to Plans, the best summaries of Einstein's Madrid lectures.[28] The division among authors of articles appearing in various organs of the religious press identifies precisely the frontier separating intransigents from those who welcomed participation in civil discourse. In particular, the Catholic commentary on relativity in the 1920s reveals the distance separating Catholics of the old school from the new generation of clerical commentators, better educated and socialized under the conditions of civil discourse and academic freedom in vogue from around 1900. Father Huertas noted that Einstein had attracted support from all groups in Spain and had not been the "object of either political *philias* or *phobias*."[29]

In considering the Spanish reception of relativity we must ask what there was in the common experience of so many men of the Catholic right (Terradas, Plans, Rafael, or the engineers Pérez del Pulgar and Emilio Herrera) which made relativity so attractive to them. First, there was a social factor: to accept relativity was a way to embrace modern science without appearing to oppose traditional Catholic values (as had not been possible in the case of Darwinism). Second, there appears to have been a contextual factor related to the Catholic education of such persons. The Neo-Scholastic tradition had opposed Kantian notions of time and space in that Kant – decried by Rafael as the patriarch of modernist philosophy – had identified space and time as *a priori* categories of the intellect. Ingenuous Catholic critics of relativity, confusing *absolute* with *real* space and time, supposed Einstein's position to support that of Kant. "Nothing more false," Rafael declared :

there is nothing in common between the ideas of the patriarch of modernist philosophy and those of Einstein, but rather the impossibility of *directly* perceiving space and time as absolute in themselves, an impossibility which is not contrary to scholastic philosophy which sustained, against Newton, that space and time, as we conceive them, are not necessary, eternal and immediate entities, independent of God, however great his immensity (*sensorium*) and eternity (as Leibnitz reproved) might be, but simply entities of reason or the ideal, with their real basis in the existence of extensive, permanent and successive beings. This not only is not in contradiction with modern theories, insofar as these are the clear fruit of experience and represent a true advance, but they are sufficiently more in conformity with them than what has generally been admitted up to now.[30]

Einstein, in other words, had demonstrated the correct relationship between mind and reality.

Although Rafael was a Jesuit and Terradas a layman, the latter was just as avid a believer, according to Rafael's necrological sketch.[31] Terradas had a rather carefully worked out theological justification for general relativity, stressing that its philosophical ramifications proceed directly from absolute differential calculus, whose "philosophical reper- cussions were totally unforseen." If space and time are not real entities, but only fictitious ones "whose real basis is in the existence of extensive and changeable bodies" as many Thomist philosophers (e.g., Suárez) have held, then there is no metaphysical nor physical objection to admitting that the real properties of both depend on those bodies. If one further admits that there are as many true times as there are real movements, then "all the difficulties of a philosophical order which might be raised against the application of absolute differential calculus to the study of movement disappear as if by magic." These ideas, he states, are charac- teristic of the objectivist philosophers, especially the scholastics:

to admit the application of absolute differential calculus the subjectivists must overcome two difficulties, either because, like Descartes, they locate the essence of bodies in their extensions, or because they deny the absolute reality of extensive beings, as did Berkeley, Hume and Kant: (1) the explanation of the cause of variety in space-time and (2) the non-arbitrariness of the predicted results.[32]

Turning to cosmology, Terradas compares Einstein's notion of space as finite and closed – "but time is infinite, and the universe, cylindrical" – with de Sitter's conception, wherein not only space, but also time, was finite. He concludes: "Evidently Einstein's solution (possibily without his realizing it) is in greater conformity with the idea of eternity in Catholic philosophy."[33]

In embracing relativity, therefore, conservative Catholic scientists could do away with Newtonian absolutes by associating them with Kantian *a priori* categories and oppose modernist philosophy at the same time. This was not a point that was ever explicitly stressed but it helps to locate the Catholic relativists in the intellectual spectrum between anti-Kantian traditionalists and secular modernists. It can also be taken as a given that the vast majority of Catholic relativists were anti-positivists *ex hypothesi* and would therefore have been attracted to Einstein's brand of neo-idealism.

Finally, there was a generational factor at play. Catholic members of the "Generation of 1914", no less than their more secular-minded colleagues, were anxious to shake off the burden of immobility associ- ated with nineteenth-century elites. The ideologues of Primo de Rivera's

dictatorship (1923–30, a period in which much of the relativity discussion took place) stressed the relativism of political structures and
invoked Einstein in order to justify their rejection of the sham parliamentary democracy of the Restoration.[34] Paradoxically, civil discourse continued during the Dictatorship, in spite of the regime's
promotion of clerical obscurantism. Primo's attempt to pack the Junta
para Ampliación de Estudios with conservative clerics failed because
the appointees continued to share the consensus that strove for ideological neutrality in science. With the disintegration of the Dictatorship and
the installation of the Second Republic in 1931 the political stakes were
raised and that consensus began to break down. A number of conservatives who spoke in favor of Einstein in 1923 now opposed him: *El
Debate* was critical of his appointment to a chair at Madrid in 1933 and
in 1936, in an article published the same week the Spanish Civil broke
out, Ricardo Royo-Villanova who as rector of the University of Zaragoza during Einstein's visit urged that the blackboards used by the
physicist to illustrate his lectures be preserved for posterity, now argued
that relativity was "an intellectualist creation lacking all scientific value."[35]
Not all the right deserted Einstein, by any means, but the relationship
between conditions of civil discourse and the receptivity of ideologues to
specific ideas can be clearly established. The Franco period brought a
return to pre-1900 conditions; with the passing of Terradas and the
conservative relativists of his generation by 1950 the way was open for
all kinds of antirelativist nonesense, like the counter-theory of Julio
Palacios which was widely believed in late-Franco Spain owing to the
great prestige of its author.[36]

5. SPANISH ENGINEERS AND RELATIVITY

Cabrera, Plans and Terradas were the authors of numerous treatises on
relativity aimed not at the general reading public but at those with
scientific education. Even when they wrote in a "popular" medium –
Terradas' fifty-page account of relativity in the Espasa-Calpe Encyclopedia[37] is an example – they did not have a general audience in mind, but
one rather with mathematical education and demonstrated interest in
relativity: the engineering community. Engineers were the primary
"consumers" of relativity in Spain and, along with physicians, were the
most conspicious members of a group that I call the "scientific middle
class".[38] This group included all persons with scientific education,

including pharmacists, secondary school science teachers, industrial physicists and chemists, science writers and so forth. They are important to our story because in Spain there weren't enough research scientists in the first two and a half decades of this century to fill the ranks of professional scientific societies, such as the Sociedad Española de Física y Química, or general societies such as the Asociación Española para el Progreso de la Ciencia. In such societies members of the "scientific middle class" outnumbered research scientists on the order of 3–2 or 3–1, depending on how rigorous a standard is applied to the definition of "scientist". Members of the scientific middle class not only provided the rank-and-file of scientific societies, but also formed the bulk of the readership of scientific books and journals. The scientific establishment would not have been able to function without this important class of "consumers" of science. This sociological context is particularly apparent in the reception of scientific ideas. The most militant Spanish Darwinists of the 1870s and 80s, for example, tended to be medical doctors and secondary school science teachers. Similarly, the locus of the relativity debate was in the engineering schools rather than in the faculties of science.

As early as 1915 the professor of applied optics at the Escola dels Enginyers [Industrial Engineering School] of Barcelona, J. Mañas i Bonví, included an ample discussion of relativity in his textbook, *Optica aplicada*. A special supplement discusses special relativity in the context of the history of ether theory and stating that Einstein had denied the existence of the ether "as a fixed medium for the transmission of electromagnetic actions." At the same school, in the 1920s, Ferran (Fernando) Tallada taught rational mechanics from a relativistic perspective, as did his counterpart in the Industrial Engineering School of Madrid, Carlos Mataix Aracil. In 1923, Mataix wrote a special supplement to his textbook, *Mecánica racional*, introducing students to relativity in classical mechanics, special relativity, simultaneity, the universe of Minkowski, the Lorentz transformations, local time, longitudinal contraction, mass and energy and general relativity (in that order).[39]

The best documented course on relativity given in an engineering school during this period was Enric de Rafael's "Notions of Classical and Relativistic Physics" offered at the Catholic Institute of Arts and Industries, a Jesuit engineering school in Madrid, in 1921–22. In pondering a topic for his elective course, Rafael, opining that living science was both more agreeable and more instructive than dead science, chose relativity. Because of the deficient physics preparation of his students, this meant that Rafael, a mathematician, had to first review the elements of Newtonian mechanics, to which he devoted the first third of his course; the second trimester was devoted to the special theory, the third to the general. For reading, Rafael lists a number of volumes in German and English, but probably Plans' *Nociones fundamentales de la mecánica relativista* was the operative text. Rafael indicated that he would not go into the complex arguments that philosophers of science raised about relativity and, in effect, warned the students not to believe what general philosophers might say about Einstein's theories.[40] Here he doubtless had in mind the typical run of antirelativity tracts put out by theologian-scientists who were incapable of following either the physics or the philosophy of relativity. Students in this course were doubtless among those to whom Einstein delivered an impromptu lecture on the finite nature of the universe during his 1923 visit to Madrid.[41] Rafael's year-long course devoted wholly to relativity was untypical, however. More common were short courses, such as that given at the Highway Engineering School (Escuela de Caminos) in the spring of 1923 by the engineer Pedro Lucia Ordóñez, a survey of special and general relativity (and of the mathematics required to understand them) attended by "an abundant group of civil engineers and of students and professors of the School, who departed well satisfied by [the lectures]."[42] Relativity was diffused in engineering schools less in monographic courses than by its inclusion in the normal curriculum, particular in courses on rational mechanics, as indicated above.

Interestingly, philosophical doubts about relativity were voiced not by mathematicians or physicists but by Catholic engineers; indeed the scant philosophical debate that relativity stimulated in Spain in the 1920s was conducted almost wholly by engineers. The sticking point was both the physical meaning of c as the limit of velocities and the philosophical implications thereof. For some, like the aeronautical engineer Emilio Herrera (1879–1967), there was an inability to deal with the lack of free parameters in Einstein's formulation of special relativity. To Herrera the scientist's intuition was constrained by the growing number of universal constants, of which c was only the latest in a series which included Avogadro's number, Planck's constant and others. This feeling of constraint was offensive first to his religious values ("It is an unsustainable error to suppose ourselves in possession of numbers which limit and regulate creation.") as well as to his personal neo-Cartesian cosmology (in which Einsteinian space-time was valid in our four-dimensional universe, but not necessarily for those of higher dimensions which he presumed to exist). Herrera was enthusiastic about general relativity, since his own cosmological notions had (he thought) anticipated certain of Einstein's formulations such as the curvature of space. With the passage of years Herrera overcame his scruples about c and defended relativity against Julio Palacios' countertheory which he believed brought ridicule upon Spanish science.[43] José A. Pérez del Pulgar, director of the Catholic Institute represented another line of thought, that which could not disentangle the dynamics of relativity from its kinematics. He accepted the mechanics of Einstein-Minkowski as "the true one" but balked at accepting c as the limit of velocities. He denied that there was maximum velocity in nature while praising Einstein's gravitation theory as the best part of his work.[44] Both Herrera and Pérez del Pulgar's collaborator, Vicente Burgaleta, a railway engineer, participated in a polemic over a "Relativity Paradox" from *Nature*, which presented two rigid intersecting triangles, the point of whose intersection seemed, under specified conditions, capable of exceeding the speed of light. Herrera appealed to Sir Arthur Eddington for a solution which, when received, satisfied no one.[45] There is no doubt that, because of the role played by mathematicians in the reception of relativity and their overriding interest in general relativity, the physical meaning of the special theory was not well understood.

Notice of relativity reached the general public mainly through the efforts of engineers, some of whom were particularly gifted as popularizers. Manuel Velasco de Pando, a Sevillian industrialist and graduate of the Industrial Engineering School of Bilbao lectured on relativity in Seville and Bilbao throughout the 1920s. According to one listener, who before hearing Velasco had doubted that he knew enough mathematics to understand Einstein, "it was a revelation to me to hear Sr. Velasco deliver, first in the Academy of Letters and later in the Athenaeum of Seville . . . a number of lectures on Einstein's theory of relativity, which were attended by such a throng of people that, especially in the Athenaeum, the lecture hall and adjoining rooms were filled and there were even listeners on the stairway."[46]

Popular pamphlets, sold at kiosks and railroad stations, were also written by engineers. Civil engineers Salvador Corbella Alvarez and José Ochoa y Benjumea each produced such a pamphlet designed to explicate Einstein's theories without the use of mathematics.[47]

The constant coverage of relativity in the early twenties in journals such as *Madrid Científico*, a general-science journal written and read by engineers of the capital, and *Ibérica*, a Jesuit popular science journal with a similar readership, together with teaching of relativity in various formats in the engineering schools bespeaks unusually high interest in Einstein by this socially conservative group. What did engineers have to gain by their noisy espousal of relativity? First, relativity provided the first major public platform on which conservative Catholics could openly espouse the cause of modern science. Having been unable to accept the new Darwinian biology (and having suffered ostracism by progressive intellectuals because of it), conservatives could now recoup the prestige accruing to those donning the halo of modern science. In the particular context of the "scientific middle class", engineers could now appear as the intellectual equals to their main rivals the physicians, whose spokemen had been, since the 1870s, Darwinian and liberal. Relativity provided an unusual opportunity for engineers to acquire intellectual prestige. The Einstein myth held relativity to be incomprehensible and did not distinguish between special and general theories. In fact, special relativity was not difficult to understand and many undergraduate engineering students had learned its rudiments in their text books. The sense of having comprehended the incomprehensible both built esprit and conveyed prestige, which was subsequently heightened when large numbers of engineers studied absolute differential calculus and gained thereby at least a minimal understanding of the more difficult general theory.

Finally we must stress the pivotal role played by engineers in the formal reception of relativity in Spain. Through their membership in both the exact and physical sections of the Royal Academy of Sciences their backing for the two prizes eventually won by Plans can be seen as an expression of demand by engineers not only for more information about relativity but for the tools which would enable them to understand it. Subsequently they took the leading role in popularizing and promoting Einsteinian ideas for a lay audience, having already integrated them into the curricula of their schools. During Einstein's visit they were conspicuous by their attendance at his lectures to the point where the public was able to identify the unity of interest between professional engineering and pure science. Engineers had, since the late nineteenth century adopted the cause of science as an ideological cornerstone of their drive for professional recognition. Relativity was an obvious and convenient symbol for that aspect of their professional credo and they seized upon it enthusiastically.

6. RELATIVITY AND THE SPANISH INTELLIGENTSIA: ORTEGA Y GASSET

In a country such as Spain where the scientific establishment was very small and well integrated within an intellectual world which was likewise small and where everyone seemed to know everyone else, the conductivity of scientific ideas was heightened. Ideas spread rapidly and diffused easily across disciplinary boundaries as well as those defined by interest or education. Scientists from many disciplines other than the three most directly concerned pronounced on relativity. Chemists, as I have mentioned, were troubled enough about relativity's supposed connotations of energeticism to have combatted it; physiologists familiar with Mach's notions of physiological time detected in relativity an amplification of this same theme. Relativity was discussed in circles of affinity which included scientists and non-scientists. Where the circle of affinity is semi-institutionalized, as in the Spanish *tertulia*, conductivity of ideas is heightened, and channeled in detectable ways. The important tertulia of the philosopher José Ortega y Gasset (1883–1955), for example, included Blas Cabrera, but no mathematicians. Therefore Ortega's ideas on relativity emanated mainly from the discussion of the special theory and he was not much concerned with general relativity.

Ortega made two important contributions to the Spanish discussion of relativity. The first, his famous essay on the historical significance of relativity, written in 1922, had its kernel in a carefully worded riposte to the typical allegation of Catholic anti-relativists that relativity was subjectivist and that it denied the existence of absolutes. Implicitly addressing anti-relativists, Ortega first criticizes a current of European thought he called "utopianism". This value, characterized in "the enormous zeal for dominating the real" was, moreover, "specific to Europeans" and was a byproduct of a European rationalism which insisted upon absolutes, in politics as well as in physics. The need for absolutes is at the root of scientific reductionism (Ortega does not identify it as such), which he ridicules by alluding to Jacques Loeb's exaggerated use of tropisms to explain animal and human behavior.[48]

Against the background of utopic/absolutist thought, Einstein was a "breath of fresh air". Elaborating on the Kantian distinction between pure reason and sense data, he associates a physics which is purely geometric with Kantian pure reason and, by implication, with absolutism and utopianism. An example is the Michelson-Morley experiment. Lorentz's solution was utopic, that of the old rationalism. But Einstein "inverts the inveterate relationship which existed between reason and observation", by insisting that geometry yield to observation.[49] Einsteinian physics was absolutist, he explains, because it holds that physical laws are true whatever the system of reference used. But it was not absolute in the aprioristic sense of the old rationalism. In classical physics our knowledge was relative because we could never achieve knowledge of the absolute. In Einstein's physics our knowledge is absolute; reality is relative.[50]

During Einstein's visit to Madrid, Ortega delivered a short speech introducing the guest at the Residencia de Estudiantes.[51] It is an eloquent statement on the role of physics in western culture and on Einstein's place in the history of physical thought. Physics, he begins, is the characteristic, idiosyncratic discipline of western culture: a synthesis of thought and action which combined theory with practical dominion over things. In the same line as the great figures of this tradition – Copernicus, Galileo, Kepler, and Newton – came Einstein. In order to describe Einstein's revolution, Ortega did not think it necessary to dissect Einstein's science itself. (He believed it difficult to grasp, "although not as difficult to comprehend as is said.") There is an easier route, "which is to fix the significance of Einstein's intellectual act, to inquire into the intellectual faculties which mostly characterize him." In this light, one finds that Einstein's achievement lay in "a new kind of experience."

For Kant, Ortega explained, knowledge was a product of two factors: sense data and *a priori* thought. The history of modern physics can be described in terms of an oscillation between those two poles. In Descartes' system, the idea predominated over observation and experiment, inasmuch as the physical image of the world would be constructed on a purely geometric basis. Kant, while recognizing the importance of both factors, "really conceded to experience the role of observing how the laws of geometry are fulfilled, so that decisive evidence continued to rest on reason." Einstein turns this state of affairs completely around:

Einstein represents the opposite point of view. For him, what is strictly rational cannot arbitrate physical things, because mathematics is a formal science, not a science of things. Reason builds a repertory of ordinal concepts; but it is the experiment which selects the applicable order. Euclidean space is an order; experience, not theory, must decide whether or not it is applicable to the world. Einstein, then, symbolizes a shift towards empiricism.

To say, as many did, that Einstein merely confirmed Kantian doctrine on the subjectivity of time and space was simply wrong. Ortega's view of relativistic physics as empirical is almost unique among Spanish commentators, especially in view of the common misunderstanding of the role of mathematics in its formulation (antirelativist scientists and scientist/theologians portrayed it as excessively abstract). Although Ortega does not here allude to the experimental proofs of Einstein's theories, his argument was perhaps the clearest articulation in Spain of the 1920s of the widespread feeling that such experimental confirmation legitimized a new system of physical thought, a sentiment which underlay much of the popular enthusiasm for relativity.

Ortega was initially drawn to relativity because it appeared to provide scientific substantiation for his own notion of "perspectivism", formulated in 1916. The perspectivist argument, whose analysis of observation is at least in a broad sense analagous to relativistic concepts, was really no more than a plea for cultural relativism. Although his 1922 essay has now passed into the canon of relativistic philosophy,[52] it was harshly criticized at the time by scientists and philosophers alike. Miquel (Miguel) Masriera, a physical chemist who had studied with Hermann Weyl, asserted that Ortega had missed the main point about Einstein's critique of observation:

What Ortega has said about how relativity is essentially objectivist, anti-Kantian or vitalist is, as they say today, pure dilettantism. Relativity is chiefly characterized by not stating how objects are, but how we see them. Clearly, it establishes an absolute value, in the sense of the independence of each system of observation, and this is the so-called "interval" between phenomena. But this value can never have philosophical objectivity, for it has no tangible existence, but rather the character of a necessary abstraction, or, more concretely, of a mathematical indeterminant.[53]

For the conservative Catholic philosopher José Pemartín, Ortega's 1922 essay, in advancing the argument for perspectivism, had disfigured "not only the sense of Einstein's renovation, but even the nature of classical mechanics, which he erroneously presents as something utopian, a pure entity of reason, arbitrarily imposed on reality by rationalist obfuscation." Ortega distorts classical physics by asserting that Galileo and Newton made the universe Euclidean simply because reason so dictated. But the truth, says Pemartín, is that they saw in Euclidean geometry a close fit with experimental data. Leverrier's discovery of Neptune with a pencil, later confirmed by observation, is a kind of a Euclidean anticipation of Einstein's non-Euclidean feats. Also false was Ortega's interpretation of Einstein's achievement: that he had inverted the relationship between reason and observation and stood Kant on his head. Pemartín quotes Einstein himself to the end that time and space in a Gaussian four-dimensional system lose all physical reality. (There is a conceptual stand-off here: since neither Ortega nor Pemartín grasped the core of Einstein's critique of physical observation, there was no point in arguing what was real or not real in Einstein's system.) In a footnote, Ortega had alluded to the latent anticausality of relativity theory. Pemartín thought it nonsensical to stress Einstein's antirationalism and, without alluding to Ortega's contraposition of Einstein and Descartes, noted the profound Cartesian roots of the former.[54] (Pemartín, it should be noted here, was the ideologue of Primo de Rivera's regime who cited Einstein approvingly as justifying a "relativistic" approach to politics.)

Ortega was virtually the only Spanish philosopher to make an extended commentary on relativity in the 1920s, although his disciple Ramiro Ledesma wrote a polemical review of Hans Dreisch's antirelativist book from an enthusiastically pro-Einstein position (an interesting episode in view of Ledesma's future role as a founder of Spanish fascism). Other philosophers of the 1920s – Miguel de Unamuno and Eugeni d'Ors, for example – mentioned Einstein only in passing.

7. EINSTEIN'S VISIT AND THE QUICKENING OF DIFFUSION

Relativity diffused, as geographers would have it: downward through the urban hierarchy, from centers of large population to smaller ones and from the best educated to the less well educated.[55] Einstein's visit to Barcelona, Madrid and Zaragoza in February-March 1923 crystalized discussion of relativity at various "popular" levels. Here I will discuss only three: the issue of incomprehensibility and the displacement of the "gens du monde"; the penetration of less well educated groups through the scientific middle class's information network; and, last, how Einstein's visit revealed a number of significant shifts in the popular image of science.

7.1. Incomprehensibility and the Displacement of the Gens du Monde

Michel Biezunski has described how, in the case of France, the *gens du monde*, the typical arbiters of bourgeois culture exhibited resentment at their displacement as cultural arbiters owing to their inability to understand Einstein's ideas.[56] A similar process unfolded in Spain and was amply articulated during Einstein's visit. Most of the discussion of relativity's comprehensibility originated with literati. One common note which they struck was that of embarrassment, as in the Catalan writer Josep Maria (José María) Sagarra's open confession: "I attended Einstein's lectures knowing that I would understand little of his explanations, half afraid of playing the ridiculous role of falling asleep. I went to the lectures without saying a word to anyone, as if I'd be ashamed . . ." Rafagas, a Zaragozan columnist, was even more blunt: "We went to hear him . . . in the certainty of not being able to understand; we felt ourselves to be as wretched as the lowest of household bugs." Inability to understand "truths born yesterday," filled him with anguish. Yet, "if anything consoles us, it is the spectacle presented by many persons interested, intrigued by the scholar's theories. They waved their arms about like blind people. But doesn't this same gesticulation indicate that they seek light?" The most evocative self-deprecatory image to appear in the comprehensibility discussion was that of the "slave at the sermon" (*el negro al sermón*). Joan (Juan) Colomines Maseras used this image to describe audience reaction to Einstein's lectures in Barcelona: "We must confess that many of the listeners got from the lecturer's explanation what the slave got from the sermon, as demonstrated by the looks of exhaustion which surprised us on a great number of familiar faces, and sighs of liberation escaped from many breasts upon hearing the closing words."[57]

Other commentators were just as frank but more petulant. In the view of Carles (Carlos) Soldevila, one of the leading voices of the Catalan intellectual bourgeoisie, Einstein's popularity was explained by the public's blind faith in science. What did Einstein do? "The immense majority of the people of Barcelona are totally ignorant of it. The rest are divided among those who said that Einstein was a great genius, inventor of an enormously curious and transcendant theory, and the miniscule group of specialists who had read the theory and understood it." Indeed, it was difficult to find a similar case of celebrity in all the history of science. The press and popularizing books had brought his name before the public. And so, in sum, "Einstein is famous because a few hundred mathematicians have believed him worthy of so being." The rest of the public must take it on faith alone, because "the theory of relativity, in spite of all attempts to popularize it, is something that the good burghers who walk along the streets reading the paper can never understand." For Soldevila, then, there was a connection between Einstein's popularity, "almost divine", the incomprehensibility of his theory (which made its acceptance an act of faith), and a kind of conspiracy on the part of scientists to bring about this result. A similar interpretation was made by the humorist Julio Camba who observed that everyone admired Einstein but few, himself included, knew why. He supposed that the inventors of absolute differential calculus had jobbed their invention to Einstein just to ensure it would be used (and, it followed, so that no one else but they would be able to understand him).[58]

Other commentators tried to analyze the perceived failure of popularization. Lucanor, columnist for the Madrid daily *La Epoca*, noted that the "cultural middle class" had not achieved "that intelligible vision" necessary to translate the new physics for the average reader. There is a clear need "to socialize these conquests, [but that] was not Einstein's task but that of popularization." Spaniards, Lucanor duly notes, once had a great popularizer, José de Echegaray, and the French still had one in the person of Henri Bergson. Lacking such figures, "the public of Madrid listened to Einstein bereft of any teacher or guide,"[59] Echegaray, winner of the Nobel Prize in literature and a conservative politician as well as mathematician (he preceded Pedro Carrasco in the chair of mathematical physics at the University of Madrid) was, of course, more than just a good popularizer of science: he was the quintessential Castilian *gens du monde* whose word on science could be believed because of his social status. His departure was symbolic of his class's loss of control over a segment of culture, pure science, whose significance it had only recently begun to appreciate.

This kind of argument implied that the role filled by Echegaray would, henceforth, have to be filled by two separate classes of the bourgeois intelligentsia. Engineers were more than willing to fill the role on the scientific side, at the very moment they had turned the tide of public sentiment in their generation-long battle to win recognition as a profession alongside medicine, law and others long established. There would now have to be a "cultural middle class" (Lucanor) and a "scientific middle class" (the argument presented in this essay for the social and cultural role of the engineering community).

A minority not only took exception to the incomprehensibility argument but denied the validity of the very terms of that debate. To the philosopher Rafael Selfa Mora declarations by intellectuals that they had, or had not, understood Einstein were largely self-serving:

As for those who say they have understood Einstein, unless they are among the exceptionally privileged group in Spain who are learned, we doubt their comprehension. On the other hand, those who say that the number who can listen to him is limited to half a dozen also make us doubt their competence. Both groups disguise their thoughts or dress them with pedantry; the first, so self-possessed that they believe they can transmit to us the knowledge they have been able to capture from the complexity of higher understanding which the theory of relativity supposes; the second, because they either include themselves among the half dozen, excluding the rest, or they disguise their sin by saying that what is difficult to master in totality is absolutely incomprehensible, and thus they justify the passivity of their intellects or the incapacity of their understanding.

The majority of intellectuals have not been able to understand wholly the scientific exposition of the professor; but most have been able to listen to him and understand him. A theory only accessible to eight or ten human minds, given the current progress of science, would have little or no value.[60]

7.2. Popular Access to Scientific Ideas

Here we look at the lower levels of the educational hierarchy in order to make a distinction between formal, articulated access to scientific ideas at the popular level and informal, disarticulated diffusion, on the other hand. In both cases, the information loss is so enormous that the net result is practically the same. But in analyzing the processes of diffusion, we learn something of its differing social contexts.

What distinguishes popular reception from that of the scientific middle class is that, in the former, there is no organized or formal, articulated access to scientific ideas. Such access requires, first, a specific level of scientific education necessary to place the new facts in perspective and, second, some institutional supports whereby the ex-science student can relate to authority figures (e.g., former professors) in order to validate or invalidate the new ideas. I have in mind articles in popular engineering magazines, public lectures, newspaper reports or interviews, and so forth. Freeform reading or discussion of popularizing literature by the "man in the street", on the other hand, must lead, in the vast majority of cases, to a subjective, impressionistic, and disarticulated perception of the idea in question.

We know that relativity was discussed in small towns. A provincial columnist observed apropos of Einstein's visit:

Virtually no one understands his famous theories; but everyone is talking about relativity and the artistic physicist. The coffee-house strategists . . . move lumps of sugar around to give an idea of what an "inertial" system is. The press fills column after absurd column, horribly distorting the relativistic harmonies. Cartoonists and humorists squander their charm and good humor on Einsteinian curves and space. The role of the generalizer, never lacking in any latitude, tends to make relativity the panacea of universal knowledge which, at the same time as it resolves problems of the interatomic world, poses the gravest questions about bullfighting.[61] Even in village *tertulias*, where the vicar, the doctor and the mayor play cards in the back room of a grubby pharmacy, they speak of its portent, one believing it a corroboration of some theologian's conjecture, another holding it to be no more than a hoax based on the application of useless calculations to what a forgotten and insignificant book by an old professor of physics said on such and such a page of some chapter.[62]

Consider now these fictitious participants of the discussion in the grubby village pharmacy. The vicar, a political reactionary, has enjoyed an anti-relativity piece in a theological journal, written by a theologian attempting to defend "traditional" values; he opines that Einstein is incomprehensible and has subverted good philosophy and good science with mathematics. The physician, who uses Roberto Novoa Santos' pathology textbook in his daily practice, follows this authority in believing that the time of a patient's reaction to the thump of a neurological hammer bears a "relative" relation to "real" time. The mayor, an alliophile liberal, follows the newspaper in presuming relativity to be a triumph of progress; his authority is a politician, Luis Araquistain, not a scientist. The pharmacist has been assured by Eugenio Piñerúa's series in *La Farmacia Española* that matter is deprived of its classical meaning by Einstein's theory.[63] When a customer enters the shop and declares that "Light has weight" (the favorite relativity catch-phrase of Spanish cartoonists), the village experts will respond in consonance with the most congruent authority. Here we observe relativity diffused through an intellectual field; the variety of social and economic realities implied by Sánchez Peguero's description are likewise refracted through the intellectual field.[64]

At the very lowest level of the "scientific middle class" one could acquire some notions of hypergeometry, but without the ability to

interpret them in the context of the new physics. In the journal published by the Aides and Auxiliaries of the Corps of State Civil Engineers, a professional association of geometrical practitioners who
worked as aides to civil engineers or surveyors, we find a curious tribute
to Einstein signed by Pablo Pulido, an aide in the land survey office of
Cuenca. Pulido's notes, reflecting his attendance at Einstein's lectures in
Madrid, constitute a peculiar mixture of references: non-Euclidean
geometry, Emilio Herrera's ideas on hyperspace, and four-dimensional
notions gleaned from popular theosophy.[65] In all of these cases, the end
product of the diffusion process is incomprehension, but each of our
characters believes he has comprehended. Such persons have sharply
different views of the relation of science to society than a Soldevila or a
Sagarra who knew they had *not* comprehended.

7.3. Einstein and the Image of Science in Spain

Einstein's visit stimulated changes in the popular image of science or
else legitimized shifts already underway before his arrival. It had long
been part of the ideological baggage of Spanish conservatism to criticize
science as a fetish of the left, and this line surfaced under the stimulus of
Einstein's visit, particularly on the occasion of his address to anarchists
in Barcelona. Anarchists, in their union education programs, had made
no secret of their messianic view of science as the salvation of the
working class, and Einstein's address reinforced this image. Yet, if civil
discourse was to have effect, such a view of science had to be undermined. Such views were raised during Einstein's visit by the far right,
but the visible enthusiasm of engineers and other demonstrably conservative voices demonstrated the isolation of those who still insisted on
making science an ideological weapon. Another notion entertained by
the old right was that science had to be utilitarian or it had no social
value. Einstein's presence stimulated an open discussion on the value of
pure scientific research, as opposed to that with utilitarian or practical
objectives only. Science writers were at pains to point out that science
was ideation and that the public ought not to be overly concerned with
practical results emanating from Einstein's theories. In this context,
Mariano Poto drew a comparison with Darwinism: "Darwinian evolution was [once] no more than pure ideation. But with the passage of
time we have seen it applied to the transmutation of living species; and
nowadays don't researchers seek to acquire in the laboratories a model

to demonstrate the route that nature followed in the evolution of the elements? . . . . Darwin revolutionized man's soul; Einstein has achieved much more."[66]

Many made the point that the practical minded were unable to fathom what Einstein's theories held for them. For some, the very absence of utility was praiseworthy. Comparing public interest in relativity with that in King Tut's Tomb, José María Salaverría declared that people pursued such knowledge precisely because there was no practical benefit to be had from it. Salaverría had been present at a tea offered to Einstein by the Marqueses de Villavieja, attended by members of the nobility, scientists and intellectuals – civil discourse in action. In Salaverría's view, that which united the guests was exactly "that cult of the useless." This was a sardonic reference, but not a derogatory one. He went on to point out that Einstein was not the kind of scientist to traffic with industrialists, in the manner of Nobel or Edison. Indeed, there was a surfeit of practicality in the world: "We have a practical science, a practical morality, virtually a practical religion (pragmatism). If culture is not practical, it isn't culture," the columnist concluded. For him, the distinguishing feature of Einstein's wisdom was that it was not practical and did not produce money. The fact that it was not widely understood simply underscored that relativity brought to mind the highest moments of western civilization, when the greatest truths of science and religion were forbidden to the masses. Others, like the mathematician and popularizer Francisco Vera, pointed out that one simply had to recast one's conception of utility; relativity did indeed have utility, but of a purely scientific kind which opened up new fields of mathematical and philosophical speculation.[67] These kinds of arguments, which were fully congruent with the objectives of engineers, provided a kind of popular legitimation, originating with the non-scientific intelligentsia, for the instauration of pure scientific research in Spain, in a society which had previously held no such general value.

8. CONCLUSION

In conclusion I would like to comment on what the contextual approach to reception studies offers first to the study of science and second to the study of society. If in studying the Spanish case, we had limited our discussion only to the scientific reception of relativity we would have perceived one of the two central motors of reception, the strength and energy of Spanish mathematics, only to have missed the other, namely the engineering community's "demand" for the introduction of relativity. Einstein was a symbol of science's prestige: to espouse his ideas was to associate oneself with a high cultural value for science. To master his ideas was to accrue the prestige itself. What does the process tell us about Spanish society? We know that Spanish engineers, in their drive for professional status, adopted the ethos of science as an ideological principle. Their enthusiastic support of Einstein, when many of the other connotations of his persona (his liberalism, his Jewishness, his distaste for the conventional) were negative, demonstrates the depth of their commitment to that ethos.

Most of the elements that affected the scientific reception of relativity positively or negatively were present in the broader society as well. On the "popular" level, the pervasive, positive influence of civil discourse in science, which had been in action for nearly a quarter of a century at the time of Einstein's visit, had created a context in which pure research was viewed positively in public opinion. The drag of traditionalism, seen in the non-responsiveness of educational structures (that is, in the tension between the "gentlemanly" and the professional models of higher education), the narrowness and rigidity of the literary intelligentsia, and the residual anti-intellectualism of upper and lower classes alike, appears as a drone against which the clear new melody of the modern scientific ethos was heard. The Spain that received relativity was a society caught up in far-reaching process of value change associated with modernization. The mastery of the forces of nature that Einstein represented was a convenient symbol of the nation's aspirations to modernity and which therefore held great appeal throughout the society.

NOTES

[1] The present essay is based on my book, *Einstein in Spain* (Princeton, Princeton University Press, 1988) and on the more detailed Spanish version, *Einstein y los españoles* (Madrid, Alianza, 1986).

[2] Esteban Terradas, 'Teorías modernas acerca de la emisión de la luz,' in Asociación Española para el Progreso de la Ciencia, Zaragoza Congress, 2nd section, *Ciencias físico-químicas* (Zaragoza, 1908), pp. 1–21; and 'Sobre la emisión de radicaciónes por cuerpos fijos o en movimiento,' *Memorias*, Real Academia de Ciencias y Artes, 3rd epoch (Barcelona, 1909). Blas Cabrera, 'La teoría de los electrones y la constitución de la materia.' In this essay I provide the given names of Catalan personages in that language, with the Castilian equivalent following in parentheses.

[3] Cabrera, 'Principios fundamentales de análisis vectorial en el espacio de tres dimensiones y en el universo de Minkowski,' *Revista de la Real Academia de Ciencias Exactas*, **11** (1912–13), 326–344, and following; Terradas, 'Sobre'l principi de la relativitat,' *Arxius de l'Institut de Ciències*, **1** (2) (1912), 84–94; Antoni Roca, 'Incidència del pensament d'Einstein a Catalunya,' in *Centenari de la naixença d'Albert Einstein* (Barcelona, Institut d'Estudis Catalans, 1981), pp. 165–184, on pp. 170–172.

[4] Cabrera, *¿Qué es la electricidad?* (Madrid, Residencia de Estudiantes, 1917), pp. 173–176.

[5] Enrique de Rafael, 'De relatividad (Apuntes con ocasión de las conferencias de E. Terradas en el Institut),' *Ibérica*, **15** (1921), 218–22, 376–379.

[6] Cabrera, *Momento actual de la física* (Madrid, Real Academia de Ciencias Exactas, 1921), p. 8.

[7] For an example of Comas' semi-Newtonian emissive theory of light, see 'Nueva teoría emisiva de la luz y la energía radiante en general,' *Scientia*, **36** (1924), 375–382, and discussion in Glick, *Einstein in Spain*, chapter 5.

[8] Antonio Lafuente, 'La relatividad y Einstein en España,' *Mundo Científico*, **2** (1982), 584–591, on p. 588.

[9] Julio Rey Pastor, 'Federico Enriques,' clipping from *La Nación* (Buenos Aires), 1923. I discuss the relationship between Spanish and Italian mathematicians in *Einstein in Spain*, chapter 1, and 'Einstein, Rey Pastor y la promoción de la ciencia en España,' in *Actas. I Simposio sobre Julio Rey Pastor*, Luis Español González, ed. (Logroño, Colegio Universitario de La Rioja, 1985), pp. 79–90, on pp. 80–85.

[10] Giorgio Israel, 'Julio Rey Pastor e la matematica italiana: analisi di alcune conessioni,' in *Actas* (n. 9, above), pp. 105–117, on pp. 114–115.

[11] On Lorente de Nó's work in Italy, see *Revista Matemática Hispano-Americana*, **1** (1919), 224. Lorente de Nó was the translator of Einstein's popular book, *Über die spezielle und die allgemeine Relativitätstheorie*, published was 'Sobre la teoría de la relatividad especial y general,' *Revista Matemática Hispano-Americana*, **3** (1921–22), 194–199 and following and later issued as a book under the title, *La teoría de la relatividad al alcance de todos* (Madrid, Biblioteca Scientia, 1923; 3rd ed., 1925). Puig Adam's thesis was published as 'Resolución de algunos problemas elementales en mecánica relativista restringida,' *Revista de la Real Academia de Ciencias Exactas*, **20** (1922), 161–216. Peña was the author of an appendix on relatividad included in the Spanish translation of William Watson's popular physics textbook: 'Bosquejo de la teoría de la Relatividad,' in Watson, *Curso de física* (Barcelona, Labor, 1925), pp. 867–886.

[12] On Plans' superiority as a teacher of relativity, see my necrology of his student, 'In memoriam: Tomás Rodríguez Bachiller,' *Dynamis*, **2** (1982), 403–409, on p. 405. By applying Paul Appell's analogy between the form of equilibrium of a cord and the trajectory of a light ray, Plans derived a formula for the diffusion of light in a gravitational field with an equation that was, he said, simpler, more natural and logical than that of Einstein (the deflection was the angle of the asymptotes of the hyperbola formed by the ray passing through the field); 'Nota sobre la forma de los rayos luminosos en el cuerpo de un centro gravitatorio según la teoría de Einstein, *Anales de la Sociedad Española de Física y Química*, **18**, pt. 1 (1920), 367–373.

[13] In his writings on relativity, Plans always stressed experimental results. In a 1920 article, he noted with respect to general relativity, that one could ask nothing more of a theory than to have its predictions confirmed by observation and, with the hindsight of

several more years, he added that "the great success of Einstein's theory and its relativistic mechanics" was the correct prediction of the deformation of the perihelion of Mercury; 'Algunas ideas sobre la relatividad, *Ibérica*, **13** (1920), 380; and 'Bosquejo histórico y estado actual de la mecánica celeste,' *ibid.*, **23** (1925), 111. Throughout the 1920s Plans provided a running critique, always from the perspective of an Einsteinian stalwart, of the continuing series of ether-drag experiments by Dayton Miller and others for the readers of *Ibérica*. See, for example, 'El experimento de Miller y la teoría de la relatividad,' *Ibérica*, **27** (1927), 169–171, and 'Nuevas repeticiones del experimento de Michelson,' *ibid.*, **28** (1927), 94–95.

[14] 'Curso Levi-Civita,' *Ibérica*, **15** (1921), 98–99. In this additional talk he clarified certain concepts raised in the first lecture on the stability of movement. Levi-Civita's lectures were published in Catalan as *Questions de Mecánica clàssica i relativista* (Barcelona, Institut d'Estudis Catalans, 1922).

[15] On the banquet for Levi-Civita, see *El Sol* (Madrid), February 2, 1921.

[16] Plans, 'Proceso histórico del cálculo diferencial absoluto y su importancia actual,' *Asociación Española para el Progreso de las Ciencias, Congreso de Oporto* (Madrid, 1921), **I**, 23–43 (quotations on pp. 24 and 41).

[17] Plans, *Nociones de cálculo diferencial absoluto y sus aplicaciones* (Madrid, Real Academia de Ciencias Exactas, 1924), p. 5.

[18] Lorente de Nó in *Revista Matemática Hispano-Americana*, **7** (1925), 206; José María Orts in *Ibérica*, **24** (1925), 335.

[19] F. Navarro Borras, 'Don José María Plans y Freyre,' *Anales de la Universidad de Madrid, Ciencias*, **3** (1934), 242.

[20] See, for example, Rey Pastor's declaration to that effect in an interview with Ramiro Ledesma Ramos, 'El matemático Rey Pastor,' *La Gaceta Literaria*, **2**, No. 30 (March 15, 1925), p. 1.

[21] Jeffrey Crelinsten, 'William Wallace Campbell and the "Einstein Problem": An Observational Astronomer Confronts the Theory of Relativity,' *Historical Studies in the Physical Sciences*, **14** (1983), 88.

[22] I discuss the problem of science and civil discourse in *Einstein in Spain*, introduction, and in 'Ciencia y discurso civil en España, 1868–1960,' *Revista de Occidente*, in press.

[23] Glick, 'Tomás Rodríguez Bachiller' (n. 12, above), p. 408.

[24] Terradas, in *La Veu de Catalunya*, March 22, 1934, cited by Navarro Borras, 'José María Plans,' (n. 19, above), p. 231.

[25] Arthur Eddington, *Espacio, tiempo, gravitación* (Madrid, Calpe, 1922). Letter of Plans to Terradas, January 1, 1922, Terradas Archives, Institut d'Estudis Catalans, Barcelona. *El Sol*, August 1, 1922.

[26] Enrique de Rafael, 'La teoría de la relatividad,': *Razón y Fe*, **64** (1922), 344–359; José Antonio Pérez del Pulgar, 'El valor filosófico del relativismo: Einstein y Santo Tomás,' *ibid.*, **78** (1927), 503–511; Eustaquio Ugarte de Ercilla, 'Exposición y refutación de la relatividad,' *ibid.*, **73** (1925), 426–428.

[27] Benjamín Navarro, 'La relatividad,' *Revista Calasancia*, **10** (1922), 38–47; Ataulfo Huertas, 'La relatividad de Einstein,' *ibid.*, **11** (1923), 241–254, 290–309; 369–384; José María Goicoechea, 'Crítica de las teorías de Einstein,' *ibid.*, **11** (1923), 563–585.

[28] In *El Debate*: Enrique de Benito, 'Notas de un oyente profano' March 6, 1923 [favorable]; Bruno Ibeas, 'El einsteinianismo y la venida de Einstein,' March 7 [sceptical; rare example of a priest defending "positive" concept of time and space against Ein-

steinian idealism]; Manuel Graña, 'Aspectos de la relatividad,' March 14 [favorable; attracted by idealist blurring of boundary between spirit and matter]. Summaries of Einstein's lectures by Tomás Rodríguez Bachiller, March 4, 6 and 8; English translation in my *Einstein in Spain*, Appendix III.

[29] Huertas, 'La relatividad de Einstein,' (n. 27, above), p. 241.

[30] Enrique de Rafael, 'De relatividad,' *Ibérica*, **15** (1921), 91, n. 1 (emphasis that of Rafael). In his course notes, however, he indicates that both Einstein and Weyl had tied their theories to subjective, a priori Kantian forms; 'Nociones de mecánica clásica y relativista (Conferencias semanales en el ICAI en el curso 1921–1922), *Anales de la Asociación de Ingenieros del Instituto Católico de Artes e Industrias* [hereinafter cited as *Anales ICAI*], **1** (1922), 20–26 and following, on p. 187.

[31] Enrique de Rafael, 'Juventud y formación científica de Terradas,' in *Discursos pronunciados en la sesión necrológica en honor de . . . Esteban Terradas e Illa* (Madrid, Real Academia de Ciencias Exactas, 1951), p.10.

[32] [E. Terradas], 'Relatividad,' *Encicplopedia Ilustrada Universal* (Madrid, Espasa-Calpe), vol. 50 [1923], 455–512, on pp. 458–459.

[33] *Ibid.*, p. 459.

[34] José Pemartín argued that the relativism of the twentieth century was preferable, in politics as in science, to the absolute and categorical philosophies of the preceding century. See Shlomo Ben Ami, *Fascism from Above: The Dictatorship of Primo de Rivera in Spain, 1923–1930* (Oxford, Clarendon Press, 1983), p. 183.

[35] Ricardo Royo-Villanova, 'La crisis de la ciencia,' *El Siglo Médico*, **98** (1936), 58–70, on pp. 61–62.

[36] Julio Palacios, *Relatividad: Una nueva teoría* (Madrid, Espasa-Calpe, 1960). The heart of Palacios' approach was to alter the construction of clocks to eliminate the contraction of time; see discussion by Henri Arzeliès, *Relativistic Kinematics* (Oxford, Pergamon, 1966), pp. 190–191.

[37] See n. 32, above.

[38] See the extended discussion of this class in *Einstein in Spain*, chapter 6.

[39] J. Mañas y Bonví, *Optica aplicada* (Barcelona, 1915), 'Suplemento a la página 427,' p. viii. Carlos Mataix Aracil, *Primeras nociones de mecánica relativista* (Madrid, Koehler, 1923).

[40] Enrique de Rafael, 'Nociones de Mecánica clásica y relativista,' (n. 30, above), p. 21.

[41] Einstein lectured informally to engineering students at the headquarters of the Association of Engineers, which was the alumni association of the Instituto Católico, on March 8, 1923. See the accounts in the Madrid dailies *ABC*, *El Noticiero* and *El Debate* for March 9 and a Spanish version of Einstein's remarks in the latter.

[42] *Revista de Obras Públicas*, **71** (1923), 32; *Anuario de la Escuela Especial de Ingenieros de Caminos, Curso de* 1922–1923 (Madrid, 1923), p. 60.

[43] Emilio Herrera, *Algunas consideraciones sobre la teoría de la relatividad de Einstein* (Madrid, Imprenta del Memorial de Ingenieros, 1922), quotation on p. 24. On Herrera's cosmology and notions on the curvature of space, see 'Relación de hipergeometría con la mecánica celeste,' *Memorial de Ingenieros del Ejército*, 5th epoch, **33** (1916), 371–388; **34** (1917), 221–235; and 'L'Univers de Descartes,' *Le Génie Civil*, **140** (1963), 280–282, and following. On his critique of Palacios, see my article 'Emilio Herrera and Spanish Technology,' in Herrera, *Flying: Memoirs of a Spanish Aeronaut* (Albuquerque, University of New Mexico Press, 1984), pp. 173–215, on pp. 206–208.

[44] José A. Pérez del Pulgar and Vicente Burgaleta, 'Observaciones sobre la mecánica de Einstein-Minkowski,' *Anales ICAI*, **2** (1923), 480–494; **3** (1924), 485–496; and Pérez del Pulgar, 'Portée philosophique de la théorie de la relativité,' *Archives de Philosophie*, **3** (1925), 106–140.

[45] The "paradox" appeared in *Nature*, **110** (1922), 844, and the Spanish polemic in *Madrid Científico*, **30** (1933), 33–35, 52, and 67.

[46] *Palabras pronunciadas por el Excmo. Sr. Gobernador Militar de Sevilla, D. Antonio Fernández Barreto* (Sevilla, Cámara de Comercio, Industria y Navegación, 1929), p. 5; cited by Antonio Lafuente, *Introducción de relatividad especial en España*, unpub. licentiate thesis, University of Barcelona, 1978.

[47] Salvador Corbella Alvarez, *La teoría de la relatividad de Einstein al alcance de todos* (Barcelona, España en Africa, 1921); José Ochoa y Benumea, *El espacio y el tiempo desde Newton a Einstein* (Barcelona, Bazar Ritz, 1924).

[48] José Ortega y Gasset, 'El sentido histórico de la teoría de Einstein, in *El tema del nuestro tiempo*, 18th ed. (Madrid, Revista de Occidente, 1976), pp. 149–168, on pp. 149–151.

[49] *Ibid.*, pp. 152–154.

[50] *Ibid.*, pp. 142–143.

[51] Ortega, 'Mesura a Einstein,' in *ibid.*, pp. 189–193.

[52] E.g., it is included in the anthology, *Relativity Theory: Its Origins and Impact on Modern Thought*, L. Pearce Williams, ed. (New York, Wiley, 1968).

[53] Miguel Masriera, 'El valor del relativismo,' *La Vanguardia* (Barcelona), February 4, 1925.

[54] José Pemartín, 'La física y el espíritu,' *Acción Española*, **3** (1932), 595–604, **4** (1933), 27–37, and following, on **4**: 144–146.

[55] In practice, of course, ideas do not follow a strict educational, social or demographic path but rather diffuse through overlapping domains of discourse. Groups of affinity, like the *tertulia*, in which specific ideas are discussed, may, for example, contain persons of different social or educational backgrounds.

[56] Michel Biezunski, *La diffusion de la théorie de la relativité en France*, unpub. doctoral diss., University of Paris, 1981, pp. 83, 88.

[57] Josep Maria Sagarra, 'Einstein,' *La Publicitat* (Barcelona), March 4, 1923; Rafagas (pseudonym), 'Lecciones de humildad,' *El Heraldo de Aragón* (Zaragoza), March 14; Juan Colominas Maseras, 'Einstein en Barcelona,' *El Pueblo* (Valencia), March 2.

[58] Carles Soldevila, 'La popularitat d'Einstein,' *La Publicitat*, February 25, 1923; Julio Camba, 'Los admiradores de Einstein,' *El Sol*, March 6.

[59] Lucanor (pseudonym), 'Después de oír a Einstein,' *La Epoca*, March 16, 1923.

[60] Rafael Selfa Mora, 'La sed intelectual,' *El Luchador* (Alicante), March 14, 1923.

[61] The allusion is probably to Otto Kaestner, *El toreo científico* [Scientific bullfighting], Fernando de Ormaza, tr. (Madrid, 1923).

[62] C. Sánchez Peguero, 'Un aspecto minimal de la relatividad,' *El Noticiero* (Barcelona), March 13, 1923. Sánchez Peguero also uses the slave at the sermon image in this article.

[63] (Vicar:) Teodoro Rodríguez, 'Relatividad, modernismo, matematicismo,' *La Ciudad de Dios*, **135** (1923), 42–67 and following; (doctor:) Roberto Novoa Santos, *Physis y psyquis* (Santiago de Compostela, 1922), pp. 32–36; (mayor:) Luis Araquistain, 'Einstein o la razón estremecida,' in *El arca de Noé* (Valencia, Sempere, 1926), pp. 89–94

(originally published in *El Sol* during Einstein's visit); (pharmacist:) Eugenio Piñerúa, 'Nociones acerca de la Teoría de la Relatividad,' *La Farmacia Española,* **55** (1923), 241–243 and following, on p. 276.

[64] Pierre Bourdieu, 'Intellectual Fields and Creative Project,' in Michael F.D. Young, ed., *Knowledge and Control* (London, Collier-Macmillan, 1971), pp. 161–188, on p. 185.

[65] Pablo Pulido, 'Einstein en España: Algunos apuntes relacionados con la teoría de la relatividad,' *El Auxiliar de la Ingeniería y la Arquitectura* (Madrid), **3**, (1923), 84–86.

[66] Mariano Poto, 'Einstein y su teoría,' *El Liberal* (Madrid), March 1, 1923.

[67] José María Salaverría, 'Las originalidades einsteinianas,' *ABC*, March 10, 1923; Francisco Vera, 'La tercera conferencia de Einstein: Consecuencia relativistas,' *El Liberal*, March 8.

Boston University

V.P. VIZGIN AND G.E. GORELIK

THE RECEPTION OF THE THEORY OF RELATIVITY IN
RUSSIA AND THE USSR*

The creation and confirmation of the theory of relativity was one of the main events of the revolution in the natural sciences in the first third of the 20th century. The revolutionary process is not only an act of scientific discovery which is fixed in one or a series of publications. It is also a process of the reception and assimilation of discovery which unfolds in the "space" of the entire scientific community and stretches over years and even decades. Science in its essence is international. But the "internationalization" of scientific ideas and theories which arise in one or several countries is at times a complex and lengthy phenomenon. The study of such a kind of phenomenon at once clarifies the condition of science in an individual country, and the peculiarities of its scientific and cultural development.

The biography of the theory of relativity in Russia is especially interesting because of the fact that in its first three decades (1905–1941) Russia left the physical "provinces" (distinguished in all only by a few achievements of world class) and became one of the leading physical "powers". In the history of Russian science, 1917 – the year of the victorious revolution – is always considered as a significant boundary. In our case, 1917 forms the border also because of the fact that it is possible really to talk about the reception of the general theory of relativity (GTR) in Russia only in the post-revolutionary period. The special theory of relativity (STR) by that time actually had already become the property of the Russian physical community.

Thus our work is naturally divided into two parts: 1) the reception of the STR in pre-revolutionary Russia (1905–1917); and 2) the reception of the GTR in the USSR (1917–1941). Toward the end of the 1930s several important events in the GTR and its development occurred which are definitive mark of the period being examined.

The limited length of this article requires us to leave out a number of important details and extensive argumentation.

The literature which is devoted to the history and reception of RT in

* Translated by Paul R. Josephson, Program in Science, Technology and Society, MIT

Thomas F. Glick (ed.), The Comparative Reception of Relativity, 265–326.
© 1987 *by D. Reidel Publishing Company.*

Russia is quite small. Especially useful for this work were the collective work on the history of the natural sciences in Russia [190] (the section which touches upon the TR written by O.A. Lezhneva), materials on the life and works of A.A. Friedmann which are collected in [193], works of K. Kh. Delokarov [203], I. Ia. Itenberg [203], and V. Ia. Frenkel' [206, 208, 210]. The personal reminiscences of the mathematician B.L. Laptev (who finished the mathematics department of Kazan' University in the mid-1920s), and A.I. Ansel'm and M.A. Korets (scientists at Leningrad Physico-Technical Institute in the 1930s), to whom the authors express their thanks, were also used.

The authors are also grateful to A.T. Grigorian, I. Iu. Kobzarev, O.A. Lezhneva, L.S. Polak, V. Ia. Frenkel', B.E. Iavelov, and A.K. Iankovskii for useful and stimulating discussions.

I. THE STR IN RUSSIA (1905–1917)

1. Russian Physics at the Beginning of the 20th Century

Research on physics was concentrated in Moscow and Petersburg mainly in universities and in physics institutes created within them at the turn of the century; to a smaller degree in provincial capital city technical institutions and universities in Kiev, Khar'kov, Odessa, Kazan, and so on [182, 183, 190, 194]. On the whole, physics in Russia lagged behind physics of the leading European countries, although there were achievements on a world level, especially in the region of experiment; as one example, it is sufficient to name the experiments of P.N. Lebedev (1866–1912) on the pressure of light and the experiments of A.A. Eichenwald (1864–1944) on the electrodynamics of moving bodies. Both of these physicists worked in Moscow where already in the 1890s the outstanding Russian physicist of the second half of the 19th century, A.G. Stoletov (1839–1896), worked.

After the death of Stoletov in 1896, the main figure in the area of physics in Moscow University was one of the greatest Russian theoreticians of the period, N.A. Umov (1846–1915), who made a fundamental contribution to studies on the motion of energy. Lebedev worked at Moscow University from 1891 (from 1900 as full Professor), Eichenwald from 1906 through 1911. It was in Moscow that one of the first Russian physics schools was formed under Lebedev's leadership. In 1911, however, a great loss was inflicted upon Moscow physics as a result of the departure from the University of almost all its great physicists (in sign of protest against the arbitrariness of the Government).[1] Scientific life moved into physical laboratories which were organized with private funds; there, after the death of Lebedev, his students continued research.

The scientific successes of the leaders of Petersburg physics, professors of the university I.I. Borgman (1849–1914) and O.D. Khvol'son (1852–1934), were obviously more modest than the achievements of the Muscovites. But they raised the level of teaching (both were brilliant lecturers and created original and detailed courses of physics[2]), and also played a leading role in the Russian Physico-Chemical Society (RFKhO) – the main organization which united Russian physicists and published the major physics journal (*Zhurnal russkogo fiziko-khimicheskogo obshchestva, chast' fizikicheskaia*, hereafter *ZhRFKhO*). Only upon the return from Germany of two young physicists, A.F. Ioffe (1880–1960) and D.S. Rozhdestvenskii (1876–1940), in the middle of the first decade of the 20th century and the arrival of P.S. Ehrenfest in 1907 did scientific life in Petersburg enliven. The first two created a new impulse to the development of the experiment which was oriented toward the newest achievements in physics, while the latter, who would live in Petersburg until 1912, organized a theoretical seminar from which sprang the Leningrad school of theoreticians, which would occupy a secure position in world physics [204, 205, 210].

A considerable number of Stoletov's and Lebedev's students worked in the provincial universities in conjunction with their own pupils as well as with students completing Petersburg University. Often they were great scholars: in Kiev, N.N. Schiller (1848–1910) and later L.I. Kordysh (1878–1932); in Khar'kov, N.D. Pil'chikov (1857–1908) and A.P. Gruzintsev (1851–1917), and somewhat later D.A. Rozhanskii (1882–1936); in Odessa N.A. Umov (until 1893), F.N. Shvedov (1840–1905), followed by N.P. Kasterin (1869–1949); and in Kazan D.A. Goldhammer (1860–1922), and V.A. Ul'ianin (1863–1934).

At the First International Congress of Physicists in Paris in 1900 the Russian delegation consisted of forty-nine people, although according to various estimates the number of actively working physicists in Russia barely exceeded one hundred. The scientific revolution in Russian physics brought to the fore problems of the physics of electrons, electrodynamics of moving bodies, radioactivity, the study of spectra, thermodynamics and statistical physics. The newest theoretical successes, connected with quanta and the theory of relativity, began to attract special attention toward the end of the first decade.

2. "Pre-relativism" in Russia

The reception of one or another scientific conception in an individual country depends on the contribution to the preconditions of that conception which scholars of the given country make. The special theory of relativity was the natural result of the development of Maxwell's theory of the electromagnetic field, the electron theory and the electrodynamics of moving bodies. Maxwellian electrodynamics flourished in Russia in the last third of the 19th century [190, 224]; the earlier mentioned courses of Borgman and Khvol'son, in particular, are evidence.[3] The three most important achievements of Russian physics at the end of the 19th and the beginning of the 20th centuries (already mentioned in connection with the names of the Muscovites Umov, Lebedev and Eichenwald) give an idea of the Russian contribution to the preconditions of STR.

The work of Umov on the localization and motion of energy (1874) and the development of this conception by J. Poynting and others actually constituted an important part of the theory of the electromagnetic field and anticipated the relativistic generalization of the law of conservation of energy [200]. The problem of light pressure which became the main theme of the scientific creativity of Lebedev is also connected to this circle of questions. In 1899–1901 Lebedev experimentally recorded and measured the pressure of light on solid bodies and, at the end of the first decade, on gases [190, 194]. The pressure of light also tied Maxwell's theory of field with the theory of relativity. "The fact of light pressure demonstrated [by P.N. Lebedev]," the outstanding Soviet physicist S.I. Vavilov wrote, "facilitated the concretization of that indissoluble tie between mass and energy, which in all its breadth was clarified by the theory of relativity." [170: 166]. It is not accidental that text books and monographs on the TR beginning with the book of M.v. Laue [44], as a rule, devote space to the discussion of light pressure. The problem of the "ether drift" occupied Lebedev as well: at the end of the 1890s and beginning of the 1900s he attempted to utilize light pressure, the Doppler effect, and an experiment similar to the Rouland experiment in order to solve the question of the motion of the earth relative to the ether.

The analogous experiments of Eichenwald on the registration of the magnetic field of convection currents and displacement currents (1903–1901), which became the experimental classic of the electrodynamics of moving bodies [5, 7, 15, 19] are better known. Together with the experiments of Michelson-Morley, Rowland, Roentgen, Trouton-Noble and others, they formed the experimental foundation of the STR; in the dispute between supporters of the conception of the motionless ether (Lorentz) and of ether drag (Hertz) they provided some of the most powerful evidence in support of the former.

Soon after the works of Minkowski, first within the framework of STR, and then later the GTR, the deep tie between TR and non-Euclidean geometry was clarified. Unexpectedly, the problematics of non-Euclidean geometry turned out to be topical in the TR. In the motherland of N.I. Lobachevskii (1792–1856), especially in his native Kazan University, the traditions of the "Copernicus of geometry" were vital, in particular a fixed interest in the problem of "Geometry and Reality." The Kazan mathematicians F.M. Suvorov (1845–1911), A.V. Vasil'ev (1853–1929), A.P. Kotel'nikov (1856–1944), D.M. Sintsov (1867–1946), D.N. Zeiliger (1864–1936) and others not only were developing non-Euclidean geometry, but also wrote about the conditionality of the geometry of real space by laws of physics, the connection between motion and geometric structure, and the possible realization of non-Euclidean geometry in nature,[4] and they studied mechanics in non-Euclidean spaces.

On the eve of the theory of relativity they were active propagandists of Lobachevskii's ideas and of non-Euclidean geometry in general; in 1893 the collection of the classical works of E. Beltrami, B. Riemann, H. Helmholtz, including H. Poincaré's "On the Foundations of Geometry," [3] were published, and in 1895 the N.I. Lobachevskii prize was established, the first laureates of which were S. Lie (1897), W. Killing (1900), and D. Hilbert (1904). Along with the Kazan geometers, the Moscow mathematicians B.K. Mlodzeevskii, D.F. Egorov (1869–1931) and especially V.F. Kagan (1869–1953) (at first in Odessa and then in Moscow) were occupied with analagous problems, Kagan being the author of a detailed work on the foundations of geometry, the second volume of which was published in 1907 and was devoted to detailed analysis of classical works of Riemann, Beltrami, F. Klein, Lie, Helmholtz and others [13].

Thus, the Russian contribution to the experimental, theoretical and

mathematical preconditions of TR were sufficient so that Russian scholars could understand its significance and be included in the development
of the new theory. Not coincidentally the beginning of the 1910s
Lebedev and Umov correctly evaluated the STR (the latter himself
carried out important research on STR and actively propagandized it),
and the Kazan mathematicians immediately responded to the work of
Minkowski, and were the first in Russia to have translated his famous
paper *Raum und Zeit* [30, 45, 84].

3. Relativistic Ideas through the Prism of the Electromagnetic Worldview (1906–09)

The facts and conclusions of the electron theory and the electrodynamics of moving bodies which seemed at that time revolutionary and
which received subsequently a clear interpretation from the point of
view of STR (the negative result of experiments on the detection of the
"ether drift," the "Lorentz-Fitzgerald contraction," the dependence of
the mass of electrons on their speed, the conception of electromagnetic
mass, and so on) were looked at by many soon after the discovery of
STR as unambiguous confirmation of the electromagnetic picture of the
universe. The first stage of a scientific revolution in physics consisted
precisely in the transition from the classical mechanical worldview to the
electromagnetic. This is how V.I. Lenin evaluated the character of the
scientific revolution in physics in his book *Materialism and Empiriocri-
ticism* [17] which was published in 1909.[5]

In a speech devoted to the anniversary of Benjamin Franklin, the
young Petersburg physicist K.K. Baumgart (1860–1963) spoke at the
beginning of 1906 about the triumph of the electromagnetic worldview
in its radical variant, that is in the variant of M. Abraham and W.
Kaufmann [9]. During this talk the names of Lorentz and Poincaré (let
alone Einstein) were not mentioned and, in the character of the most
revolutionary conclusion of the new approach, the electromagnetic
interpretation of the mass of the electron and its dependence upon its
speed were discussed.[6] In that very year, D.S. Rozhdestvenskii, who
was coming to a more moderate variant of the electromagnetic worldview in the spirit of Lorentz and Poincaré, presented a thorough review
of the electrodynamics of moving bodies [11]: an analysis of the expanse
of experiments on the problem of the "ether drift" including the
experiments of Michelson-Morley and Eichenwald, the theoretical con-

structions of Lorentz, connected with the hypothesis of contraction, the critical observations of Poincaré, and so on. Having agreed with Lorentz about the immobility of the ether he also came to the point of view of E. Cohn who considered that the ether, as a result, turns out to be a superfluous concept.[7]

Already in the following year (that is, 1907), Baumgart in a review "On Electromagnetic Mass," wrote about the theory of Einstein (having in mind STR) [12]. Having described the well-known discussion between supporters of the theories of Abraham, on the one hand, and Lorentz-Einstein, on the other[8] (the author writes about the "theory of Einstein which practically coincides with that of Lorentz" [12: 317]), he compared these theories and considered the merits and drawbacks of each. He saw the advantage of the Lorentz-Einstein theory in the fact that it supported the negative result of the Michelson-Morely experiment and it was based on a very general principle, namely the principle of relativity which is not fulfilled in the theory of Abraham. Baumgart himself was inclined to prefer Abraham's theory, inasmuch as "it is consistently and rigorously electrical" [12: 318].[9]

In 1907–09 the reader might encounter on the pages of the *ZhRFKhO* more than a few pronouncements in the spirit of one or another variant of the electromagnetic worldview. Moreover in several of them the opposition between "Abraham and Lorentz-Einstein" is veiled, in others it is mentioned, but the theories of Lorentz and Einstein, as a rule, are not differentiated. The Petersburg physicist V.K. Lebedinskii (1868–1937) writes about the substitution of electromagnetic *Weltanschauung* for the mechanical one [14: 332]; still another Petersburger, the later well-known theoretician V.R. Bursian (1886–1945), provides an in-depth review of researches connected with the Michelson-Morley experiment, and in this plan he considers the theory of Lorentz to be the main theoretical achievement (explicit references to Einstein and the STR are absent) [16]; Eichenwald once again describes in detail his experiments which had become particularly timely),[10] and having rejected Hertz' electrodynamics of moving bodies, subscribes to the theory of Lorentz.

Very likely, only O.D. Khvol'son in a speech entitled "Successes of Physics in 1908" (delivered in February 1909) [18], clearly talked about TR as one of the greatest theoretical achievements of the recent years. The most important aspect of this he saw in the radical transformation of the bases of classical mechanics.[11] But all the same, Khvol'son, who earlier than many had greeted the STR with enthusiasm (he wrote "the principle of relativity has been introduced into science by Einstein with great success" [18: 137]), seemingly considered that behind it (i. e. STR) stood the electromagnetic worldview: "Instead of that *mechanico-monism* to which science strove as its desired ideal, at the present time in the distant and dark expanses the hazy outlines of a new ideal in science, takes shape – *electro-monism*" [18: 192]. Soon, however, Khvol'son became a convinced supporter of STR and its energetic propagandist.

4. The Leaders of Russian Physics on the RT (1909–1914)

The attitude toward RT for a series of leading Russian physicists was generally the same as their attitude toward the peculiar phenomenology of the electromagnetic worldview. Borgman, Eichenwald and Goldhammer all took an ethero-electromagnetic position.

Borgman, by the way, was very interested in RT. The series "New Ideas in Physics" appeared under his editorship in 1912. The collection "The Principle of Relativity" contained translations of articles by Einstein (apparently this is the first Russian translation of the founder of the theory of relativity[12]), P. Frank, G. Lewis and R. Tolman, J. Classen and others on STR [58]. In fact, the volume "Ether and Matter," in which Borgman, trying to observe objectivity published programmatic articles both by supporters of the ether (J.J. Thomson and P. Lenard) and of its opponents (M. Planck and N. Campbell), was devoted as well to the problems of STR and the ether [57]. At the second Mendeleev Congress, held in December 1911 in Petersburg, Borgman referred to the TR which touches "the very foundation of theoretical physics and arouses unusual interest," but called for judicious care in relation to its all too bold conclusions and expressed hope for the preservation of the conception of the ether: "The idea of the ether directed the investigations of all great researchers in the area of physics. . . . Surely, in the future it will serve us as well." [31: 17][13]

Eichenwald, whose experiments were well known to all concerned with the problem of the electrodynamics of moving bodies and the STR, was an even more convinced follower of the electromagnetic worldview and the ether.[14] Eichenwald, apart from his experiments, wrote about the theories of Hertz, Lorentz and Cohn, but along these lines there is not one word about Einstein nor about the STR in a review of 1908–09.

He gave preference to Lorentz' treatment of the electrodynamics of moving bodies. Moreover, already in the 1920s he suggested that "the experiment of Michelson-Morley might give another result if it were repeated in a vacuum and in the last years of his work in Moscow he planned such an experiment."[178: 170] In other words, he even then maintained all hopes for the detection of the "ether drift" and likewise of direct proof of the existence of the ether.

D.A. Goldhammer, professor of Kazan University, who also played a significant role in the Russian physical community, although he was, in the final analysis, an opponent of the TR, repeatedly and very clearly made pronouncements concerning this theory, being carried away by the courage, wit, and mathematical depth of the theoretical constructions of Einstein and Minkowski. Occasionally, it can be shown that he shared their views, but he concluded his eloquent speeches with a return to the ether.[15] Goldhammer, first examining the STR as a good phenomenological theory of ether phenomena, later came to a rejection of Lorentzian, and correspondingly, of relativistic electrodynamics of moving bodies and returned to a variant of the theory of Hertz. In an article published in 1916 he wrote that the STR "introduces far too expansive change in our conceptions of space and time and leads to other difficulties the significance of which, apparently, is too great [97: 1]."

Three other authorities of Russian physics, the Muscovites Lebedev and Umov, and also the Petersburger Khvol'son became supporters of the STR, although their positive evaluations (Lebedev) or active support (Umov and Khvol'son) date to the beginning of the second decade. In the course of several years following the creation of RT, Lebedev was intensely occupied with the experimental evidence of light pressure on gases; and only towards 1910 were these experiments successfully concluded, and he could come forward with a review of the newest scientific achievements which contained an evaluation of RT. Lebedev noted the connection between the new theory and experiments on the detection of the motion of the earth or of other bodies with respect to the ether, and the circumstance that the STR inverts the problem since on the bases of these experiments the principle of relativity is postulated and the necessary conditions for its fulfillment are determined. As a result it was necessary to modify the laws of physics, and the very idea of the ether turns out to be superfluous.[16] The early interest of Lebedev in the experimental detection of the "ether drift," the organic tie of the conception of light pressure with relativistic teachings on energy and impulse, and the relativistic thematics of theoretical research of his student P. S. Epstein (see below) – all this indicates the connection of this outstanding Russian physicist with the TR and his predisposition to its acceptance.

N.A. Umov who already firmly held positions on the electromagnetic conception of nature in 1905,[17] in 1910 enthusiastically joined in the development of the TR and its popularization. In 1910–1912 he published two articles devoted to the axiomatics and the mathematically rigorous conclusion of the Lorentz transformations [33, 49, 86]. An analogous conclusion, based on the analysis of conditions of the invariance of the wave equation was utilized by Laue in his first monograph on STR [44] and significantly later by V.A. Fok (1898–1974) who alluded to Umov as his predecessor [186].

At the Second Mendeelev Congress in December 1911 Umov presented a long paper in which he spoke in detail about the theory of relativity which he considered to be the main theoretical achievement of recent times [48]. During this talk he clearly differentiated the "electromagnetic" approach of Lorentz and the more universal, general physical approach of Einstein and Minkowski.[18] Behind the "relativism" of STR he saw its theoretico-invariant essence and consequently he valued highly its four dimensional formulation presented by Minkowski.[19] And although Umov was not a "consistent relativist" and at times expressed himself in an "ether-electromagnetic" spirit, at this Congress he opposed the "antirelativist" Goldhammer. By the way, in Umov's paper we meet certainly for the first time in Russian literature a discussion of the question of the equivalence of inertial and gravitational mass and the bending of the light of stars around the sun which was predicted by Einstein on the basis of the principle of equivalence.

O.D. Khvol'son played a major role in the dissemination of the idea of the TR in Russia. Already in 1909 he spoke of it as an outstanding success of physical theory.[18] In 1912 Khovl'son wrote a thorough and clear review of STR which was published in three editions [66, 67] including his *Kurs fiziki* [91]. This account successfully combined rigor and consistency with attention to the principal physical and philosophical aspects of the theory. In addition to this, a comprehensive list of literature in which there were more than twenty references to the works of Russian scholars (including Umov, Bursian, Kordysh, Gruzintsev, V.S. Ignatowsky (1875–1943), K.N. Shaposhnikov and others) was appended to the work. Noting the electrodynamic sources of the theory, he emphasized its general physical significance inasmuch as it "represents the foundation of a new doctrine, first and foremost about space and time" [91: 333]. (Hereafter we will cite the second edition of *Kurs*); and he compared it with the heliocentric revolution, although the scale of the relativistic revolution seemed to him even more impressive.[20] Khvol'son spoke about the "incomprehensible strangeness" of the conclusions of the STR which complicated its reception and therefore devoted special attention to the analysis of its physical bases. Although in regard to the question of the ether he sooner joined its opponents (". . . it is impossible to grant the existence of the ether if you accept the RT in all its entirety" [91: 357]), he saw in the electromagnetic foundation of nature the deep reason of relativistic correlations of mechanics.[21] In addition to the first textbook account of the STR the first popular science book on the TR belongs to Khvol'son (1914) [79]. Under his editorship important scientific and popular science books were published which facilitated the acceptance of the TR in Russia, for example, the book of A. Michelson (1912) [63, 64][22] and the brochure of M. La Rosa (1914) [87]. The first account of the TR in Russian encyclopedias also belongs to Khvol'son (1916) [101].[23] Having noted that in the preceding ten years the literature on the TR exceeded five hundred titles, he emphasized its revolutionary, nonclassical character, and difficulties with its understanding and study tied to that.

5. Russian "relativists"

Of course, Umov and Khvol'son certainly may be ranked completely among the relativists, all the more so since their scientific authority and lecturing and popularizing talent to a great degree facilitated the attraction of the scientific community to the RT. But in the first public discussion of the problems of the RT which occurred on December 30, 1909, at a special meetings of the physics section at the 12th Congress of Russian Naturalists and Physicians [23], neither Umov nor Khvol'son took part. The main figure at this gathering was P.S. Ehrenfest who had come from Vienna to Petersburg in 1907 and who worked there for five years [210]. He organized a theoretical seminar which became the center of attraction for young Petersburg physicists, and which gave birth to a Leningrad school of theoretical physics. The school was closely connected with that of Ioffe which was formed and fully manifested in the Soviet period. Besides Ioffe, who became the close personal friend of Ehrenfest, young and talented Petersburg physicists (Rozhdestvenskii, Baumgart, Bursian, L.D. Isakov (1884–1942), A.A. Dobiash (1876–1932), G.G. Weihardt (?–1919), Iu. I. Krutkov(1890–1952) and others) and mathematicians (S.N. Bernstein (1880–1968), Ia. D. Tamarkin (1888–1945), A.A. Friedmann (1888–1925) and others) attended the seminar. After the departure of Ehrenfest for Leiden the seminar was conducted by his students Bursian and Weihardt and, after 1915, the subsequently famous Soviet physicists N.N. Semenov (1896–) and I.V. Obreimov (1894–1981). Ehrenfest took an active part in the work of the Russian Physico-Chemical Society, first of all in the ZhRFKhO.

The first works of Ehrenfest on the TR date to 1906–07 and are devoted to the agreement of various models of the electron with it [201]. Works on relativistic mechanics of deformed media and absolutely rigid bodies, included works by Ehrenfest himself, and the works also of two other participants in the mentioned relativistic discussion, W.S. Ignatowsky, a graduate student of Petersburg University, and the student of Lebedev, P.S. Epstein (1883–1966), who were stimulated by these problems. Ehrenfest called into question the very possibility of the existence of concept of an absolutely rigid body in STR through his remarkable thought experiment with the rigidly rotating disk (the "Ehrenfest disk") [20]. On the other hand, the "paradox of Ehrenfest" led to the idea of the possibility of the destruction of Euclidean metric correlations in non-inertial frames of reference, having played a role in the genesis of GTR [214, 216]. On the pages primarily of German journals a broad polemic arose on the problem of a relativistic absolutely rigid body, in which, in addition to the German physicists G. Herglotz, F. Noether, M. Born, and M. Laue, Ehrenfest himself and W.S. Ignatowsky took active part.

Ignatowsky[24] was the honorary secretary of the "relativistic" meeting, and Goldhammer was its honorary chairman [23]. Three papers were delivered: Ignatowsky's "The Derivation of the Invariance of the Speed of Light from the Principle of Relativity," Ehrenfest's "The Principle of Relativity and the Rigid Body," and Epstein's "Theory of Relativity and the Structure of the Electron." Beside these three, Eichenwald, Schiller, the mathematician A.V. Vasil'ev, the mechanician I.V. Meshcherskii (1859–1935), and others took part in the discussions. Ehrenfest's talk met with great success. On the day after the meeting Ioffe told his wife, V.A. Kravtsova, "The Congress at the start was boring, but then interest grew and it achieved an apogee with Ehrenfest's talk which was unusually successful both in content and in the effect which its sincerity and enthusiasm produced . . . [210: 45]" The student of Ehrenfest L.D. Isakov wrote in his review: "The meeting of December 30 evoked great interest due to the critical question of the principle of relativity in the sphere of which all papers revolved [61: 48.]" Ehrenfest spoke not so much about the achievements of the STR and its paradoxical conclusions, as about the difficulties of the theory connected with the concept of the absolutely rigid body, with departure beyond the limits of uniform and rectalinear motion and with the thought experiment with the rotating disk [35].

Ignatowsky examined the question of the derivation of the analog of the Lorentz transformation which may be based only on the principle of relativity (that is without utilization of the postulate of the invariance of the speed of light). Henceforth, he published a series of articles in German journals on the axiomatics of the STR [24, 26, 27], in particular, parallel with P. Frank and H. Rothe [50] he gave the group-theoretical derivation of the Lorentz transformation to which W. Pauli devoted an entire page in his encyclopedia article.[25] If you add to this the cycle of the works by Ignatowsky on relativistic mechanics of continuous medium [41, 42] and of the absolutely rigid body [25, 39], for which you may find references in the same well-known monographic review of Pauli, then Ignatowsky may rightly be called one of the first Russian relativists. Two circumstances surely influenced Ignatowsky's "relativistic" orientation: his study and subsequent work in Germany (from 1906 to 1914 in Giessen and Berlin), where STR found fertile soil and by 1909 had gone through extensive development, and his clearly expressed interest mathematical physics (in 1909–10, his two volume textbook on applications of vector analysis in theoretical physics, which contained a large amount of material on electrodynamics and also references to tensors, was published) [28].

In Epstein's paper, various models of the electron were compared with the demands of STR and with experiments similar to Kaufmann's. Although the model to which the speaker gave preference, namely A. Righi's, soon lost its meaning, the utilization of STR as the basic theoretical criterion of selection was a method characteristic for a "relativistic" theoretician. In the "relativistic" discussion of 30 December 1909, questions which touched upon the bases of STR were broached. "In the discussions," Isakov wrote in the already cited review, " . . . all of the complexities and difficulties in understanding the questions touched upon were expressed. I.V. Meshcherskii (a professor of Petersburg University, and one of the founders of mechanics of variable mass) did not wish to see in the basic equations of TR anything besides the usual formal transformations of variables; on this point, A.V. Vasil'ev correctly noted that in this case we are not returning to earlier variables . . . W.S. Ignatowsky referred to the defined physical essence of the transformations of RT, and P.S. Ehrenfest noted, as an especially characteristic sign, the mixture of spacial and temporal coordinates in the formulations of the transformations, and hence the change of the definition of simultaneity [29: 49]."

Two years later at the Second Mendeleev Congress (in December 1911), the discussion on the problems of TR was continued in papers by Umov, Ehrenfest, and Goldhammer. The latter, having defended the conception of the ether and classical notions of space and time, polemicized with Umov who "attempted to make as much as possible more accessible in order to understand those new concepts which acceptance of the theory of relativity entailed [61: 27]." Ehrenfest, recognizing the merits of STR, placed it in opposition to Ritz's theory, which was more systematic from his point of view and "which characterizes the process of radiation like Newtonian theory of emission and which thus reduces the electrodynamical principle of relativity to the mechanical [61: 28]."

In spite of the critical direction of several works and addresses of Ehrenfest on RT, he was one of the main centers of "relativistic" activity in Russian from 1907–1912. In 1911, an early review [55] dealing with the first monograph on STR written by M.v. Laue [44] appeared. The author of the review was Ehrenfest, and he considered that Laue's book indicated the broad recognition and maturity of the TR.[26] Unfortunately, Ehrenfest, and still earlier Ignatowsky and Epstein, left Russia (at the end of 1912 Ehrenfest became professor at Leiden University; Ignatowsky returned to Russia only in 1914, and Epstein on the advice of his teacher, P.N. Lebedev, left in 1910 for Munich to study with Sommerfeld, and never returned to Russia). Epstein soon published a detailed theoretical work on relativistic mechanics [54] which was mentioned repeatedly in Pauli's review, both in connection with the key problem of relativistic statics, and in a plan for a rigorous basis to an approach to relativistic mechanics, independent of electrodynamics, in the spirit of G. Lewis and R. Tolman.

In these years (1909–11) the first relativists appeared in other university cities of Russia as well: in Kazan – mathematicians (see the following section), in Kiev – L.I. Kordysh who had worked in the Kiev Politechnical Institute and Kiev University; and in Khar'kov-the professor of Khar'kov University A.P. Gruzintsev. Kordysh, subsequently one of the leading Kiev theoreticians [157], at the beginning of 1910 produced a simple derivation of the Lorentz transformations [38]. He was one of the first in Russia to become familiar with the GTR (see below). Gruzintsev, who, by the way, considered that with respect to STR one should manifest judicious care and not exaggerate its physical significance, presented a more systematic account of STR [43].[27]

We will also mention the series of works by the teacher (and later professor) of Ivanovo-Voznesensk Politechnical Institute and Iaroslavl' Pedagogical Institute, K.N. Shaposhnikov, which were published in *ZhRFKhO* and in German journals [68–71, 80, 81] in 1912–1915, and which were devoted mainly to relativistic mechanics and thermodynamics. In these works, several derivations of well-known relativistic correlations were defined more precisely; Shaposhnikov did not succeed in producing new results.

And thus, it may be said that around 1910 Russia had more than a few major physicists who defended the position of the TR: in Moscow, Umov, Epstein, and Lebedev; in Petersburg – Kvol'son, Ehrenfest, and several of his students, and also Ignatowsky – who was more closely connected in this period with German scientific community; in Kiev – Kordysh; in Kharkov' – Gruzintsev; in Kazan and Odessa, the mathematicians A.V. Vasil'ev, V.F. Kagan (1869–1953), and others. Such influential Russian physicists as Borgman, Goldhammer and Eichenwald, who were closer to the position of Lorentz and who did not consider it possible to reject the concept of the ether, nonetheless also did much for the dissemination of the ideas of TR.

6. Beyond the Limits of the Scientific Society of Physicists: Mathematicians and Mechanicians

As has already been mentioned, the mathematician A.V. Vasil'ev and the mechanician I.V. Meshcherskii participated in the "relativistic" meeting in December 1909. During this discussion, the latter, judging by his own observations, did not see any new physical content in the Lorentz transformations, and therefore, spoke against the TR. Vasil'ev however defended the theory, stressing its physical novelty, and was supported by the "relativists," Ehrenfest and Ignatowsky. In this debate the polarity of the relationship of the mathematicians and mechanicians to the RT was reflected.

In general, mathematicians in Russia were interested in STR and came out in support of it. Of course, mathematicians were constrained by the "chains" of generally accepted physical concepts to a lesser degree than physicists and mechanicians. Therefore, the radical conclusions of STR disturbed them less, and its mathematical order and the possibility of utilizing new geometric structures was more attractive to them.

An important additional factor promoting Russian mathematicians' attraction to the RT was, as already noted, the tradition of N.I. Lobachevskii. Not coincidentally, Kazan geometers already in 1909 held a special meeting devoted to the four-dimensional formulation of STR [31]. At this meeting, D.N. Zeiliger, a specialist on linear geometry and helical calculus, presented a paper. Approximately sixty people attended this meeting; the prominent Kazan' mathematicians A.P. Kotel'nikov, N.I. Porfir'ev and others participated in the discussion on the paper. The well-known paper of Minkowski, *Raum und Zeit*, was first published in the *Izvestiia* of the Kazan Physico-Mathematics Society in 1910 in a translation by A.V. Vasil'ev, the leader of Kazan' geometers who in 1907 moved to Petersburg [30].[28] In 1910, at yet another meeting of the Kazan Physico-Mathematics Society, mathematicians and physicists jointly discussed the theory of relativity [32].

A.V. Vasil'ev was an active propagandist not only of non-Euclidean geometry and the tradition of Lobachevskii in Russia, but also the theory of relativity, which he examined as the natural development of the ideas of Lobachevskii. He wrote about this both in a book about Lobachevskii (1914) [85], and in a brochure, *Matematika* (1916) [95].[29] And consequently Kazan geometers manifested great interest in the RT and contributed significantly to its development and teaching (see below).

In Novorossiisk University (in Odessa), V.F. Kagan, the author of a major work on the fundamentals of geometry (the second volume of which, published in 1907, contains an account of the bases of Riemannian geometry and a discussion of the problem of the relation of geometry to reality), and I.M. Zanchevskii (1861–1929), a specialist on applied mathematics and translator of the renowned Columbia University lectures by Max Planck on theoretical physics [47] which contained a brilliant account of the STR, became interested in RT. Kagan attentively followed work on TR, but only at the beginning of the 1920s became occupied directly with this theory, and at that time gave one of the first lecture courses on RT [196].

Mechanicians on the whole turned out to be more conservative with respect to RT. Such authorities on Russian mechanics such as N.E. Zhukovskii (1847–1921) and Ia. I. Grdina (1871–1931) took a rather negative position. The engineer I.E. Orlov, and L.B. Slepian (1889–1959), the student of the electrotechnician V.F. Mitkevich (1872–1951), were also authors of anti-relativist articles in *ZhRFKhO*. N.P. Kasterin, the student of Stoletov, known for his research in the area of acoustics, also came out against RT, siding with the mechanicians. The sceptical relationship of K.E. Tsiolkovskii (1857–1935) to the theory of RT is also well-known [129: 43]. Surely, specialists in the area of mechanics to a larger degree than physics were on the whole inclined to absolutize ideas and concepts based on classical mechanics, but the paradoxical conclusions of the TR were more difficult to accept for researchers with an engineering, practical character.

Beginning in 1912, Grdina, the founder of mechanics of living organisms, a major specialist in the area of theoretical mechanics, and professor of the Ekaterinoslavl' Higher Mining School, came out with a series of articles directed against RT [199]. He considered that the experimental basis of STR was too poor and its conclusions too paradoxical "to treat more attentively such explanations of the negative result of the Michelson experiment, which do not lead, necessarily, to the demolishing and complications of the bases and the conclusions of the physical sciences, answering at the same time other data of experiment and observations." [98: 1] His own approach was in accordance with the work of Goldhammer on the electrodynamics of moving bodies [97] and based on an understanding of the ether which postulates combination of the ideas of the ether drag in the first moments with the immobility of the ether at great distances from the sources of radiation. In spite of criticism – the critical review of the work of Grdina by G.N. Rautian (1889–1963) [90], a subsequently well-known Soviet optician, was fully correct and precise – the latter continued to come out against the RT in the 1920s [118].

Zhukovskii was not an active and systematic opponent of RT. In 1916, for example, he spoke rather highly of Umov's derivations of the

Lorentz transformation, having called them the "best mathematical interpretation of the principle of relativity [98: 436]," but in the beginning of 1918 he delivered a speech "Old Mechanics in the New Physics [101]," in which he defended classico-mechanical ideas in opposition to relativistic ones.[30] In this talk, he referred in particular to Kasterin who had developed an ethero-hydrodynamical model of the electromagnetic field [103: 260].

The engineer Orlov tried to prove the equivalence of STR with Ritz' theory and, likewise, the reducibility of the TR to classical mechanics [89]. Ehrenfest, by means of ingenious thought experiments refuted the constructions of Orlov [92]. Slepian considered the relativistic arguments and derivations to be contrary to common scientific sense and to scientifico-materialist ideas, and called for a return to reliable conceptions of space and time which were characteristic of classical mechanics [95, 100].

The relativistic corrections which were characteristic of the STR were too small to consider in astronomy. Therefore, the majority of astronomers, including Russian, did not manifest great interest in STR. There were, however, exceptions: K. Schwarzschild and W. de Sitter in the West and B.P. Gerasimovich (1881–1937) in Russia. The outstanding Soviet astrophysicist, later director of Pulkovo Observatory, Gerasimovich in 1913 published a work on the relativistic theory of the aberration of light [76]. At this time he graduated from Khark'kov University. Gerasimovich regarded the general theoretical significance of STR highly, and believed that the theoretical bases of astronomy should be reconstructed in connection with the new theory, for the confirmation of which, on the other hand, the cosmos is the most suitable laboratory.[31] Later this evaluation was completely justified.

II. THE RECEPTION OF THE GENERAL THEORY OF RELAVITY IN THE USSR.

7. The First Years of Soviet Russia: The Socio-Scientific Background

The general theory of relativity (hereafter, GTR) was born in a country separated from the majority of European countries including Russia by the fronts of the world war. However, scientific communication was not entirely interrupted by the war, primarily due to scientists of neutral Holland. W. de Sitter expounded the ideas of the GTR, of Einstein's cosmology, and his own results in a series of articles which were published in an English journal [102] immediately after Einstein's foundation-laying works; as is known, this was the primary condition for the organization of Eddington's expedition in 1919 which was accomplished with the triumphant confirmation of the predictions of the GTR. In 1915 an article by Lorentz was published in *ZhRFKhO* [93]. This was apparently the first scientific article in Russian devoted to the GTR, which was then being created. It is possible to think that its appearance is indebted to P.S. Ehrenfest who in 1912 moved from Russia to Holland on the invitation of Lorentz and became his successor in the Physics Department in Leiden (a research note of Ehrenfest, received the very day that Lorentz' article was, appeared immediately following Lorentz'). Ehrenfest, as is known, played an outstanding role in the sharp rise in the level of theoretical physics in Russia both before and after the 1917 Revolution [210], in particular in the assimilation of the STR in pre-Revolutionary Russia (see Part I).

The events connected with the Revolution of 1917, the Civil War and foreign intervention all the more isolated Russia from the outside world. The character of this isolation changed after the establishment of peace between Soviet Russia and Germany and after the Revolution in Germany in 1918. Among internees who had returned to Russia from Germany was V.K. Frederiks, who played an important role in the dissemination of the GTR in Russia. In Berlin Russian language publishing houses were springing up. In 1921 one of them published Einstein's brochure [110] with a preface of the author to the Russian edition:

More than any other time during these present troubled times one should concern oneself with everything that can bring people of different nations and languages together. From this point of view it is especially important to facilitate a lively exchange of artistic and scientific works even during these current difficult conditions. Therefore I am especially pleased that my little book is being published in Russian, all the more since Mr. Itel'son, whom I respect, serves as a guarantee of an excellent translation. The author has been repeatedly abused because he calls his pamphlets "popular." Therefore, it would be correct if the Russian reader, who during the study of the book encounters difficulties, would not be annoyed with himself or Mr. Itel'son. The true guilty party is none other than the author.

A. Einstein
Berlin. 10.IX. 1920

This was one of the very first works of Einstein on the GTR which was published in Russian.

The powerful effect of the events of civil history in Soviet Russia are also characteristic of the history of the reception of the GTR during the first decade of Soviet power.

At the beginning of the 1920s, great interest in the TR flared up in broad circles of society as in other countries. The reasons for such interest in part coincide with reasons which also effected other countries: "apparent" and "well-known" properties of space and time were subjected to revolutionary changes; the radical change in the world of scientific ideas had especially peaceful meaning against the background of the recently, violently and sanguinely changing world of people; events connected with the confirmation of the predictions of the GTR in a world which had grown tired of the nationalism and chauvinism raging during war had a clearly international character. Besides that, the atmosphere of general social upheaval in Revolutionary Russia, which was consonant with the revolution in physics, and the correlation of forces in scientific society which was changed by social processes in favor of its younger members, had great significance.

All of these reasons led to the publication of a large number of translated and original popular books on the TR (in the 1920s around thirty books were published), and articles and talks of Einstein were published frequently. His name became well known in Russia, far beyond the limits of the circle of physics. In 1922 V.I. Lenin called Einstein one of the "great reformers of the natural sciences."[32] In 1922 at the suggestion of A.F. Ioffe, P.P. Lazarev (1878–1942), and V.A. Steklov (1844–1926), he was selected as a foreign member of the Russian Academy of Sciences, and in 1926 an honorary Academician of the Academy of Sciences of the USSR [208]. The ideas of RT (or those that were taken for them) excited the imagination of a poet,[33] and landed in the works of an ethnographer [116] and a religious thinker [113].

And nevertheless, the informational isolation of Russia from world science was felt until the beginning of the 1920s. In December 1919 a telegram was sent from Rozhdestvenskii, Krutkov and Frederiks from Petrograd to Lorentz and Ehrenfest in Leiden: it specifically reported the absence of literature since the beginning of 1917, and contained a request to acquaint them with the physical works of the Amsterdam Academy [106]. In June 1920, as soon as the possibility of postal communication with the West was reestablished (through Estonia),

Ioffe sent his first letter to Ehrenfest after a six year break in which he wrote, in particular, "Now our main misfortune is the complete absence of foreign literature which we have been lacking since the beginning of 1917. And my first and main request to you is to send us the journals and main books on physics [192]." This letter was received by Ehrenfest after a two month delay. However, the situation with respect to literature changed only after research trips to the West in 1921 by a number of prominent Russian scholars (A.A. Arkhangel'skii (? –1926), V.M. Chulanovskii (1889–1969), A.F. Ioffe, A.N. Krylov, D.S. Rozhdestvenskii, O.D. Khvol'son) to purchase scientific literature and instruments: the quantity of literature acquired weighed tons [192].

Nevertheless, in 1922 an appeal "To German Physicists" was published in *Zeitschrift für Physik* [115] with a call to help physicists in Russia overcome the informational famine and collect journals and offprints of the preceding years for them.

8. A.A. Friedmann – Founder of Non-stationary Cosmology

All the more remarkable is that fact that only twenty-five pages later, in that very volume of *Zeitschrift für Physik*, A.A. Friedmann's article which initiated contemporary cosmology – the theory of the expanding universe – was published. This work of the Soviet scholar became the most important result in the area of GTR after the work of Einstein in 1915–17. It is well-known that at first Einstein considered the calculations of Friedmann to be incorrect, and only after a letter from Friedmann and conversations with Krutkov [206] did Einstein acknowledge Friedmann's results "to be correct and to shed new light". [120] And afterwards, when the work of Friedmann began to be hidden in the West behind the essentially equivalent work of Lemaître which was published five years later, Einstein more than once indicated that Friedmann had initiated the theory of the expanding universe [152: 350; 180: 599].

In the creative work of Friedmann works on RT at first glance seem rather unexpected. Basically his work was devoted to theoretical hydromechanics and dynamic meteorology, and not long before his death (from abdominal typhus) he was appointed the Director of the Geophysical Observatory. But the many facts of his biography when taken as a whole speak to the fact that Friedmann (1888–1925) was in no way an incidental figure in the Russian history of the GTR.

For Friedmann, the rare combination of the approach of a mathematician who strives toward logical rigor, and the passion of a naturalist who bravely moves toward contact with the most vital reality, was characteristic. The first scientific works of Friedmann were devoted to the theory of numbers. He finished the Mathematics Department at Petersburg University in 1910. His teacher was the outstanding Russian mathematician, V.A. Steklov. Friedmann was professor of mechanics at Perm' University (1918–1920) and professor of mathematics at Petrograd University (1920–1924) [193].

In World War I Friedmann joined the army as a volunteer, served in aviation, mastered the profession of pilot-observer and under hostile fire "experimentally" studied aeroballistic properties of bombs dropped from the air, and compiled ballistical tables. Later he participated practically in the organization of production of aviation instruments. And, finally, in 1925, Professor Friedmann participated in the flight of a balloon to the record height of 7400 meters. At that time, such flights involved great risk and demanded great courage.

Because of the small number of teachers at Perm University, Friedmann, in addition to mechanics, was required to teach courses on differential geometry and general physics. It was necessary for him, in his own words, to prepare a great deal for these courses, but it is easy to understand that this prepared him for his future work on GTR. In a letter to Ehrenfest of August 6, 1920, Friedmann wrote that he was occupied (besides dynamic meteorology, hydrodynamics of compressible fluids, and a series of questions which touched upon aviational technology) with the axiomatics of the STR and "I very much want to study the large [the general] principle of relativity, but do not have the time [206: 16]." It is natural to think that courses of physics and differential geometry which were under his belt, psychologically and technically facilitated the assimilation of the Einstein's theory of gravity. Evidently, V.K. Frederiks, who from May 1919 became a professor of Petrograd University, played no small stimulating role in this process.

At the outbreak of World War I Frederiks (1885–1943) was in Göttingen working with Hilbert. He was interned, but owing to the protection of the famous mathematician he was able to remain his assistant. Thus, the birth of the GTR (with Hilbert's active participation, as is known) apparently occurred before his eyes [206]. Therefore, it is not surprising that when he returned to Russia in 1918, as soon as intergovernmental relations in Europe permitted, Frederiks began to promote the ideas of the GTR. In 1921 his review "The General Theory of Relativity of Einstein [109]" appeared in *UFN*. Immediately after Fredericks' article a paper appeared by G.S. Landsberg (1890–1957) [108] on the results of Eddington's expedition on the verification of the bending of the lightrays in the field of the sun which had been predicted by Einstein. (A characteristic sign of the times: the author indicates that the corresponding English journal is not available in Russia and therefore he is basing this report on Freundlich's German article). These were the first works in Russia which set forth the GTR and its experimental confirmation.[34]

According to the testimony of V.A. Fok, Friedmann and Frederiks "were the first who familiarized Russian physicists, who worked in Petrograd, with the theory of gravity recently created by Einstein. This was at the very beginning of the 1920s when the blockade of Soviet Russia had just been broken, and scientific literature from abroad began to arrive. In the Physical Institute of the University a seminar gathered where among other things papers on the Einstein's theory were delivered. The participants of the seminar were professors and students of the senior class (and at that time there were very few). The basic speakers on the TR were V.K. Frederiks and A.A. Friedmann, but sometimes also Iu. A. Krutkov, V.R. Bursian and others spoke. The talks of Frederiks and Friedmann I remember clearly. The style of these talks was different: Frederiks deeply understood the physical side of the theory, but did not like the mathematical computations; Friedmann stressed not physics, but mathematics. He strived for mathematical rigor and attributed great importance to the full and exact formulation of the initial preconditions. The discussions which arose between Frederiks and Friedmann were very interesting [191: 399]."

The assimilation of GTR by Friedmann was very intense and to a large extent fruitful. Together with Frederiks he undertook the serious work, *Osnovy teorii otnositel'nosti*, in which it was intended to set forth "sufficiently rigorously from the logical point of view" the bases of tensor analysis, multi-dimensional geometry, electrodynamics, and the special and general principle of relativity [122: 3]" Only the first part of this book was published (in 1924), the manuscript of which was ready in 1922 [122]. In 1923, the popular book of A.A. Friedmann, *Mir kak prostranstvo i vremia* [119], which was devoted to the GTR and oriented toward the rather prepared reader (it was destined for a philosophical journal) was published. In 1924 the article of Friedmann appeared (together with the well-known Dutch geometer J. Schouten) which considered several degenerate cases of general linear connections which, in particular, generalize the Weyl displacement and, the authors ventured, "perhaps will find application in physics [125]."

Finally, the most important work of Friedmann in the area of the GTR was the cosmological non-stationary model which now bears his name. This model was set forth in two articles in *Zeitschrift für Physik* in 1922 and 1924 [114, 123].

According to the testimony of V.A. Fok, the approach of a mathematician predominates in the attitude of Friedmann toward the TR:[35] "Friedmann often said that his cause was to show the possible solutions of the equations of Einstein, and then let physicists do what they wanted with these solutions [191: 402]."

The mathematicity of Friedmann was manifested in his own cosmological works. Mathematically speaking, he too formally regarded the point singularity of his solution. Counting off time from that moment, and introducing for easily understood reasons the expression "time which has passed since the creation of the universe [193: 236]," he could evoke criticism even from non-militant atheists. In addition to this, he in fact placed a period at the beginning of cosmological expansion without any kind of reservation, and not a question mark, to which physicist should arrive, recognizing that in the situation where the density of matter goes to infinity, the question arises about the applicability of the original equations and the entire corresponding conceptual apparatus[36]. For the mathematician it is entirely natural to examine the solution in its entire area of definition, while it is obvious no kind of limitations are formulated.

Only a mathematician could also impassively indicate a solution with negative density as one of the possible solutions.

The mathematicity of Friedmann may be found in his powerful sympathies toward the unified theory of Weyl (expressed in [119]), and his sympathy to the "problem of axiomatiziation of physics," proclaimed by Hilbert, in the aggregate with some kind of underestimation of the independent role of quantum ideas [122: 26] (here it is possible to detect the influence of Frederiks, who was Hilbert's assistant in the period of the creation of Hilbert's so-called unified theory).

It is possible to think that the mathematicity of Friedmann could have facilitated the appearance of his cosmological work. It was easier for a mathematician to see that Einstein's cosmological solution is too degenerate a case: that by not destroying the spatial symmetry, it is natural to anticipate a solution which is dependent on time. Besides, it was easier for a mathematician to resist the hypnosis of the recognized physical genius of Einstein and to abandon his physical ideas in favor of the static model – the "slowness of the motion of the stars."

However, the fact that Friedmann was not a mathematician in "pure form" is without question. To begin with, the very area toward which all his basic mathematical interests were directed – the mechanics of fluid and gas – has a large physical component. Friedmann also had strictly physical inclinations. While still in his student years he participated in the work of Ehrenfest's seminar in Petersburg. The conditions of the time (which were determined by war and revolution), professorship, the insufficiencies of scientific and teaching personnel – all this stimulated the expansion of his scientific interests. In addition to this, he took upon himself the obligation of "mathematical gendarme [206: 17]" in the area of quantum ideas, which on the eve of the appearance of quantum mechanics was in a turbulent state.

And, if he did not have physical interests and knowledge (which developed as already noted, not only as a result of "self improvement," but also owing to the influence of social conditions), Friedmann hardly would have turned in general to the TR – this region was too far from his basic interests. The "physicist" in Friedmann also appeared at the very end of his first article on cosmology when he calculated the "period of the universe"; with reservations which somehow compensated the non-lawful (in any case for a real mathematician) logic, he accepted for the mass of the universe the magnitude $5 \times 10^{21} M_\odot$ (De Sitter produced such a magnitude for Einstein's static model), and with the cosmological constant being equal to zero, he produced the "period of the universe" of 10^{10} years.

In sum, this was the scholar whom the history of science, with the participation of social conditions, prepared and led to the theoretical discovery of one of the most remarkable facts which characterize the contemporary physical picture of the universe: the discovery of the expansion of the universe.

9. The Teaching of the TR

In Soviet Russia, special courses on the theory of relativity were offered in institutions of higher education from the very beginning of the 1920s.

In Petrograd, Ia. I. Frenkel' (1894–1952) gave one of the first courses on TR in the Spring of 1922 (in the Petrograd Polytechnical Institute in the Physico-Mechanical Department created by A.F. Ioffe). Frenkel' had the possibility to become familiar with TR earlier than other Russian physicists. In 1918–1920 he taught in the Tavrian University in the Crimea (which was first occupied by German troops, and then by Wrangel's army), where German journals were regularly delivered [206].

Frenkel's book ([118], the "first 'non-popular' guide to the TR [in Russian]"), which is based on his course, is very distinctive. Although it contains a sufficiently complete account of the STR and GTR (including the cosmological model of Einstein), the earmark of the book, however, was not in demonstration of the revolutionary changes in the concepts of space and time, but, rather in showing the deep native stance of the theory of relativity with the electrodynamics of Maxwell. The book begins with a broad electrodynamic introduction which with skillful reasoning (of a mathematical character) leads to the relativistic correlation $E = Mc^2$. The author of the book is an obvious supporter of the electromagnetic picture of the world, albeit a rather distinctive picture. For him, the very possibility of the universalization of the properties of physical reality, which are detected in the region of electromagnetic phenomena for all physical phenomena are a consequence of the unified electromechanical nature of all physical phenomena (including gravitation!). Precisely this universality of electromagnetism leads to the space-time universality of the TR. The author even considers that it is more correct to call the TR electromechanics, and the GTR – gravitational electromechanics (inasmuch as all physics is simply the mechanics of electrons and of the electromagnetic field).

It does not follow, however, to think that Frenkel's picture of the world was too attached to 19th century concepts. For example, he decisively and without vacillation rejects the concept of the ether as having outlived its usefulness (to which then, after Einstein's "conciliatory" speech [in Leiden] in 1920, attention was attracted anew, and with which even several followers of the theory of relativity did not wish to part). He went even further, contending that space and time are only "empty abstractions, forms of the interrelation of material bodies." But it is necessary to say that the methodological reasonings of the author do not crush the account of the mathematical apparatus of STR and GTR. In particular, for Frenkel' the geometrization of gravity is the effective, macroscopic, non-linear manifestation of electromagnetism which re-

sponds to the universality of the properties of space-time: "The properties of space and time are conditioned by electromagnetic forces and are
manifested in the form of gravitational forces [118: 279]." But this,
by the way, does not hinder the description of Riemannian bases of
GTR. The decisiveness with which Frenkel' rejects the unified theory of
Weyl makes an impression. He rejects it, in the first place qualitatively,
because of the fact that in it the "dualistic nature of matter which
consists of positive and negative electrons" was not reflected and, in the
second place, because of the non-correspondence of Weyl's equations of
motion with equations of Lorentz: "The theory of Weyl is presented to
us, therefore, as no more than a mathematical diversion – quite elegant,
but devoid of any kind of physical meaning [118: 278]."

The book of Frederiks and Friedmann, *Osnovy teorii otnositel'nosti*
[122], had an entirely different character. And the first part of the book,
the only part published, fully provides the possibility of seeing this. This
is a thorough, detailed account of the TR which is based on a very solid
mathematical foundation of the geometry of general linear connection
in the manifold of arbitrary dimensionality and the theory of groups.
The starting point for the authors turns out to be the geometry of
space-time (they do not believe in an electromagnetic global picture of
the world, as is noted directly in the "Introduction"). The title of the
only part published, "Tensor Analysis," does not fully correspond with
its contents. Besides tensor algebra and tensor analysis (with extensive
coverage of Riemannian and Weylian geometry, including integral
theorems) there is in the book a large section (25 pages) "On groups of
Transformations." Here, on the material of the transformation of
coordinates, the "concept of groups of transformations, a concept which
plays such a fundamental role in contemporary mathematics, and which,
apparently, is beginning to penetrate into the region of theoretical
physics," is introduced. The group character of the set of general
continuous transformations of coordinates is shown, the concept of Lie
groups is discussed, and the most important concrete groups of transformations of coordinates in the space of arbitrary dimensionality are
examined (apparently, taking into account a five-dimensional generalization of GTR): including affine, projective, orthogonal, and homothe-
tic groups. And, finally, the Lorentz group is examined, utilizing the
space-time diagrams of Minkowski for illustration, and establishing the
correspondence of geometric and kinematic conceptions (the group of
Galileo is introduced as the degenerate case, which corresponds to the

speed of light $c = \infty$). Likewise, the geometric contents of STR are described.

Incidentally, Frederiks had already taught the TR in Petrograd University in 1921. His survey on the GTR in *UFN* [109] gives the ideas of the general relativistic part of these lectures.

Courses on TR were available not only to students of Petrograd. In Moscow University I.E. Tamm (1895–1971) taught a course on RT in 1923/24 in the Physics Department, and P.S. Urysohn (1898–1924), known for his outstanding results in general topology, taught another in the Mathematics Division; in Kazan University, the mathematician N.N. Parfent'ev (1877–1943), and in Odessa V.F. Kagan taught courses on RT.

10. The Relationship of Mathematics and Astronomers to RT

In general mathematicians regarded TR with understood enthusiasm. Indeed, new mathematical ideas and methods entered into physics, the mathematical component in theoretical physics grew sharply, and several unanticipated physical applications of mathematical constructions were detected, which had appeared quite recently. Of course, both national-historical and personal-biographical circumstances had an effect. Among mathematicians the tradition of Lobachevskii with its attention to the interrelationship of geometry and physics was strong. This tradition was especially vital among mathematicians of Kazan University where the activity of Lobachevskii (see Part I) was centered. In 1923 the book of the prominent representatives of the Kazan mathematics school A.V. Vasil'ev, *Prostranstvo, vremiia i dvizhenie. Istoricheskie osnovy teorii otnositel'nosti* [177], appeared in which the ideas of STR and GTR were set forth with great attention to their philosophical, mathematical and physical pre-history.

The successor of Vasil'ev in Kazan, N.N. Parfent'ev, taught not only a special course on RT for students in the middle of the 1920s, but also gave lectures accessible to a wide audience. The Kazan' school of geometry maintained longterm and fruitful contacts with the TR.[37] The 100th Anniversary of Lobachevskian Geometry, which was observed in 1926, and which was accompanied by the publication of a collection of articles and by the preparation for publication of the works of Lobachevskii (on the initiative and under the editorship of A.P. Kotel'nikov and V.F. Kagan), was a noteworthy event in the scientific life of the country and attracted additional attention to the role of the works of Lobachevskii in the genesis of Riemannian geometry, set forth in the language of GTR.

On the other hand, for P.S. Urysohn who belonged to the Moscow mathematical school, interests in TR were determined apparently by personal-biographical reasons – by physical interests clearly expressed already from adolescent years (experimental research on x-ray radiation was the first scientific work of the creator of the topological theory of dimensionality. It is possible to observe the interconnection of the mathematical works of Urysohn and his interests in TR [218]). In addition to the two semester courses at Moscow University ("The Mathematic Bases of the Principle of Relativity" and "The Mathematical Bases of RT"), for four hours weekly in 1923/24, Urysohn together with the other creator of the Soviet school of topology and his friend, P.S. Aleksandrov (1896–1983), delivered a cycle of four public lectures in January 1923 entitled, "On the Mathematical Cognition of the Universe in the Light of TR."

Not only mathematicians accepted the TR. As is well-known, the relationship of world astronomy (in any case, astrophysics), to the TR was basically positive (apparently, one of the main reasons was the fact that only the relativistic correlation $E = Mc^2$ provided hope for uncovering the physical sources of stellar energy). An example of such an approach is *Vselennaia pri svete teorii otnositel'nosti* [126], a book of B.P. Gerasimovich (see Part I), who in the 1930s was the Director of the Pulkovskaia Observatory. (It is interesting to note that this book was approved by Glavpolitprosvet for *sovpartshkoly* and communist *vuzy*[38]). In Gerasimovich's book, from the astronomical point of view STR, GTR, cosmology and experimental verifications of GTR are recounted with enthusiasm; in particular, the author refers to his own propositions to verify the effect of the shift of the periastr on close binary stars, where this effect should be more powerful than for Mercury. At the same time another prominent Soviet astronomer, V.G. Fesenkov (1889–1972), in an article devoted to astronomical confirmation of RT, reached a more cautious conclusion: "The principle of relativity does not stand in contradiction to any kind of well-established phenomena, but to maintain that it is verified in practice, to a lesser degree, is premature [127: 216]." In an encyclopedia article in 1931, Fesenkov wrote, "The RT at present cannot be verified completely and undoubtedly with the help of astronomical observations.

Nonetheless, not one of the well-known phenomena contradict it [149: p. 362]."

11. Philosophical Discussions on the Theory of Relativity: The Problem of the Ether

The dissemination of the ideas of TR and their reception was accompanied by an intense ideological struggle which took different forms in different countries. In the USSR this struggle took the form of philosophical discussions [203] which continued until the 1950s (in essence until the celebrated conference on philosophical questions in the natural sciences in 1958 [158]), but they were most heated and effective in the 1920s. The most prominent and active opponents of TR – the physicist and professor of Moscow University A.K. Timiriazev (1880–1955), and the electrotechnician and Academician V.F. Mitkevich-began with the usual physical and quasi-physical objections toward the TR (based on a classical understanding of the ether), but then, in view of the lack of success of their criticism, began to attract argumentation of a philosophical and even political character. A number of philosophers and philosophizing physicists (A.A. Maksimov (1891–1976), I.E. Orlov and Z.A. Tseitlin), who were distinguished by superficial familiarity with the theory of relativity, devotion to mechanism and dogmatic understanding of the philosophy of Marxism, joined their side. These authors rejected the TR (as anti-Materialistic and all too speculative) in the name of dialectical materialism, but in fact because of a limited understanding of classical physics, a mechanistic world view, and a notorious "common sense." For weapons they took the authority of J.J. Thomson and P. Lenard, the doubtful results on the ether drift of Dayton Miller (whose article was republished by Timiriazev in a philosophical journal), interpretations of TR by philosophers of various orientations, and chance comments of physicists which were taken out of context.

Among physicist-professionals attacks of this kind were not seriously perceived, and were decisively rebuffed in the sphere of philosophy and scientific publication both from physicists – Ioffe, Tamm, Vavilov, and Fok [155, 160, 171, 172], and from philosophers (S. Iu. Semkovskii (1882 –), A.A. Gol'tsman, B.M. Gessen (1883–1938). Among philosophical works devoted to RT, the book of Semkovskii [121], which is distinguished by a deep understanding of the physical content of the TR and the penetrating analysis of it from the point of view of the philosophy of Marxism, stands out especially. Already in the 1920s the dialectical materialist understanding of TR which corresponds to the contemporary viewpoint had been elaborated. It was clearly realized that a scientific theory cannot be contradicted genuinely by a scientific philosophy, that the question of the truth of a physical theory is located entirely within the competence of physics, and at the same time it was clarified that the TR itself confirms the general ideas of Marxist philosophy in a remarkable way: the dialectical interconnection of the concepts of the absolute and the relative, the concepts of space, time and motion; the purely philosophical understanding of matter, which is not burdened by a mechanistic world view or any kind of physical attributes, but which records only objective reality which is given in the sensations of man. And if the activity of opponents of TR did not cease at that, then first of all it is because new scientific ideas, in the words of Planck, "are victorious not because they convince their opponents to recognize their correctness, but for the most part because these opponents gradually die out, and the following generation assimilates the truth at once." One should not, however, underestimate the great effort required of an individual of adult age to assimilate the ideas of the TR.

One of the main causes for the rejection of the TR was, as in other countries, the status of the understanding of the ether. For physicist-theoreticians, in whose area of professional interests electrodynamics and the theory of relativity entered, there were no special problems, inasmuch as the concept of the ether simply was unnecessary; it did not work. But for the broader circle of followers of the TR and even among physicists the problem of the ether remained for some time rather troublesome.

Thus, for example, Boris Gessen, (who with E.V. Shpol'skii (1892–1975) was editor of *UFN*, and together with A.F. Ioffe was editor of the Physics Section of the *Great Soviet Encyclopedia* (*BSE*)), was head of the Physics Department of Moscow University and is known especially for his socio-economic analysis of the genesis of Newton's physics), studied the history and methodology of physics in his book of 1928 [134] and in his article "Efir" in *BSE* (1931). In spite of his correct evaluation of STR and positive Marxist philosophical evaluation, he saw one of the central problems of physics of that time in the problem of the ether. In this way, he in part (without the requisite understanding) leaned on the position of Einstein of the 1920s in regard to the ether, which in essence transferred several functions of the ether to space-time. However, from the text of Gessen it is clear that the main root of his bent for the ether was found same in 19th century physics.[39]

The evolution of the relationship to the ether of one of the most prominent Soviet physicist-experimentalist, S.I. Vavilov (1891–1951) (from 1945 the President of the Academy of Sciences), provides an even more interesting example. His book of 1928 *Eksperimental'nye osnovaniia teorii otnositel'nosti* signifies an important stage in the process of the taking root of the TR in the USSR. This was a thorough review of all of the most important aspects of the empirical status of STR and GTR. The general conclusion of the book was that "there is not one fact known which contradicts the GTR; on the contrary, in all cases within the limits of exactness tolerated by experiment, the results of the TR are confirmed," and although "the theory of Einstein is still not complete (the role of the electromagnetic field remains unclarified, as are the nature of the electron, several cosmological derivations, the connection between quantum phenomena, and so on), the kernel of the theory stands on a very firm experimental foundation. Such theories grow, become perfected, and do not perish." [133: 163; 15].

Vavilov was not only an outstanding experimentalist (in particular, under his direction Vavilov-Cherenkov radiation was discovered), but he was also a well-known historian of science (he published in Russian Newton's *Opticks* and his biography). Psychologically and pedagogically it was important that in his book [133] the TR is presented as already having occupied its place in the history of science and having been organically connected with its predecessors. This goal was served by the epigraphs from Newton which preface all eight sections of the book (they also demonstrate that "many postulates and results of the TR would not seem to be entirely unanticipated or unacceptable even to the creator of classical physics," [133: 5] and by the reminder that the initial idea for GTR – the principle of equivalence with its now seeming triviality – was born in the struggle of Galileo with Aristotelian physics.

And in that, despite the historic merits of the concept of the ether before physics, Vavilov unconditionally ascertains that "the hypothesis of the ether (at least its most developed variants) contradicts the principle of relativity in the real world, not agreeing with a series of experiments in moving systems [133: 12]." However, the relationship of Vavilov to the ether was not always thus. At the beginning of the 1920s, for example, he attentively (if not with sympathy) regarded the ether constructs of Lenard, and in a review of a speech of Einstein in

1920 on the ether noted with relief, "Most significant is the 'lifting of the ban' on the hypothesis of the universal ether by the very author of this 'ban' which has hypnotized science for fifteen years and which undoubtedly acted as a brake on the natural development of hypotheses which are valuable for physics [107]."

The first accounts of the TR were, of course, not free from the all too powerful pressure of new ideas which were advanced by the creators of TR: the conventionalism of Poincaré, Einstein's adherence to the idea of complete relationality of space-time and the understanding of GTR as a theory of arbitrarily moving frames of reference, the axiomatic ideal of Hilbert. The somewhat exaggerated quality was natural in describing the ideas of GTR in the period when these ideas had just penetrated into the consciousness of the physics community (provisions and clarification of the kinship between the old and the new at first could even hinder the reception). But in fact, the method "from one extreme to another" – if not better, is then in any case the most common method of assimilation of new ideas. The essential expansion of the class of frames of reference being examined in physics, the systems of coordinates, induced talk about the complete arbitrariness of arithmetization of space-time.

In the book of Friedmann, *Mir kak prostranstvo i vremia* such a complete arbitrariness in the selection of the system of coordinates contradicted the supposed smoothess of coordinates (their agreement with the topology of space-time). The equivalence of all possible frames of reference, taken in this book to the notorious equivalence of the Ptolemaic and Copernican systems, had only limited (narrowly pedagogical) meaning, when really only "points of view" and not dynamic systems are kept in mind. Friedmann himself understood beautifully the significance of the Copernican revolution.[40]

In the 1930s the understanding of the GTR and its possibilities became more clear [146, 166], although as is known, to this day several fundamental positions of GTR evoke discussion.

12. The Development of the Ideas of the TR in the 1930s

Toward the end of the 1920s physicists in the USSR not only became fully familiar with the TR[41], but even began actively to participate in its development. They succeeded in taking part in the elaboration of a program of unified field theories (Tamm, Fok, Mandel', Rumer); in

1929, articles by Fok and Ivanenko appeared on general relativistic generalization of spinor equations of Dirac.[42]

The problem of the synthesis of STR and quantum mechanics was especially urgent for the TR at the turn of the 1920s–1930s, as were the astrophysical and cosmological supplements of the TR. The work of L.D. Landau (1908–1968) with R. Peierls was connected with the first problem [151] (in this work, which played an important stimulating role, the correctness of the concept of a "field in a point" in relativistic quantum theory was subjected to doubt), as were the ideas of V.A. Ambartsumian (1908–) and D.D. Ivanenko (1904–) on quantization and discreteness of space [145] (the main reason for the dying out of these ideas was the difficulty of their relativistic formulation [147]).

In 1931 M.P. Bronshtein (1906–1938) published a thorough review of relativistic cosmology in *UFN* [146]. In this review (which was very timely in connection with the discovery in 1929 of the Hubble expansion of the universe) light is shed skillfully on the astronomical and physical bases of cosmology (with an introduction to the GTR), the static models of Einstein and De Sitter; and most importantly, the non-stationary model of Friedmann with the supplement of LeMaître were examined. In the review the "Russian mathematician," the work of whom was "half-forgotten," was given his due. The main difficulty of cosmology – the lack of correspondence between the Hubble growth of the universe and cosmological data – was correctly pointed out. The article of Bronshtein was supplied with an editorial forward and comments which were characteristic for that time. They were written, apparently, by Gessen, and although they acknowledge the possibility of general relativistic formulation of the cosmological problem, they contain comments which are connected to the philosophical-ontological overload of elements of the mathematical model of Friedmann (such as the radius of the universe and the initial singularity of the solution of Friedmann).

In the beginnings of the 1930s cosmology and fundamental astrophysics attracted physicist-theoreticians in connection with the problem of beta-decay and the construction of relativistic quantum theory; in this, there were hopes of discovering the real physical phenomena which belong to the essentially quantum-relativistic or to the *ch*-region, and, simultaneously, of establishing the source of stellar energy (according to Bohr's views). The interior of the star seemed to be a suitable place where *ch*-physics should act, where (according to the suggestion of

Bohr) in processes which violate the law of the conservation of energy, the stellar energy may arise and where, because of those still unknown ch-laws on a fundamental, microscopic level the irreversibility of physical processes may be manifested [154]. At that time great expectations were aroused in connection with the future ch-theory – the theoretical derivation of the constant of fine structure, the explanation of atomism of charge and mass. In connection with all this, the aspiration to examine the possibility of stable stellar structure under maximally general suppositions was understandable. In 1932 Landau produced a remarkable result in this direction – the existence of a maximum for the mass of stable configuration for which gravitation is balanced by the pressure of the degenerated gas of fermions (electrons) [153]. This signified that the cooled star of sufficiently large mass (after the depletion of energy sources) does not have a state of equilibrium. (The limit of stellar mass was independently discovered by S. Chandrasekhar and carries now his name). Then this result was perceived primarily as the unavoidability of the ch-stage in the evolution of a sufficiently massive star, as the existence of the important real area of physical phenomena which were required in ch-theory [162]. At the beginning of the 1930s it was also considered that nuclear physics and the physics of cosmic rays required ch-theory.

Experimental physics of the characteristic (energetic, space-time) scales visible at that time did not demand connection to ch-theory of gravitation. But the principal necessity of the synthesis of the full theory of relativity (i.e. GTR) with quantum theory was perceived by Einstein already in 1916. The important features of such a synthesis were revealed in the first in-depth research of the problem of quantization of gravitation which M.P. Bronshtein undertook (his doctoral dissertation in 1935 and articles [168]). He constructed a quantum theory of gravitation in approximation of a weak field, having produced, as its consequence, the law of gravitation of Newton and the Einsteinian formula for gravitational radiation. But its main result is the principal irreconciliability of GTR and quantum theory without fundamental reconstruction of the conceptual apparatus, without the substitution of the usual concepts of space and time "by terms which are far more fundamental and are devoid of obviousness." Actually, Bronshtein discovered Planckian limits of applicability of GTR and quantum theory of fields. This result maintains its significance to this day and indicates one of the main obstacles on the path of the creation of a unified theory of all fundamental interactions. Bronshtein came to this result, having shown that the method found by Bohr and Rosenfeld to legitimize the concept "field in a point" for ch-theory (which is subjected to doubt by the work of Landau and Peierls [151]) is not realizable in cGh-theory [219].

In 1937 the important work of Bronshtein concerning the physical nature of the cosmological red shift [169] was published. At that time the hypothesis was advanced (in particular, by the well-known physicist W. Heitler) about the fact that Hubble's red shift in spectra of distant galaxies was called forth not by the expansion of the universe (that is, as a result of the Doppler Effect), but by the "reddening" of photons, which are traveling through intergalactic distances, as a result of spontaneous decay of photons to lower energies. This hypothesis arose within the framework of contemporary quantum electrodynamics ("theory of radiation") which did not have, at that time, consistent formulation because of difficulties of taking into account vacuum effects. Bronshtein rejected the possibility of such an explanation of the cosmological red shift first, on the basis of rather general and elegant considerations connected with relativistic invariance[43], and second on the basis of direct and at that time complicated quantum-electrodynamical calculation; thus the fact of the expansion of the universe received reliable theoretical grounding. This was, apparently, the first work to stress is the interaction between the physics of elementary particles and cosmology. This interaction has, of course, become a typical attribute of the physics of our time.

There are two other works of Soviet physicists completed at the end of the 1930s which relate to the TR and characterize its position in the USSR. A deep understanding of STR and electrodynamics helped I.E. Tamm and I.M. Frank (1908–) in 1937 give a theoretical description of Vavilov-Cherenkov radiation (discovered in 1934) as radiation of electrons which are moving in a medium with speed exceeding that of light; this work was awarded the Nobel Prize [173, 222]. In 1939 an article by V.A. Fok which was devoted to the question of the motion of finite (astigmatic) bodies in the GTR was published [175]. This work (together with an analogous work by Einstein, Infeld, and Hoffmann) has great significance for the deriviation of general relativistic corrections in celestial mechanics and for correct realization of the correspondence between GR and Newtonian mechanics.

The scientific achievements of Soviet physics in the 1930s reflect the fact that the RT was fully rooted in it, although attacks on the RT and its active supporters (Ioffe, Vavilov, Tamm, Frenkel', Fok and others) from the side of philosophical opponents in no way ceased. The situation with respect to relativistic cosmology was much more complex. Ideas arising within its framework in the second half of the 1930s of the possible finiteness of space and the expansion of the universe actually did not have open defenders. This was connected not only with the ideological pressure of the supporters of mechanicism but also with empirical bases of cosmology which were scant at that time, and with the very specific formulation of the cosmological problem for physics. In particular, V.A. Fok skeptically evaluated the models of relativistic cosmology in his article [176] in which he also wrote that Einstein is "one of the greatest scientists of the present time whose name is famous and valued by every educated individual," that his theory of gravity "may be considered established as solidly as the Newtonian, the generalization of which it is," and that "the name of Einstein is equal in splendor to the name of Isaac Newton."

Ideas connected with the TR became rather customary even among the broad public (this is reflected, for example, in the novel of M.A. Bulgakov, *Master and Margarita*, where Woland utilizes for his purposes the fifth dimension). In the central newspaper, *Izvestiia*, articles of Rumer and Gerasimovich were published which were devoted to RT and cosmology [163, 164], and Bronshtein gave a lecture on RT over the radio.

Among advanced courses on RT the lectures of L.I. Mandel'shtam (1879–1944) at Moscow University in 1933/34, in which on the basis of a large amount of historical material the TR was presented as the unavoidable result of the development of physics, are noteworthy for their depth [202]. The publication of a series of translations served well the goals of physics education: Einstein [167], the well-known book of Eddington [161], Sommerfeld's classical collection of works on RT [165], supplemented by the 1906 article of Poincaré on STR (the collection was furnished with biographies of Lorentz, Poincaré, Einstein and Minkowski, and also with detailed historico-scientific commentaries prepared by Frederiks and Ivanenko under whose editorship the volume appeared. Simultaneously, in the most important scientific popular journal of that time an article of Frederiks was published which was devoted to the history and perspectives of the TR [166].

The end of the period under discussion may be connected with the appearance in 1941 of the book of L.D. Landau and E.M. Lifshits

(1915–) [179], the first Soviet textbook on TR, which gained international recognition and which basically maintains its ideological-methodological content in following editions and serves to this day as an effective short course introducing STR and GTR.

CONCLUSION

In spite of the general lag of Russian physics in comparison with West European, especially in the area of theoretical physics, the contribution of Russian scholars to experimental (Lebedev, Eichenwald), theoretical (Umov), and mathematical (the tradition of Lobachevskii) precursors of the TR were fully adequate so that they could evaluate its significance and be included in the development of the new theory.

In the beginning (1906–1909), the reception of STR was connected with its understanding within the framework of the electromagnetic picture of the world; there was no differentiation between the STR and the theory of Lorentz (in the majority of cases physicists referred to the theory of Lorentz-Einstein as opposed to the theory of Abraham, the latter in agreement with the most radical variant of the electromagnetic conception of physics).

During the following two to three years (1909–1912) the positions of the leading Russian physicists were defined fairly clearly with respect to the TR: 1) the recognition of the theory in the variant of Einstein, Planck, Minkowski, the understanding of its non-classical character (Umov, Khvol'son; the more moderate appraisal was Lebedev's although he too accepted the relativistic conclusion about the collapse of the concept of the ether); and 2) the aspiration to preserve the conception of the ether and classical ideas about space and time, the interpretation of the theory through the prism of the electromagnetic picture of the world (Borgman, Eichenwald, Goldhammer).

The positive relationship to the RT from the side of leaders of Russian physics (such as Lebedev, Umov, Khvol'son; others, for example Borgman and Goldhammer were also interested in the TR and promoted it) facilitated the acceptance of the RT and the dissemination of its ideas. Also important was the activity of Ehrenfest who worked in Russia from 1907 until 1912 and who influenced the formation of a Russian school of theoretical physics. Finally, the authority of German physics and the very close contacts with it before 1914 (many Russian physicists received education or trained in Germany, the leading Ger man journals willingly printed works of Russian "relativists") played an important role in the reception of RT in Tsarist Russia. At the two most important scientific forums of this period (the 12th Congress of Russian Naturalists and Physicists in 1909–1910 and the 2nd Mendeleev Congress in 1911) the enormous interest of scientific society in TR was clearly manifested, and the two positions in relation to it which were mentioned earlier were clarified.

The basic contribution to the TR which was made by Russian physicists in this period was connected with elaborations and modifications of derivations of the Lorentz transformations, problems of relativistic mechanics (especially of a continuous medium and relativistic theory of rigid bodies).

Owing to the tradition of Lobachevskii, Russian geometers manifested heightened interest in the STR and also played an appreciable role in the diffusion of its ideas. But specialists in the area of mechanics, engineers or physicists of a more practical or "mechanical" direction turned out to be more conservative in their evaluation of the TR, remaining supporters of the ether and Newtonian conceptions of space, time and motion.

The particularities of the reception of the STR in Russia to a certain degree influenced the character of the reception of the TR (at first, GTR) in the USSR. The basic contribution to the popularization, teaching, and development of GTR in the 1920s was made by Leningrad physicists who were connected to the school of Ioffe, Rozhdestvenskii, and closely tied to the theoretical tradition of Ehrenfest (Krutkov, Friedmann, Frederiks, Frenkel', Fok and others). The tradition of Lobachevskii, now enriched by the TR, continued to thrive and develop: mathematicians popularized the ideas of RT and taught it in universities. These discussions of the TR increasingly acquired a philosophical character connected with the problem of the ether and the relationship of the theory to the electromagnetic picture of the world.

Difficult social conditions caused by the civil war and foreign intervention did not hinder the penetration of GTR in Russia. As in other countries, in Soviet Russia in the first half of the 1920s a real relativistic "boom" occurred: dozens of popular books were published on RT and the name of Einstein, selected in 1922 to the Russian Academy of Sciences, became almost common place. The loud sounds of revolutionary ideas of the TR corresponded to the general atmosphere of revolutionary rebuilding of society.

In spite of the powerful informational famine and the difficult conditions of life, Russia gave birth to one of the most important achievements in the area of GTR after the works of Einstein in 1915–1917 – the non-stationary cosmological model of Friedmann (1922). Although works on the TR may seem to be incidental in the works of Friedmann – a mathematician and hydromechanician – in contemporary Russia it was difficult to find a scholar more suitable for such an advance in the GTR. His personal qualities (a rare combination of mathematical purposefulness and vital interest in fundamental problems of physics) played a role, as did the specific social conditions (because of the insufficiency of personnel he was required to teach physics and differential geometry), and the influence of V.K. Frederiks, one of the main propangandists of GTR in Soviet Russia.

From the beginning of the 1920s special courses on the TR were given in higher educational institutions of Soviet Russia (in Petrograd-Frenkel', Frederiks, Friedmann, in Moscow – Tamm, Mandel'shtam and the mathematician Urysohn). The first text books (Frenkel', 1923; Frederiks, Friedmann, 1924) were published, in which the radical theoretical developments and divergencies and complexities in interpretation of the relativistic program were reflected.

In accordance with the tradition of Lobachevskii, Russian mathematicians met GTR with enthusiasm. Among astronomers there were both strong supporters of GTR and also those who reacted to it quite cautiously.

Discussions around the TR in the 1920s quickly acquired a philosophical character. Several philosophers (Semkovskii and Gol'tsman) and leading physicists (Vavilov, Ioffe, Tamm, Fok and Frenkel') resisted the opponents of the TR (the leaders of whom were the physicist Timiriazev and the electrotechnician Mitkevich).

From the end of the 1920s the TR became firmly rooted in Soviet physics, and the main directions of its development in the 1930s were connected with relativistic quantum theory, astrophysics and cosmology. The achievements of Soviet theoreticians speak eloquently about the position of RT: Landau (quantum relativistic limitations of physical concepts, 1931; the limit of the mass of gravitating configurations of relativistic Fermi-gas, 1932), Bronshtein (the first fundamental research of quantization of gravitation with important physical results, 1935; the relativistic quantum foundation of the Doppler nature of the cosmological red shift, 1937), Tamm and Frank (the theory of radiation of faster-than-light electrons, 1937), and Fok (the theory of motion of finite masses in GTR, 1939).

The GTR was examined in the physics community on the whole as a fundamental conquest of science which is comparable in significance to Newtonian theory, although the attitude to relativistic cosmology, especially at the end of the 1930s, was basically skeptical. Finally, the publication of the collection of classics of relativism (1935) and the textbook of Landau and Lifshits (1941) which received international recognition were important landmarks in the history of book publication on the RT.

NOTES

[1] In these years in Moscow University "groups (three to five people) specialized in physics" and "the huge physical institute stood empty." [178: 173]

[2] The first volume of the noted *Kurs fiziki* of O.D. Khvol'son was published in 1897, the last, the fourth volume, appeared in 1914 [91]. *Kurs* was immediately translated into German and French [128]. The book of I.I. Borgman, *Osnovaniia ucheniia ob elektricheskikh i magnitnykh iavleniiakh*, published in 1893, was one of the first thorough descriptions of Maxwellian electrodynamics in Russian scientific literature [2].

[3] Works of O.A. Lezhneva are devoted to the reception of the Maxwellian theory of the electromagnetic field [190, 224].

[4] A.V. Vasil'ev in a brochure *Prostranstvo i dvizhenie* (1900) asked, "Are the laws of motion which are deduced from an examination of the motion of solid bodies [and, likewise, Euclidean geometric relations] applicable to the motions of molecules which are hidden from us?" Later, he wrote about the possibility of the utilization of a geometry of a four-dimensional space and continued: "If the hypothesis of the fourth dimension may serve as the explanation for the phenomena of molecular physics, then even more can it permit the existence of other groups of motion, which differ from the Euclidean group for explanation of these phenomena [4: 11–12]." For this he cited the pronouncement of B. Riemann and Lobachevskii on the physical conditionality of geometric structure.

[5] "When the physicists say that 'matter is disappearing,'" wrote V.I. Lenin, "They mean that hitherto natural science reduced its investigations of the physical world to the three ultimate concepts – matter, electricity and ether [and they strived to reduce the latter two elements to the first – the authors]; but now only the latter two remain, for matter has succeeded in being reduced to electricity . . . [17: 274–275]." In another place he emphasized, "It is, of course, complete nonsense, that materialism has confirmed . . . absolutely a 'mechanical' and not an electromagnetic, or any other immeasureably more complex picture of the world, than *matter in motion* [17: 296]."

[6] "The question of the real or seeming character of the mass of the electron," Baumgart said, "was thus advanced first . . . Works in which electrons were considered to possess real mass immediately disappeared and ceased to appear. Opponents of Abraham and Kaufmann were not to be found [9: 71]."

[7] Rozhdestvenskii still considered it useful to maintain the concept of the ether: "As the

final result of all experiments and theories for an entire century, the certainty that the ether is motionless arose, but for this there is no direct proof." According to Rozhdest-venskii, Cohn, on the basis of his theory, came to the conclusion "that if you put a motionless ether at the basis of the theory, then for further construction of the theory the concept of the ether is entirely unnecessary; it is suffucent to suppose that the electromag-netic energy may be spread in a vacuum where there is no matter." "Why then is the ether necessary?" Rozhdestvenskii asked later. He answered: "In order to make for itself the concrete idea, picture, mechanism of the phenomenon since only such a concrete idea leads to new questions, to new riddles of nature, to the furthest expansion of our knowledge in this area [11: 80]."

[8] This discussion occurred at the Congress of German Naturalists and Physicians in Stuttgart (16–22 September 1906). Many leading Germany physicists took part in it including W. Kaufmann, M. Abraham, A. Sommerfeld, M. Planck, C. Runge, A. Bucherer, R. Gans, and others.

[9] Baumgart concluded this comparison of competing theories with the words, "The debate between supporters of the two views is still unresolved; it is necessary to wait until the further development of these views to uncover their contradictions and make it possible to conduct a decisive experiment between them. But the very participation in this debate of the most outstanding scholars indicates what importance the question of electromagnetic mass has for the theory of electricity. It may certainly be said that from the time of Maxwell new concepts, which are more important than the concept of electromagnetic mass, have not been introduced into the theory of electricity" [12: 319].

[10] The editorial board of the journal, *Jahrb. d. Rad. u. Elektronik*, in which the most complete reviews on the problems of contemporary physics were printed in those years, commissioned a review from Eichenwald in 1908, specially devoted to his experiments [15]. In 1909 the Russian version of this review appeared in *ZhRFKhO* [19].

[11] "The development of electronic theory," Khvol'son said, "led, as everyone knows, to the necessity *of rejecting the old, Newtonian mechanics*. Now it is necessary to build a new mechanics, to search for a new, reliable foundation for it. *Planck* did this in his remark-able work *on the dynamics of moving systems*. . . . Planck accepts the *principle of least action* and *the principle of relativity* which was introduced with such success into science by Einstein, and builds a new theoretical mechanics upon them [18: 137]."

[12] In 1911 the translation of two short articles by Einstein (one of which was coauthored with L. Hopf) appeared in *ZhRFKhO*: the first, on the motion of a resonator in a field of radiation [52], and the second on the theory of opalescence of fluids close to the critical state [53].

[13] Two years earlier Borgman spoke about the decline of the mechanistic *Weltanschauung* and its replacement by the electromagnetic worldview. He underlined the foundation-laying significance of the concept of the ether; however, he considered that its mechanics should undergo radical change: "Only the mechanics of the electron and the atom will open the curtain and reveal to us the anatomy of the ether [21: 107]."

[14] At the 12th Congress of Russian Naturalists and Physicians he said, "We see, thus, that the electromagnetic theory has completely changed our view of the phenomena of the external world: earlier there were attempts to explain electrical phenomena mechanically, that is through the motion of matter. Now, on the contrary, the basic mechanical phenomena, for example inertia, are derived from laws of the electromagnetic field [34: 334]."

[15] This is how he described the creation of the TR at the 2nd Mendeleev Congress: "It happened in 1905 when Albert Einstein in Bern proclaimed the so-called 'postulate of relativity,' which deeply rocked 'physical thought' and which beyond expression explosively hurled it from questions of contemporary scientific life to eternal questions of space and time [40: 176]." He clearly understood the principal difference between the approaches of Lorentz and Einstein and, on the other hand, the organic unity of STR and the four-dimensional formulation of the theory set forth by Minkowski: "The natural question arises where lies the reason for this relativity, which comprises the philosophical meaning of the theory of Einstein. Hermann Minkowski in Göttingen provided the answers to this question [40: 177]." Having begun happily, he finished sadly: "And if relativists change the standard conception of time for a new one, then this is because they *a priori* secretly rejected the ether, whereas in essence the TR gets along quite well with the ether. And from our point of view, all these different times are not real times, but only seeming, for real time is one – it is the time of the motionless ether [40: 184]." And further: "And if it is so, then the entire philosophical significance of the broad interpretation of the formulas of Einstein collapses, and we are forced to return to the ether: this is its effect, that the size of bodies changes during motion; it is its effect, that the speed of light is critical for matter, during which something occurs to matter which is unknown to us. This is the ether, finally . . . the ether which established not the real, but only the seeming connection between space and time, which figures in the formulas of Einstein and Minkowski [40: 186]."

[16] We present the pronouncement of P.N. Lebedev from this review: "All attempts to ascertain directly this relative motion [i. e. the motion of the earth relative to the ether] have not produced any positive results and have recently led to the supposition that this problem will never be resolved, that its solution in principal is impossible; this supposition received the title 'principle of relativity' and raised the opposite task: to indicate in all those discussions the conditions on account of which uniform motion of matter with respect to the ether cannot be detected by any methods. If earlier electrodynamics suggested those examples by which one could attempt to detect the motion of the earth in the ether, then now the principle of relativity demands such supplements in electrodynamics owing to which all phenomena which envelop matter . . . occur independent of its motion in the ether – we may speak consequently only about the relative motion of masses with respect to each other and then the ether and its properties, and what is more the very hypothesis of the existence of the ether in the form that it has been understood until now, is already superfluous and unnecessary [62: 379–380]."

[17] He wrote at that time: "We arrive at the conclusion that the ancient branch of natural science and its basis – mechanics – should be reconstructed on new principles – on the electromagnetic properties of the ether [8: 282]."

[18] "The derivations of Lorentz which date to 1904 and which mainly dealt with electro-optical phenomena," he said, "served as the impetus to the proclamation of the new principle announced by Albert Einstein and its remarkable generalization by Hermann Minkowski who has recently died. We are mounting the summit of contemporary physics: the principle of relativity occupies it, and its expression is so simple that its fundamental significance is not immediately observed [48: 30–31]."

[19] "The new ideas were enveloped by Minkowski in an elegant mathematical theory." According to Minkowski, "in the universe all is given; there is no past or future, it is the eternal present . . . These pictures in the region of philosophical thought should produce

a revolution greater than the shift by Copernicus of the earth from the center of the universe. Since the time of Newton more brilliant prospects for the natural sciences have never been displayed [48: 42–43]."

[20] "The revolution called forth by the replacement of the geocentric *Weltanschauung* by the heliocentric is insignificant in comparison with that which humanity is required to endure if it accepts the principle of relativity . . . During the past seven years or so the new teaching grew into a vast and remarkably harmonious scientific system from the *formal* point of view; a massive amount of literature which is growing every day is devoted to it; the region which it envelops is continuously expanding, and its influence, which destroys all traditionally rooted concepts and which compels us to carry out a radical reevaluation of all values which are utilized by physics and which comprise the fruit of the century long work of the luminaries of science, is felt by every chapter of physics [91: 333–334]."

[21] "Einstein and also Wien set forth *new electromagnetic bases of mechanics* with which the new *Weltanschauung* is connected; mechanics (Newtonian) should neither lie at the base of the world-view, nor should the explanation of phenomena, including electromagnetic be reduced to it, but, on the contrary, laws of electromagnetic phenomena should be the primary foundation, and the mechanics of matter which surrounds us should be constructed on them [91: 366]."

[22] It is interesting to note that the book of Michelson was published almost simultaneously in Russia in two editions: one under the editorship of Khvol'son, and the second under that of the great Russian optician, A.L. Gershun [63, 64]. Rozhdestvenskii wrote in a review of both of these editions. having in mind the famous experiments of Michelson-Morley: "The TR, one of the most burning questions of contemporary physics, is based on them. So important are the results of the researches of Michelson, the significance of which is connected to the question of the recognition or rejection of the ether, that this is apparent in particular from the comments of the editors of the two translations. . . . One of them, O.D. Khvol'son, considers the ether to be the remnant of the old tool of thought which has been tossed aside, [while] the sympathies of the other (A.L. Gershun) undoubtedly are inclined toward the recognition of the ether . . . [65: 203]." The translation of the article of Michelson and Morley which contains the description of the classical experiment, had already been published in *ZhRFKhO* in 1888 [1].

[23] "This principle of relativity changes the basic concepts upon which physics is built in a fundamental way. There is almost no area of physics in which the principle of relativity has not brought about a substantial revolution. Most characteristic of it are, first, the completely new views of space and time, and second the striking paradoxicalness of several of its results. Lengthy efforts on it itself and continuous work are required in order to assimilate the bases of this new teaching; but incomparably more difficult than adjusting to the paradoxes, is to adapt thoughts and methods of reasoning to them, having removed those to which we have been accustomed from our youth [101: 922]."

[24] Vladimir Sergeevich Ignatowsky (also W. von Ignatowsky) graduated Petersburg University. In 1906–08 he studied in the University in Giessen where in 1909 he defended his doctoral dissertation; from 1911 until 1914 he taught in the Higher Technical School in Berlin. After the October Revolution in Russia he taught in different higher educational institutions in Leningrad and worked in the State Optical Institute, mainly in the area of geometric optics and the theory of optical systems. From 1932 he was a corresponding

member of the Academy of Sciences of the USSR. But even before the Revolution Ignatowsky frequently travelled to Russia and participated in the scientific life of the Russian community of physicists [177].

[25] Pauli wrote, in particular, "The simple construction of the formulas (1) [i.e. the Lorentz transformations] makes natural the question of the possibility of the derivation from general group-theoretical concepts without the demand of invariance of the equation (2) [that is the equation of the light cone]. The work of Ignatowsky and Frank and Rothe show to what degree this is possible [220: 27]."

[26] "This book presents an exhaustive account of all the literature existing to this time, of course delving in detail into the works of Lorentz, Einstein, and Planck and Minkowski. Moreover, the chapter on the dynamics contains several interesting new results, in particular, several of the basic formulas of mechanics of elastic bodies in the TR, which in this theory promises to become the main problem of the near future. The account stands out because of its extreme carefulness and exactness . . . Laue's book, certainly, will remain the most distinctive textbook of the TR for the entire time that the theory remains on its present course [55: 246]."

[27] He wrote that the supporters of STR regard the invariance of the equations of electrodynamics with respect to the Lorentz transformations "as the manifestation of the special general principle as the principle of relativity . . ., but, it seems to us," Gruzintsev continued, "that there are still no fully grounded data for this since the one negative result of the experiment of Michelson is . . . still insufficient; there are major, if not objections, then, in any event, bewilderment and doubts with the experimental aspect of the case, and even with the principle aspects . . . In any case, it is necessary to await a more detailed and complete examination of the question [38: 20]."

[28] The translation of A.V. Vasil'ev was subsequently published twice more: in a separate brochure in 1911 in Kazan [45] and in the collection "Printsip otnositel'nosti v matematike" (the fifth volume in the series New Ideas in Mathematics) in 1914 [84]. Yet another translation of Minkovski's famous paper was published in Petersburg in 1911; the translator was I.V. Iashunskii [46].

[29] Having given high praise to STR and its four-dimensional formulation (as a theory of invariants of the Lorentz group), Vasil'ev in the last of the works cited enumerates the most important achievements of the previous years in the area of TR which are connected with the utilization of Lobachevskiian geometry in STR and "the new theory of Einstein" (apparently, he had in mind the GTR or its first draft in the form of the theory of Einstein-Grossmann). He also writes about D. Hilbert's talk "The Foundations of Physics," where already the general covariant equations of the gravitational field occur in correct form [96: 54].

[30] "But I cannot agree with the attribution of such a modest role to the old mechanics," the outstanding mechanician said in a speech. "I am convinced that the problems of great light speeds and the basic problems of the electromagnetic theory will be solved with the help of the old mechanics of Galileo and Newton [103: 259]." And further: "These equations [i.e. Maxwell's] encompass all light phenomena, the electromagnetic waves of Hertz, and in the fine analysis of Abraham give a complete theory of the motion of the electron. For me the importance of the works of Einstein is doubtful in this area which was thoroughly investigated by Abraham on the basis of Maxwell's equations and classical mechanics [103: p. 259]." The address concluded with the question: "Is not the analogy of

Faraday tubes with vortices of non-compressible fluid [this analogy was developed by N.P. Kasterin] an indication of that path along which the mechanics of the ether may be constructed, and is the role of the old mechanics really lost in the new physics? [103: 260]."
[31] "The creators of the new theory," wrote Gerasimovich, having in mind the TR, "in research experiments which can confirm their basic positions, somehow completely ignored the area of celestial phenomena. In addition, it is already clear *a priori* that one should turn precisely to astronomy for decisive data." Emphasizing that astronomy has to do with great distances, times and magnitudes of mass, he draws the conclusion: "Therefore, the influence of corrections of the highest order in standard formulas and laws, which the physicist vainly searches to find in the results of his experiments, in general, should become significantly more noticeable in a comparison of astronomical observations with theory [76: 184]."
[32] In 1922 a review of Einstein's book [110] was published in *Pod znamenem marksizma*. The author of the review, Moscow University professor A.K. Timiriazev, criticized the TR from a mechanistic position, considering it incomprehensible, contradictory to common sense, idealistic through and through, and moreover, manifestly a mistaken theory [112].

Timiriazev's review attracted the attention of V.I. Lenin. However, he, as follows from his article, "On the Significance of Militant Materialism" [111], which was published in the following number of the Journal, on the contrary, considered the TR to be an outstanding scientific achievement. He would otherwise not have ranked Einstein as "a great reformer of natural science": "If Timiriazev had stipulated that the great mass of representatives of the bourgeois intelligentsia of all countries had grasped the theory of Einstein, who himself, in the words of Timiriazev, did not carry out an active campaign against the bases of materialism, then this relates not to Einstein alone, but to an entire group, if not to the majority of great reformers of natural science beginning from the end of the 19th century." [111: pg. 29] Lenin believed that one should criticize not the TR but the fashionable directions of bourgeois philosophy "which now strive to grasp onto Einstein . . . [111: 25]." We note that in the personal library of Lenin there were several books on TR including three different Soviet editions of Einstein's book *O spetsial'noi i obshchei teorii otnositel'nost* [198: 57].

Regarding Timiriazev, even in the long run he continued to oppose the TR [203, 213]. He wrote a large quantity of anti-relativistic articles [174], and in his 1933 book [156] he summed up all the "minuses" of the TR which, in his words, "all the enemies of materialism seized upon, and which, as a rule, materialists come out against . . . [156: 390]." It is worth noting that the book was supplied with a publication forward in which the decline of the methodological position of Timiriazev was mentioned: "The author considers all positions which are defended by him to be correct, inspite of the appraisal they received from the point of view of Marxist criticism, in spite of the Party decisions which concern philosophical discussion and which point to mechanism as the main danger in the given stage [of development of the USSR]." [156: 3] The views of Timiriazev were subjected to decisive criticism both of physicists and of several philosophers [121].
[33] Roman Jakobson, who had just returned to Moscow in 1920 from abroad, acquainted V.V. Maiakovskii with the TR and the discussions raging around it. This story aroused an unusual enthusiasm in the poet, and he spoke about his intention to send Einstein a greetings radiogram "To the science of the future from the art of the future [207]". At the beginning of his play "Bania" the "futurist brain of Einstein" is referred to.

[34] It is possible to get some idea about Einstein's theory of gravity from the articles of the professor of Kiev University L.I. Kordysh [105]. The author begins with methodological comments about the necessity of the sythesis of the "elementary TR" and the theory of gravitation from "proportion": (the Law of Coulomb)/(the theory of electromagnetic field of Maxwell) = (the law of Newton)/(the theory of the gravitational field). Then he briefly discusses the theory of gravitation of Abraham and Nordström and then extensively but not systematically sets forth GTR, referring to Einstein's article of 1916 setting forth GTR in detail.

[35] The famous Soviet mathematician V.I. Smirnov (1887–1974), the friend of Friedmann in gymnasium and university, entitled an article about him "An Outstanding Geophysicist and Mathematician" (*Priroda*, 1963 (11), pp. 93–96).

[36] Einstein, for example, in connection with the initial singularity in the cosmological solution wrote, "During great densities of field and matter, the equations of the field and even the variables which enter them should lose meaning [180: 611]." It is worth pointing out that this pronouncement dates to 1945.

[37] In particular, A.Z. Petrov, who is famous for his classification of pseudo-Riemannian spaces of GTR and headed the only "Department of TR" in the country which was created at Kazan University in the 1950s, belonged to this school.

[38] Glavpolitprosvet was the organ which directed the politico-educational work in the USSR; *sovpartshkoly* and communist *vuzy* were educational institutions which trained Party and government personnel.

[39] It is necessary to say that the young theoreticians were not inclined to forgive Gessen such an understanding of the ether, in spite of his actively positive relationship to the TR. After the publication in 1931 of a volume of an encyclopedia with his article "Ether," he was sent a phototelegram (signed by Bronshtein, Gamow and Landau) with a very caustic caricature mocking his attachment to the idea of the ether; Gamow (with important inaccuracies) recalls this episode and the vocal response among physicists in his autobiography [197]. But Bronshtein's article [136] testifies to the fact that the position of these young physicists was in no way unthinkingly relativistic, but was based on the fundamental understanding of the historical role and significance of the concept of the ether.

[40] "The entire progress of mechanics, and, perhaps, the entire flowering of our culture are conditioned by the transfer of attention from the terrestrial (in particular, geocentric) coordinate system to the stellar (in particular, heliocentric) coordinate system; the struggle of Copernicus and the Copernicans, the struggle of the best minds of the epoch of the Renaissance with the already decrepit ideas of Ptolemy and the scholastics, the struggle which may briefly be designated as the struggle for the stellar as opposed to the terrestrial coordinate system, placed contemporary science and contemporary culture on a firm foundation [124: p. 262]."

[41] In 1928 I.E. Tamm in the introduction to a collected volume [135] remarks that "The TR has entered into the flesh and blood of contemporary physics, and has become so much so its irreplaceable weapon," that even during the discussion of questions connected with "the structure of matter and the theory of quanta" it is necessary to be based on separate propositions and derivations of the TR.

[42] The program of the unified geometricized field theories, the prototype of which GTR served, arose at the beginning of the 1920s [223]. Right up to the creation of quantum mechanics, in spite of the absence of real physical results and the growing number of its variants, it was quite highly appreciated in scientific society. The most prominent Soviet

theoreticians (with the exception of Ia. I. Frenkel') shared the expectations of the leaders of the program of Einstein, Weyl, and Eddington. The creation of quantum mechanics, having in a decisive way strengthened the alternative quantum theoretical program, undermined these expectations. But after the non-trivial connections between the equations of Schrödinger and the five-dimensional variant of the unified field theory were unexpectedly discovered, the interest in the field geometric program was reactivated, primarily in the five-dimensional schema. In the West the researches of H.A. Mandel' [129] and V.A. Fok [129] found a response (in the latter work, speaking to the point, Fok first introduced local gauge transformations of wave functions). Fok's starting points were the manuscript of the article of Mandel' who did not know about the works of T. Kalutsa or O. Klein on five-dimensionality, and discussions with Frederiks, who soon published a thorough examination of the connections between five-dimensionality and quantum mechanics.

By the way, against the background of successes in quantum mechanics the revitalization of the program of unified theories was not especially noticeable. Physicists, including Soviet, preferred to study quantum mechanics. To the majority it began to seem that the field geometrical program and Einstein himself had gone down a blind alley. But according to the reminiscences of A.F. Ioffe, in 1926 he tried to "knock him [Einstein] off the dead-end path [189: 227]" and "put him on the right path," that is the quantum mechanical path.

Yet another flash of interest in unified field theory arose in 1929–1930, when Einstein advanced a new variant of the theory based on geometry with absolute (or "distant") parallelism. This approach stimulated attempts at unification of relativism and quantum mechanics [137–144]. Along this path, Fok and Ivanenko produced a Riemannian generalization of the spinor equation of Dirac and, more fundamentally, developed the local-gauged conception of the electromagnetic field (based on the tetradic description of metric utilized by Einstein, but without recourse to geometry with absolute parallelism). In the spring of 1929 Tamm gave a course at Moscow University, "The Theory of Gravitation and the Electromagnetic Field of A. Einstein," in which he set forth unified theories with absolute parallelism and their intrinsic variant of unification of the theory with quantum mechanics [217: 186]. Unified theories were supposed to become one of the main subjects for discussion at the First All-Union Conference on Theoretical Physics in Kharkov (1929), at which the appearance of Einstein was expected (but he, however, could not attend) and in which Fok, Ivanenko, Mandel', Frederiks, and a former co-worker of Einstein, J. Grommer (1879–1933), who worked at that time in Minsk, and others took part.[140]

Afterwards the program of the unified field theory obviously fell on the periphery although individual works appeared [158, 159] as did a book [187] (its author, Iu.B. Rumer (1901 –) was in Germany at the turn of the 1920s and 1930s and met with Einstein several times) [209].

The relationship of Soviet physicists to unified field theory "in 1931" is characterized by the last paragraph from the corresponding section of an encyclopedia article of M.P. Bronshtein and V.K. Frederiks [148: 306]: "The difficulty which unified field theory to this day has been unable to overcome consists in the fact that besides the equation of the field it is also necessary to produce equations of motion for electrically charged bodies. One will hardly succeed to construct such equations by a geometric path since the ratio e/m (charge to mass) of bodies, which consist of a large number of atoms, may be quite different, which means the equations must contain that magnitude which is not determined uniquely from geometrical data. But from the macroscopic treatment we go over to the microscopic, where e/m may take only three defined meanings (corresponding to the electron, proton, and photon), then the problem, apparently cannot be solved by only one TR (macroscopic according to ideas and methods), but demands some kind of merging of the TR with the theory of quanta. Therefore, in the opinion of many physicists the Einsteinian program of the unified field theory surely appears incapable of being fulfilled."

[43] Independent of any kind of special suppositions on the nature of the spontaneous splitting of the photon from the special principle of relativity, for $W(v)$ – the probability of the decay of a photon of frequency v – Bronshtein derives that $vW(v)$ is the relativistic invariant. This leads to the fully defined dependence of the shift of spectral lines of frequency. However, the observed cosmological red shift is the same for different ranges of spectra (that correspond fully to its Doppler nature).

BIBLIOGRAPHY

The following abbreviations are used throughout this bibliography.*

Phys. Z.	*Physikalische Zeitschrift*
Sorena	*Sotsialisticheskaia rekonstruksiia i nauka*
UFN	*Uspekhi fizicheskikh nauk*
ZhRFKhO	*Zhurnal russkogo fiziko-khimicheskogo obshchestva*
ZhETF	*Zhurnal eksperimental'noi i teoreticheskoi fiziki*

1. Michelson, A., Morley, A., 'Dvizhenie zemli po otnosheniiu k dvizheniiu svetovogo efira [The Motion of the Earth with respect to the Motion of the Ether],' *ZhRFKhO, ch. fizicheskaia* (hereafter simply *ZhRFKhO*),1888, Vol. 20, No. 5, pp. 35–36.

2. Borgman, I.I., *Osnovaniia ucheniia ob elektricheskikh i magnitnykh iavleniiakh [Foundations of the Study of Electrical and Magnetic Phenomena]* (SPB, 1893).

3. *Ob osnovaniiakh geometrii [The Foundations of Geometry]*, (Kazan: Kazan Physico-Mathematics Society, 1893). 121 pp.

4. Vasil'ev, A.V., *Prostranstvo i dvizhenie [Space and Motion]* (Moscow, 1900). 13 pp.

5. Eichenwald, A., 'Über die magnetische Wirkung bewegter Körper im elektrostatischen Felde,' *Ann. Phys.* 1904, **13**, pp. 919–943.

6. Khvol'son, O.D., *Kurs fiziki [A Course of Physics]*, Second Edition, Vol. 2 (St. Petersburg, 1904). iv – 823 pp.

7. Eichenwald, A.A., *O magnitnom deistvii tel, dvizhushchikhsia v elektrostaticheskom pole [On the Magnetic Effect of Bodies Moving in an Electrostatic Field]*, (Moscow, 1904), 144 pp.

8. Umov, N.A., 'Evoliutsiia atoma [The Evolution of the Atom],' *Nauchnoe slovo,*

* Note: Names are given in this bibliography (and throughout the article) according to their most familiar form. Hence, Eichenwald, not Eikhenval'd; Goldhammer, not Gol'dgammer; Friedmann, not Fridman; and Ignatowsky, not Ignatovskii; but Khvol'son, Fok, and Bronshtein, not Chwolson, Fock and Bronstein.

1905 (1), p. 5, as cited in N.A. Umov, *Sobranie sochenenii [Collected Works]*, Vol. 3 (Moscow, 1916), pp. 262–284.

9. Baumgart, K.K., 'Elektronnaia teoriia, kak unitarnaia [Electronic Theory, as a Unitary Theory],' *ZhRFKhO*, 1906, **38**, (4), pp. 67–71.

10. Poincaré, H., *Nauka i gipoteza [Science and Hypothesis]*, (SPB, 1906), 238 pp.

11. Rozhdestvenskii, D.S., 'Nepodvizhnost' efira pri dvizhenii materii [The Immobility of the Ether During the Motion of Matter],' *ZhRFKhO*, 1906, **38**, (4), pp. 72–80.

12. Baumgart, K.K., 'Ob elektromagnitnoi masse [Electromagnetic Mass],' *ZhRFKhO*, 1907, **39**, (8), pp. 294–300.

13. Kagan, V.F., *Osnovaniia geometrii [The Foundations of Geometry]*, Vol. 2, 'Istoricheskii ocherk razvitiia ucheniia ob osnovaniiakh geometrii [A Historical Essay on the Development of the Study of the Foundations of Geometry],' (Odessa, 1907), iv + 558 pp.

14. Lebedinskii, V.K., 'Retsenziia na kn.: G. Vitte, *O sovremennom sostoianii mekhanicheskogo ob' 'iasneniia elektricheskikh iavlenii* [A Review of the Book by H. Witte, *Über den gegenwartigen Stand der Frage nach einer mechanischen Erklärung der elektrischen Erscheinungen* (Berlin, 1906)],' in *ZhRFKhO*, 1907, **39**, (8), pp. 331–332.

15. Eichenwald, A.A., 'Über die magnetische Wirkung der elektrischen Konvektion,' *Jahrb. d. Radioaktiv. u. Elektron*, 1908, Bd. 8, pp. 82–98.

16. Bursian, V.R., 'Opyt Maikel'sona i ego znachenie dliia teorii opticheskikh iavlenii v dvizhushchikhsia telakh [Michelson's Experiment and its Significance for the Theory of Optical Phenomena in Moving Bodies],' *ZhRFKhO*, 1909, **41**, (8), pp. 284–296.

17. Lenin, V.I., *Materializm i empiriokrititsizm [Materialism and Empiriocriticism]* (1909) in Lenin, *Polnoe sobranie socheneniia [Complete Collected Works]*, 5th Edition, Vol. 18 (entire).

18. Khvol'son, O.D., 'Uspekhi fiziki v 1908 g. [Achievements of Physics in 1908],' *ZhRFKhO*, 1909, **42**, (4), pp. 119–143.

19. Eichenwald, A.A., 'O magnitnom deistvii elektricheskoi konvektsii [The Magnetic Effect of Electrical Convection],' *ZhRFKhO*, 1909, **41**, (7), pp. 235–351.

20. Ehrenfest, P.S., 'Gleichformige Rotation starrer Körper und Relativitätstheorie,' *Phys. Z.*, 1909, **10**, p. 918. Russian translation in P.S. Erenfest, *Otnositel'nost. Kvanty. Statistika [Relativity, Quanta and Statistics]*, (Moscow, 1972), pp. 37–39.

21. Borgman, I.I., *Elektrichestvo i svet. Dnevnik XII s'ezda russkiky estestvoispytatelei i vrachei [Electricity and Light. Journal of the 12th Congress of Russian Naturalists and Physicians]*, Otd. I (Moscow, 1910), pp. 93–107.

22. Goldhammer, D.A., 'Novye idei v sovremennoi fizike. Rech', sostavlennaia dlia torzhestvennogo sobraniia kazanskogo universiteta 5 noiabria 1910 [New Ideas in Contemporary Physics. Speech Given at the Grand Meeting of Kazan University on November 5, 1910],' *Fizicheskoe obozrenie*, 1911, **12**, pp. 1–35.

23. *Programmy i protokoly obshchikh, soedinennykh i sektsionnykh zasedanii [Programs and Protocols of General, Joint and Sectional Meetings]. Dnevnik XII s'ezda russkikh estestvoispytatelei i vrachei* [See [21]], Otd. II., (Moscow, 1910). The meeting of the section of physics on December 30, 1909, is found on pp. 170–172.

24. Ignatowsky, W. von., 'Einige allgemeine Bemerkungen zum Relativitatsprinzip,' *Phys. Z.*, 1910, **11**, pp. 972–975.

25. Ignatowsky, W. von. 'Der starre Körper und Relativitätsprinzip,' *Ann. Phys.*, 1910, **33**, pp. 607–630.

26. Ignatowsky, W. von. 'Einige allgemeine Bemerkungen zum Relativitätsprinzip,' *Verhandl. Deutsch. phys. Ges.*, 1910, **12**, pp. 788–796.

27. Ignatowsky, W. von., 'Das Relativitätsprinzip,' *Arch. Math. Phys*, 1910, **17**, pp. 1–24.

28. Ignatowsky, W. von, *Die Vektoranalysis und Anwendung in der theoretischen Physik*. Vols. I and II (Leipzig and Berlin, 1909–1910). Vol. I is 120 pp., Vol. II is 128 pp.

29. Isakov L.D., 'Fizika na XII s'ezde russkikh estestvoispytatelei i vrachei [Physics at the 12th Congress of Russian Naturalists and Physicians],' *ZhRFKhO*, 1910, **42**, (2), pp. 43–49, 144–157.

30. Minkowski, H., *Prostranstvo i vremiia [Space and Time]*, Trans. A.V. Vasil'eva, *Izv. Fiz.-mat. o-va pri Imper. Kazanskom u-te.*, vtoraia seriia, 1910, **16**, (4), pp. 137–155.

31. 'Protokol 144-go zasedaniia Kazanskogo fiziko-matematicheskogo obshchestva 10 oktiabria 1909 [Protocol of the 144th Meeting of the Kazan Physico-Mathematical Society, October 10, 1909],' *Izv. Fiz.-mat. o-va pri Imper. Kazanskom u-te.*, vtoraia seriia, 1910, **16**, (4), p. 77.

32. 'Protokol 147-go zasedaniia Kazanskogo fiziko-matematischeskogo obshchestva 1 maia 1910 [Protocol of the 147th Meeting of the Kazan Physico-Mathematical Society, May 1, 1910],' *Ibid.*, p. 79.

33. Umov, N.A., 'Einheitliche Ableitung der Transformationen, die mit dem Relativitätsprinzip vertraglich sind,' *Phys. Z.*, 1910, **11**, pp. 905–908.

34. Eichenwald, A.A., *Materiia i energiia [Matter and Energy]. Dnevnik XII s'ezda russkikh estestvoispytatelei i vrachei [See* [21]], Otd. I, (Moscow: 1910), pp. 331–343.

35. Erenfest, P.S., 'Printsip otnositel'nosti [The Principle of Relativity],' *ZhRFKhO*, 1910, **42**, (1), pp. 33–38; and (2), pp. 81–87. See also P. Erenfest, *Otnositel'nost. Kvanty. Statistika [See* [20]]. (Moscow, 1972), pp. I–II.

36. Ehrenfest, P.S., 'Zu Herrn v. Ignatowskys Behandlung der Bornschen Starrheitsdefinition,' *Phys. Z.*, 1910, **11**, pp. 1127–1129.

37. Borgman, I.I., 'Poslednie uspekhi v fizike. Rech' na otkrytii otdeleniia fiziki 2-go Mendeleevskogo s'ezda 21 dekabria 1911 g. [Recent Achievements in Physics. Speech at the Open Section of Physics of the Second Mendeleev Congress on December 21, 1911],' *Dnevnik 2-go Mendeleevskogo s'ezda po obshchei i prikladnoi khimii [Journal of the Second Mendeleev Congress on General and Applied Chemistry]*, (SPB, 1911), No. 3, pp. 9–17.

38. Gruzintsev, A.P., *Preobrazovaniia Lorentsa i printsip otnositel'nosti [The Lorentz Transformations and the Principle of Relativity]*, (Khar'kov, 1911), 20 pp. (See also, *Soobshchenie khar'kovskogo mat. o-va.*, ser. 2, T. 12, No. 6.)

39. Ignatowsky, W. von, 'Zur Behandlug der Born'schen Starrheitsdefinition. Erwiderung an Herrn P. Ehrenfest,' *Phys. Z.*, 1911, Bd. 12, pp. 606–607.

40. Goldhammer, D.A., 'Vremia. Prostranstvo. Efir. Rech na obshchem sobranii 2-go Mendeleevskogo s'ezda 28 dekabriia 1911 g. [Time, Space and Ether. A Speech at the General Meeting of the Second Mendeleev Congress on December 28, 1911],' *Dnevnik 2-go Mendeleevskogo s'ezda po obshchei i prikladnoi khimii [See* [37]] (SPB, 1911), **8**, pp. 142–159. Reprinted in *ZhRFKhO*, 1912, **44**, (2), pp. 165–189.

41. Ignatowsky, W. von, 'Zur Elastitätstheorie vom Standpunkte des Relativitätsprinzips,' *Phys.Z.*, 1911, **12**, pp. 441–442.
42. Kordysh, L.I., 'Elementarnyi vyvod osnovnykh formul teorii otnositel'nosti [An Elementary Derivation of the Basic Formulae of the Theory of Relativity],' *Izv. Kievskogo politekh. in-ta., otd. inzh.mekh.*, 1911, God II, Book I, pp. 43–51.
44. Laue, M. von, *Das Relativitätsprinzip*, (Braunschweig, 1911)
45. Minkovski, H., *Prostranstvo i vremia [Space and Time]*, Transl. A.V. Vasil'ev, (Kazan, 1911), 21 pp.
46. Minkovski, H., *Prostranstvo i vremia [Space and Time]*, Transl. I.V. Iashunskii (SPB, 1911), 94 pp. With biographical notes by D. Hilbert and H. Weyl and with articles by W. Wien and P. Natorp on 'The relativity principle.'
47. Planck, M., *Vosem' lektsii, chitannykh v Columbia University in the City of New York, vesnoi 1909 g. [Eight Lectures Delivered at Columbia University in the Spring of 1909]*, Transl. I.M. Zanchevskii, (SPB, 1911), 158 pp.
48. Umov, N.A. 'Kharakternye cherty i zadachi sovremennoi estestvenno-nauchnoi mysli. Rech na obshchem sobranii 2-go mendeleevskogo s''ezda 21 dekabria 1911 g. [Characteristic Traits and Tasks of Contemporary Natural Scientific Thought. A Speech at the General Meeting of the Second Mendeleev Congress on December 21, 1911],' *Dnevnik 2-go mendeleevskogo s''ezda po obshchei i prikladnoi khimii* [see [37]], (SPB, 1911), Vyp. 5, pp. 1–21. Reprinted in *ZhRFKhO*, 1912, **44**, (6), pp. 117–144.
49. Umov, N.A., 'Ob usloviakh invariantnosti volnovogo uravneniia [The Conditions of Invariance of the Wave Equation],' *Ibid.*, Vyp. 3, pp. 23–24. Reprinted in *ZhRFKhO*, 1912, Vol. 44, No. 6, pp. 349–354.
50. Frank. Ph, Rothe, H., 'Über die Transformation der Raumzeitkoordinaten von ruhenden auf bewegte Systeme,' *Ann. Phys.*, 1911, **34**, pp. 825–855.
51. Einstein, A., 'Zum Ehrenfestischen Paradoxon,' *Phys. Z.*, 1911, **12**, pp. 509–510. Russian translation in A. Einstein, *Sobranie nauchnykh trudov [Collected Scientific Works]*, Vol. I, (Moscow, 1965), pp. 187–88.
52. Einstein, A. Hopf, L, 'Statisticheskie issledovaniia dvizheniia rezonatora v pole izlucheniia [Statistical Investigations of the Motion of a Resonator in a Field of Radiation],' *ZhRFKhO*, 1911, **43**, (2), pp. 78–79.
53. Einstein, A., 'Teoriia opalestsentsii odnorodnykh zhidkostei i zhidkikh smesei vblizi kriticheskogo sostoianiia [The Theory of Opalescence of Homogenous Liquids and Liquid Mixtures Near Critical State],' *ZhRFKhO*, 1911, **43**, (5), pp. 117–119.
54. Epstein, P.S., 'Über relativistische Statik,' *Ann. Phys.*, 1911, **36**, pp. 739–795.
55. Ehrenfest, P.S., 'M. Laue. Printsip otnositel'nosti (retsenziia) [A Review of M. Laue's *Principle of Relativity*]' in *ZhRFKhO*, 1911, **43**, (6), p. 246.
56. Ehrenfest, P.S., 'Zu Herrn v. Ignatowskys Behandlung der Bornshchen Starrheitsdefinition. II,' *Phys. Z.*, 1911, **12**, pp. 412–413.
57. Borgman, I.I., ed., *Novye idei v fizike [New Ideas in Physics]*, Sb. 2, 'Efir i materiia [Ether and Matter],' (SPB, 1912), 151 pp.
58. Borgman, I.I., ed., *Novye idei v fizike [New Ideas in Physics]*, Sb. 3, 'Printsip reliativnosti [The Principle of Relativity],' (SPB, 1912).
59. Bursian, V.R., 'Opytnye issledovaniia po voprosu o vliianii dvizheniia veshchestva

na efir [Experimental Research on the Problem of the Influence of the Motion of Matter on the Ether],' in [58], pp. 1–36.

60. Grdina, Ia. I., *K voprosu o masse elektrona [Toward the Question of the Mass of the Electron]*, (Ekaterinoslav, 1912).

61. Isakov, L.D., 'Vtoroi mendeleevskii s''ezd [The Second Mendeleev Congress],' *ZhRFKhO*, 1912, **44**, (1), pp. 26–35.

62. Lebedev, P.N., 'Uspekhi fiziki v 1911 gody [Achievements of Physics in 1911],' *Russkie vedomosti*, 1912, No. 1. Reprinted in P.N. Lebedev, *Sobranie sochinenii [Collected Works]*, (Moscow 1913), pp. 376–380.

63. Michelson, A.A., *Svetovye volny i ikh primeneniia [Light Waves and Their Applications]*, Ed. O.D. Khvol'son (Odessa, 1912).

64. Michelson, A.A., *Svetovye volny i ikh primeneniia [Light Waves and Their Applications]*, Ed. A.L. Gershuna (SPB, 1912), 220 pp.

65. Rozhdestvenskii, D.S., 'A.A. Michelson. *Svetovye volny i ikh primeneniia* (Retsenziia) [A.A. Michelson, *Light Waves and Their Applications*. A Review],' *ZhRFKhO*, 1912, **44**, (5), pp. 201–203.

66. Khvol'son, O.D., 'Printsip otnositel'nosti [The Principle of Relativity],' *ZhRFKhO*, 1912, **44**, (10), pp. 377–432; 1913, **45**, (1), p. 42.

67. Khvol'son, O.D., *Printsip otnositel'nosti [The Principle of Relativity]*, (SPB, 1912), 56 pp.

68. Shaposhnikov, K.N., 'K dinamike dvizhushchegosia tela [Toward the Dynamics of a Moving Body], *ZhRFKhO*, 1912, **44**, (2), pp. 102–104.

69. Shaposhnikov, K., 'Die Minkowskischen Bewegungsgleichungen und die Plancksche Dynamik,' *Ann. Phys.*, 1912, **38**, pp. 239–244.

70. Shaposhnikov, K., 'Über die Invarianz von P und T bei Lorentz-Transformation,' *Phys. Z.*, 1912, **13**, pp. 212–213.

71. Shaposhnikov, K., 'Zur Dynamik der bewegten Körper,' *Phys. Z.*, 1912, **13**, pp. 403–404.

72. Einstein, A., 'Printsip otnositel'nosti i ego sledstviia v sovremennoi fizike [The Principle of Relativity and its Consequences in Contemporary Physics],' in [58], pp. 62–103.

73. Ehrenfest, P.S., 'Zur Frage nach der Entberlichkeit des Lichtathers,' *Phys. Z.*, 1912, **13**, p. 317.

74. Vasil'ev, A.V., ed., *Novye idei v matematike [New Ideas in Mathematics]*, Sb. 2, 'Prostranstvo i vremia I [Space and Time],' (SPB, 1913), 146 pp. This contains translations of articles by Ernst Mach, H. Poincaré, P. Langevin, L. Geffter and others.

75. Vasil'ev, A.V., ed., *Novye idei v matematike [New Ideas in Mathematics]*, Sb. 3, "Prostranstvo i vremia II [Space and Time]," (SPB, 1913), 152 pp. This contains articles by H. Poincaré, K. Schwarzchild and others.

76. Gerasimovich, B.P., 'Aberratsiia sveta i teoriia otnositel'nosti [The Aberration of Light and the Theory of Relativity],' *Izv. russk. astron. o-va*, November 1913, No. 6, Vyp. 19, pp. 183–203.

77. Lebedinskii, V.K., 'A. Puankare v mire elektrichestva [H. Poincaré in the World of Electricity],' *ZhRFKhO*, 1913, **45**, (2), pp. 115–123.

78. Poincaré, H., *Novaia mekhanika. Evoliutsiia zakonov [The New Mechanics. The Evolution of Laws]*, Translated and with commentary and two introductory articles by G.A. Gurevich. (Moscow, 1913), 179 pp.

79. Khvol'son, O.D., *Printsip otnositel'nosti. Obshchedostupnoe izlozhenie [The Principle of Relativity. A Popular Account]*. (SPB, 1913), 59 pp.

80. Shaposhnikov, K.N., 'Printsip otnositel'nosti i dinamika tochki,' *ZhRFKhO*, 1913, **45**, (9), pp. 546–550.

81. Shaposhnikov, K., 'Zur Relativdynamik des homogenen Körpers,' *Ann. Phys.*, 1913, **42**, pp. 1572–74.

82. Erenfest, P.S., 'Krizis v gipoteze o svetovom efire [The Crisis in the Hypothesis of the Ether],' *ZhRFKhO*, 1913, **45**, (4), pp. 151–162.

83. Borgman, I.I., *Novye idei v fizike [New Ideas in Physics]*, Sb. 3, 'Printsip otnositel'-nosti [The Principle of Relativity],' (SPB, 1914), 181 pp. (This contains a lengthy bibliography of works on the TR published up to 1913.)

84. Vasil'ev, A.V., ed., *Novye idei v matematike [New Ideas in Mathematics]*, Sb. 5, 'Printsip otnositel'nosti v matematike [The Principle of Relativity in Mathematics],' (SPB, 1914), 176 pp. (This contains the paper by Minkovski, *Raum und Zeit*, and also articles of M. v. Laue, F. Klein and others.)

85. Vasil'ev, A.V., *Nikolai Ivanovich Lobachevskii* (SPB, 1914), 127 pp.

86. Ivanov, A.A., ed., *Novye idei v matematike [New Ideas in Mathematics]*, Sb. 7, 'Printsip otnositel'nosti s matematicheskoi tochki zreniia II [The Principle of Relativity from a Mathematical Point of View], (SPB, 1914), 155 pp. (This contains articles of N.A. Umov, V. Varichak, N. Brill', and others).

87. La Rosa, M., *Efir. Istoriia odnoi gipotezy [The Ether. A History of A Hypothesis]*, ed. and with clarifying notes by O.D. Khvol'son, (SPB, 1914), 91 pp.

88. Lebedinskii, V.K., 'Printsip otnositel'nosti v sovremennoi fizike [The Principle of Relativity in Contemporary Physics],' *Elektrichestvo*, 1914 (1), pp. 1–16.

89. Orlov, I.E., 'Osnovnye formuly printsipa otnositel'nosti s tochki zreniia klassiches-koi mekhaniki [The Basic Formulae of the Principle of Relativity from the Point of View of Classical Mechanics],' *ZhRFKhO*, 1914, **46**, (4), pp. 163–175.

90. Rautian, G.N., Ia. I. Grdina, 'K voprosu o printsipe otnositel'nosti (retseziia) [Toward the Problem of the Principle of Relativity (A Review)],' *ZhRFKhO*, 1914, **46**, (10), pp. 391–396.

91. Khvol'son, O.D., *Kurs fiziki [A Course of Physics]*, Vol. IV, Part 2 (SPB, 1914). (See also Kurs fiziki, Vol. V, second edition, (Berlin, 1923)), 984 pp. The Theory of Relativity is covered in chapter 5 of the third part, on pp. 333–373.

92. Ehrenfest, P.S., 'Po povodu stat'i I Orlova 'Osnovnye formuly printsipa otnositel'-nosti s tochki zreniia klassicheskoi mechaniki' [In Regard to the Article of I. Orlov 'The Basic Formulae of the Principle of Relativity from the Point of View of Classical Mechanics'],'' *ZhRFKhO*, 1914, **46**, (4), pp. 175–176.

93. Lorentz, H.A., 'Nachalo Gamil'tona v einshteinovskoi teorii tiagoteniia [The Hamilton Principle in Einstein's Theory of Gravitation],' *ZhRFKhO*, 1915, **47**, (8), pp. 516–534.

94. *Russkaia bibliografiia po estestvoznaniiu i matematike [A Russian Bibliography of Natural Science and Mathematics]*, Vol. 6 (1907), Vol. 7 (1908–09), (Petrograd, 1914); Vol. 8 (1910–1911), (1915); Vol. 9 (1912–1913), (1918).

95. Slepian, L.B., 'Osnovaniia teorii otnositel'nosti [The Foundations of RT],' *ZhRFKhO*, 1915, **47**, (9), pp. 599–635.
96. Vasil'ev, A.V., *Matematika [Mathematics]*, (Kazan', 1916), 58 pp.
97. Goldhammer, D.A., *Novaia teoriia elektromagnitnykh iavlenii v dvizhushchikhsia sredakh [A New Theory of Electromagnetic Phenomena in Moving Mediums]*, (Kazan, 1916), 52 pp.
98. Grdina, Ia. I., 'Fizicheskii ili ogranichennyi printsip otnositel'nosti [The Physical or Limited Principle of Relativity],' *ZhRFKhO*, 1916, **48**, (1), pp. 1–36.
99. Zhukovskii, N.E., 'Umov kak matematik [Umov as a Mathematician],' *Matematicheskii sbornik*, 1916, **30**. See also, N.E. Zhukovskii, *Polnoe sobranie sochineniia [Complete Collected Words]*, Vol. 9 (Moscow-Leningrad, 1937), pp. 436–438.
100. Slepian, L.B., 'Istinnyi printsip otnositel'nosti i osnovaniia formul preobrazovaniia klassicheskoi mekhaniki [The Real Principle of Relativity and the Foundations of the Formulae of the Transformations of Classical Mechanics],' *ZhRFKhO*, 1916, **48**, (4), pp. 150–157.
101. Khvol'son, O.D., 'Otnositel'nosti (printsip) [Relativity (Principle of)]' in *Novyi entsiklopedicheskii slovar' [New Encyclopedic Dictionary]*, (Formerly Brokgauz-Efron), Vol. 29, 1916, pp. 922–931.
102. de Sitter, W., 'On Einstein's Theory of Gravitation and its Astronomical Consequences. Third Paper,' *Mon. Not. R. Astr. Soc.*, 1917, **78**, pp. 3–28.
103. Zhukovskii, N.E., 'Staraia mekhanika v novoi fizike. Rech' v Moskovskom matematicheskom obshchestve 3 Marta 1918 [The Old Mechanics in the New Physics. Speech at the Moscow Mathematics Society on March 3, 1918],' in *Polnoe sobranoe sochineniia [Complete Collected Works]*, Vol. 9 (Moscow-Leningrad), pp. 245–260. This work was prepared for publication by N.P. Kasterin.
104. Kasterin, N.P., 'Sur une contradiction essentielle entre las theorie de relativité d'Einstein et l'experience ,' *Izv. ross. akad. nauk*, 1918, IV seriia, No. 2–3, pp.89–98.
105. Kordysh, L.I., 'Gravitatsiia i inertsiia [Gravitation and Inertia],' *Universitetskie izvestiia (Kiev)*, 1918, No. 3–4, pp. 1–36. 'Gravitatsionnaia teoriia difraktsii [Gravitational Theory of Diffraction],' *Ibid.*, pp. 1–20.
106. 'Radiogramma Rozhdestvenskogo, Krutkova i Frederiksa v Leiden [A Radiogram from Rozhdestvenskii, Krutkov and Frederiks to Leiden]' in *Archik AN SSSR*, F. 341, op. 2, ed. xr. 69. 1. 1. (1919). See *Istoriia i metodologiia estestvennykh nauk*, Vyp. 3 (Moscow, 1965), p. 311.
107. Vavilov, S.I., 'Referat kn. Einshtein A. Efir i printsip otnositel'nosti [A Review of Einstein's *Ether and Relativity Principle*],' *UFN*, 1921, **2**, p. 300.
108. Landsberg, F.S., 'Otklonenie sveta v gravitatsionnom pole solntsa (rezul'taty angliiskoi ekspeditsii po nabliudeniiu solnechnogo zatmeniia 1919 g.) [Aberration of Light in the Gravitational Field of the Sun (The Results of the English Expedition for Observation of the Solar Eclipse in 1919)],' *UFN*, 1921, **2**, pp. 189–93.
109. Frederiks, V.K., 'Obshchii printsip otnositel'nosti Einshteina [The GTR of Einstein],' *UFN*, 1921, **2**, pp. 162–188.
110. Einstein, A., *Teoriia otnositel'nosti (obshchedostupnoe izlozhenie) [The Theory of Relativity. A Popular Account]*, (Berlin, 1921).
111. Lenin, V.I., 'O znachenii voinstvuiushchego materializma [On the Significance of Militant Materialism],' *Pod znamenem marksizma*, 1922 (3). Also in *Polnoe sobranie*

320   V.P. VIZGIN AND G.E. GORELIK

sochineniia [Complete Collected Works], Vol. 45, pp. 23–33.

112. Timiriazev, A.K., 'A. Einshtein. *O spetsial'noi i vseobshchei teorii otnositel'nosti* (retsenziia) [A Review of Einstein's *Special and General Relativity*],' *Pod znamenem marksizma*, 1922, No. 1–2, pp. 70–73.

113. Florenskii, P.A., *Mnimosti v geometrii [Imagineriness in Geometry]*, (Moscow, 1922).

114. Friedmann, A.A., 'O krivizne prostranstva [On the Curvature of Space],' See [193: pp. 229–238]. From *Zeitschrift für Physik*, 1922, **10**, pp. 377–387.

115. 'Mitteilung an dei deutschen Physiker,' *Zeitschrift für Physik*, 1922, **10**, p.352.

116. Bogoraz-Tan, V.G., *Einshtein i religiia. Primenenie printsipa otnositel'nosti k issledovaniiu religioznykh iavlenii [Einstein and Religion. The Application of RT to Research on Religious Phenomena]*, (Moscow-Petrograd, 1923).

117. Vasil'ev, A.V., *Prostranstvo, vremia, dvizhenie. Istoricheskie osnovy teorii otnositel'nosti [Space, Time and Motion. Historical Bases of the TR]*, (Petrograd, 1923).

118. Frenkel', Ia. I., *Teoriia otnositel'nosti [The RT]*, (Petrograd, 1923).

119. Friedmann, A.A., *Mir kak prostranstrvo i vremiia [The Universe as Space and Time]*, (Petrograd, 1923). See [193, pp. 244–322].

120. Einstein, A., 'K rabote A. Fridmana "O krivizne prostranstva" [On Friedmann's Work "On the Curvature of Space"],' in *Sobranie nauchnykh trudov [Collected Scientific Works]*, Vol. 2 (Moscow, 1966), p. 119. See also *Zeitschrift für Physik*, 1923, Bd. 16, p. 228.

121. Semkovskii, S. Iu., *Teoriia otnositel'nosti i materializm [The TR and Materialism]*, (Khar'kov, 1924); *Dialekticheskii materializm i printsip otnositel'nosti [Dialectical Materialism and the Principle of Relativity]*, (Moscow-Leningrad, 1926).

122. Frederiks, V.K., Friedmann, A.A., *Osnovy teorii otnositel'nosti* (Vyp. I. *Tenzorial'noe ischislenie*) *[The Basis of the RT. Part I. Tensor Analysis]*, (Leningrad, 1924).

123. Friedmann, A.A., 'O vozmozhnosti mira s postoiannoi otritsatel'noi kriviznoi prostranstva [The Possibility of a Universe with Constant Negative Curvature of Space],' in [193: pp. 238–244]. See also *Zeitschrift für Physik*, 1924, **31**, pp. 326–333.

124. Friedmann, A.A., Loitsianskii, L.G., *Teoreticheskaia mekhanika [Theoretical Mechanics]*, (Leningrad, 1924).

125. Friedmann, A., Schouten, J.A., 'Über die Geometrie der halb – symmetrischen Übertragungen,' *Math. Zeitschrift*, 1924, **21**, pp. 211–223. See also [193: p. 379.].

126. Gerasimovich, B.P., *Vselennaia pri svete teorii otnositel'nosti [The Universe in Light of the TR]*, (Khar'kov, 1925).

127. Fesenkov, V.G., 'Astronomicheskie dokazatel'stva printsipa otnositel'nosti [Astronomical Evidence of the Principle of Relativity],' *Vestnik kommunisticheskoi akademii*, 1925, Book 13, p. 200.

128. Dobiash, A.A., 'Orest Danilovich Khvol'son,' *ZhRFKhO*, 1926, **58**, (2), pp. 87–104.

129. Mandel, H., 'Zur Herleitung der Feldgleichungen in der allgemeinen Relativitätstheorie,' *Zeitschrift für Physik*, 1926, **39**, pp. 136–145.

130. Fok, V., 'Über die invariante Form der Wellen und der Bewegungsgleichungen für einen geladenen Massenpunkt,' *Z. für Physik*, 1926, **39**, pp. 226–232.

131. Frederiks, V.K., 'Teoriia Shredingera i obshchaia teoriia otnositel'nosti [Schrödinger's Theory and the GTR],' in *Osnovaniia novoi kvantovoi mekhaniki [Foundations of New Quantum Mechanics]*, ed. A.F. Ioffe, (Moscow-Leningrad, 1927)., pp. 83–98.

132. D. Hilbert, 'Referat über die geometrischen Schriften und Abhandlungen H. Weyl's, erstattet der Physiko-Mathematischen Gesellschaft am der Universitat Kasan,' *Izvestiia fiz.-mat. o-va pri Kazanskom u-te*, 1927, Vol. 2, Seriia 3-ia, pp. 66–70.

133. Vavilov, S.I., *Eksperimental'nye osnovaniia teorii otnositel'nosti [Experimental Foundations of the TR]*, (Moscow-Leningrad, 1928).

134. Gessen, B.M., *Osnovnye idei teorii otnositel'nosti [Basic Ideas of the RT]*, (Moscow-Leningrad, 1928).

135. *Fizika. Nauka XX veka. [Physics. Science of the 20th Century]*, Vol. I, (Moscow-Leningrad, 1928), p. 6.

136. Bronshtein, M.P., 'Efir v staroi i novoi fizike [The Ether in the Old and New Physics],' *Chelovek i priroda*, 1929 (16), pp. 3–9.

137. Fok, V., Ivanenko, D., 'Geometrie quantique lineaire et deplacement parallele,' *Compt. Rend.*, 1929, **188**, pp. 1470–1472.

138. Fok, V., Ivanenko, D., 'Uber eine mögliche geometrische Deutung der relativistischen Quantentheorie,' *Zeitschrift für Physik*, 1929, **54**, pp. 798–802.

139. Fok, V., Ivanenko, D., 'Quantum geometry,' *Nature*, 1929, **123**, p. 838.

140. Ivanenko, D.D., 'Kommentarii k pis'mu Einshteina ot 11 iiulia 1929 g. k I.V. Obreimovu (publikatsiia neopublikovannoe pis'mo A. Einshteina) (Commentary on Einstein's Letter of July 11, 1929, to I.V. Obreimov],' in *Voprosy istorii estestvoznaniia i tekhniki*, 1980, Vyp. 67–68, pp. 70–71.

141. Tamm, I.E., 'O sviazi einshteinovskoi edinoi teorii polia s kvantovoi teoriei [On the Connection of Unified Field Theory with Quantum Theory],' *Zhurnal prikladnoi fiziki*, 1929 (1), p. 130, as cited in *Sobranie nauchnykh trudov [Collected Scientific Works]*, Vol. 2 (Moscow, 1975), pp. 184–87.

142. Tamm, I.E., Leontovich, M.A., 'Zamechaniia k einshteinovskoi edinoi teorii poliia [Observations on Einstein's Unified Field Theory],' as cited in *Ibid.*, pp. 191–201. See also *Zeitschrift für Physik*, 1929, **57**, p. 354.

143. Fok, V.A., 'Geometrizatsiia dirakovskoi teorii elektrona [The Geometrization of Dirac's Theory of the Electron],' in *Al'bert Einshtein i teoriia gravitatsii [Albert Einstein and the Theory of Gravitation]*, (Moscow, 1979), pp. 415–432. See also *Zeitschrift für Physik*, 1929, **57**, pp. 261–277.

144. Fok, V.A., 'Volnovoe uravnenie diraka i geometriia rimana [Dirac's Wave Equation and Riemanian Geometry],' *ZhRFKhO*, **62**, (1), pp. 133–151.

145. Ambartsumian, V., Ivanenko, D., 'Zur Frage nach Vermeidung der unendlichen Selbstruckwirkung des Elektrons,' *Zeitschrift für Physik*, 1930, **64**, pp. 563–567.

146. Bronshtein, M.P., 'Sovremennoe sostoianie reliativistskoi kosmologii [The Current State of Relativistic Cosmology],' *UFN*, 1931, **11**, pp. 124–184.

147. Bronshtein, M.P., 'Novyi krizis teorii kvant [The New Crisis in Quantum Theory],' *Nauchnoe slovo*, 1931 (1), 38–55.

148. Bronshtein, M.P., Frederiks, V.K., 'Otnositel'nosti teoriia [The RT],' in *Tekhnicheskaia entsiklopediia*, Vol. 15 (Moscow-Leningrad, 1931), pp. 352–359, 362–367.

149. Fesenkov, V.G., 'Astronomicheskie proverki teorii otnositel'nosti [Astronomical Verifications of RT],' in *Ibid.*, pp. 359–362.

150. Timiriazev, A.K., 'Otnositel'nosti teoriia i filosofiia [The RT and Philosophy],' in *Ibid.*, pp. 367–371.

151. Landau, L., Peierls, R., 'Rasprostranenie printsipa neopredelennosti na reliati-

vistskuiu kvantovuiu teoriiu [The Extension of the Principle of Indeterminacy into Relativistic Quantum Theory],' in Landau, L.D., *Sobranie trudov [Collected Works]*, Vol. 1 (Moscow, 1969), pp. 56–70. See also *Zeitschrift für Physik*, 1931, **69**, p. 56.

152. Einstein, A., 'K kosmologicheskoi probleme obshchei teorii otnositel'nosti [The Cosmological Problem of the GTR],' in *Sobranie nauchnykh trudov [Collected Scientific Works]*, Vol. 2, pp. 349–352. See also *Sitzungsber. preuss. Akad. Wiss.*, 1931, p. 235.

153. Landau, L.D., 'On the theory of stars,' *Physikalische Zeitschrift der Sowjet Union (hereafter Phys. Z. Sow.)* 1932, **1**, p. 285.

154. Bronshtein, M.P., Landau, L.D., 'Vtoroi zakon termodinamiki i vselenaia [The Second Law of Thermodynamics and the Universe],' *Phys. Z. Sow.*, 1933, **4**, pp. 114–118.

155. Tamm, I.E., 'O rabote filosofov-marksistov v oblasti fisiki [The Work of Philosopher-Marxists in the Area of Physics], *Pod znamenem marksizma*, 1933 (2), pp. 220–231.

156. Timiriazev, A.K., *Vvedenie v teoreticheskuiu fiziku [Introduction to Theoretical Physics]*, (Moscow-Leningrad, 1933), 440 pp.

157. Shturm, L. Ia., 'Leon Iosifovich Kordysh (nekrolog) [Obituary],' *UFN*, 1933, **13**, (6), pp. 970–975.

158. Mandel', G.A., 'Edinaia teoriia elektromagnitnogo i gravitatsionnogo polei. Dolkad na I mezhdunarodnoi konferentsii po tenzornoi i differentsial'noi geometrii i ee prilozheniiam [The Unified Theory of Electromagnetic and Gravitational Fields. Paper at First International Conference on Tensor and Differential Geometry and Its Applications]. Moskva 17.V – 23.V. 1934,' in *Trudy seminar po vektornomu i tenzornomu analizu s ikh prilozheniiami k geometrii, mekhanike i fizike*. Vyp. 4 (Moscow-Leningrad, 1937), pp. 62–69.

159. Rumer, Iu. B., 'O geometricheskom istolkovanii materii v obshei teorii otnositel'-nosti (1934) [The Geometric Interpretation of Matter in the GTR],' in *Ibid.*, pp. 111–116.

160. Fok, V.A., 'Za podlinno-nauchnuiu sovetskuiu knigu (o knigakh A.K. Timiriazeva i V.F. Mitkevicha) [For a Truly Scientific Soviet Book (about the books of A.K. Timiriazev and V.F. Mitkevich)],' *Sorena*, 1934 (3), pp. 132–136.

161. Eddington, A.S., *Teoriia otnositel'nosti [RT]*, (Moscow-Leningrad, 1934), 508 pp.

162. Bronshtein, M.P., *Stroenie veshchestva [The Structure of Matter]*, (Leningrad-Moscow, 1935), 244 pp.

163. Gerasimovich, B., 'Vselennaia [The Universe],' *Izvestiia TsIK*, No. 108, May 3, 1935.

164. Rumer, Iu., 'Teoriia otnositel'nosti [RT],' in *Ibid.*, No. 247, October 22, 1935, and No. 273.

165. *Printsip otnositel'nosti. Sbornik rabot klassikov reliatvizma [The Principle of Relativity. A Collection of Works of the Classics of Relativism]*, eds. V.K. Frederiks and D.D. Ivanenko, (Moscow-Leningrad, 1935).

166. V.K. Frederiks, 'Sovremennoe sostoianie voprosa o teorii otnositel'nosti (k 30-letiiu teorii otnositel'nosti) [The Current State of the Question of the TR (On the Thirtieth Anniversary of RT)],' *Sorena*, 1935 (8), pp. 3–17.

167. Einstein, A., *Osnovy teorii otnositel'nosti [Foundations of RT]*, (With appendix by M.P. Bronshtein) (Moscow-Leningrad, 1935).

168. Bronshtein, M.P., 'Quantentheorie des schwacher gravitationsfeld,' *Phys. Z. Sow.*, 1936 (9), pp. 140–157; 'Kvantovanie gravitatsionnykh voln [Quantization of Gravity Waves],' *ZhETF*, 1936, **6**, pp. 195–236. (A part of this article is found in the collection [212].)

169. Bronshtein, M.P., 'O vozmozhnosti spontannogo rassheplennia fotonov [The Possibility of Spontaneous Splitting of Photons],' *ZhETF*, 1937, **7**, pp. 335–356.

170. Vavilov, S.I., 'Pamiati P.N. Lebedeva [Remembrances of P.N. Lebedev],' *Priroda*, 1937 (5), pp. 94–96, as cited in Vavilov, *Sobranie sochineniia [Collected Works]*, Vol. 3 (Moscow, 1956), pp. 165–167.

171. Vavilov, S.I., 'Po povodu knigi akad. V.F. Mitkevicha *Osnovnye fizicheskie vozzreniia* [In Regard to Academician V.F. Mitkevich's Book *Fundamental Physical Views*],' in *Pod znamenem marksizma*, 1937 (7), pp. 56–63.

172. Ioffe, A.F., 'O polozhenii na filosofskom fronte sovetskoi fiziki [The Position on the Philosophical Front in Soviet Physics],' *Pod znamenem marksizma*, 1937 (11–12), pp. 131–143.

173. Tamm, I.E., Frank, I.M., 'Kogerentnoe izluchenie bystrogo elektrona v srede [Coherent Radition of a Fast Electron in a Medium],' *Doklady akademii nauk*, 1937, **14**, pp. 107–112.

174. Gaukhman, R.P., *Bibliografiia pechatnykh trudov fizikov MGU za period* 1917–1937 *gg.* [*A Bibliography of Printed Works of Physicists of Moscow State University*, 1917–1937], (M: 1939), 469 pp.

175. Fok, V.A., 'O dvizhenii konechnykh mass v obshchei teorii otnositel'nosti [The Motion of Finite Masses in GTR],' *ZhETF*, 1939, **9**, p. 375.

176. Fok, V.A., 'Al'bert Einshtein (po povodu 60-letiia so dnia ego rozhdeniia [Albert Einstein: On the Occassion of His 60th Birthday],' *Priroda*, 1939 (7), 95–97.

177. *Poggendorffs biographisch-literarisches Handworterbuch für Mathematik, Astronomie, Physik u.s.w.*, Bd. 6 (Berlin, 1937–1939).

178. 'Fizika v moskovskom universitete, 1755–1940 [Physics in Moscow University 1755–1940],' in *Uchenye zapiski MGU*, Iubileinaia seriia, Vyp. 52. Fizika. 1940.

179. Landau, L.D., Lifshits, E.M., *Teoriia polia [Field Theory]*, (Moscow-Leningrad, 1941); 6th Edition: (Moscow, 1973).

180. Einstein, A., 'O "kosmologicheskoi probleme" [On the "Cosmological Problem"]', in *Sobranie nauchnykh trudov [Collected Scientific Works]*, Vol. 2, pp. 597–613. (See 'On the Cosmological Problem' in the *The Meaning of Relativity*, second edition (Princeton, 1945).)

181. Gaukhman, R.P., *Materialy k bibliografii po istorii russkoi nauki. Fizika [Materials for the Bibliography of the History of Russian Science. Physics]*. (Moscow, 1948), 132 pp.

182. *Ocherki po istorii fiziki v Rossii [Essays on the History of Russian Physics]*, ed. A.K. Timiriazev, (Moscow, 1949), 132 pp.

183. Lazarev, P.P., *Ocherki istorii russkoi nauki [Essays on the History of Russian Science]*, (Moscow-Leningrad, 1950), 248 pp.

184. Fok, V.A., 'Nekotorye primeneniia idei neevklidovoi geometrii lobachevskogo k fizike [Several Applications of the Ideas of Non-Euclidean Geometry of Lobachevskii in Physics],' in Kotel'nikov, A.P., Fok, V.A., *Nekotorye primeneniia idei lobachevskogo v mekhanike i fizike [Several Applications of the Ideas of Lobachevskii in Mechanics and Physics]*, (Moscow-Leningrad, 1950), pp. 48–87.

185. Gerasimova, V.M., *Ukazatel' literatury po geometrii lobachevskogo i razvitiiu ee idea [Directory of Literature on Lobachevskian Geometry and the Development of Its Ideas]*, (Moscow, 1952), 192 pp.

186. Fok, V.A., *Teoriia prostranstva, vremeni i tiagoteniia [The Theory of Space, Time and Gravitation]*, (Moscow, 1955), 554 pp.

187. Rumer, Iu. B., *Issledovaniia 5-optike [Research on Five-Dimensional Optics]*, (Moscow, 1956).

188. *Filosofskie problemy sovremennogo estestvoznaniia. Trudy vsesoiuznogo soveshchaniia po filosofskim voprosam estestvoznaniia [Philosophical Problems of Contemporary Science. Works of the All-Union Conference on Philosophical Problems of Science]*, (Moscow, 1959).

189. Ioffe, A.F., 'Al'bert Einshtein,' *UFN*, 1960, **71**, pp. 3–7. Reprinted in A.F. Ioffe, *O fizike i fizikakh [About Physics and Physicists]*, (Leningrad, 1977), pp. 224–230.

190. *Istoriia estestvoznaniia v rossii [The History of Science in Russia]*, Vol. 2 (Moscow, 1960), 703 pp. Chapter 4 physics is written by P.S. Kudriavtsev, O.A. Lezhneva, and A.E. Medunin.

191. Fok, V.A., 'Raboty A.A. Fridmana po teorii tiagoteniia Einshteina [The Works of A.A. Friedmann on Einstein's Theory of Gravitation],' (1963), in [193: pp. 398–402].

192. Sominskii, M.S., *Abram Fedorovich Ioffe* (Moscow-Leningrad, 1965), 644 pp.

193. Friedmann, A.A., *Izbrannye trudy [Selected Works]*, (Moscow, 1966), 462 pp.

194. Kravets, T.P., *Ot n'iutona do vavilona. Ocherki i vospominaniia [From Newton to Vavilov. Essays and Reminiscences]*, (Leningrad, 1967), 592 pp.

195. Iushkevich, A.P., *Istoriia matematiki v rossii do 1917 goda [The History of Mathematics in Russia to 1917]*, (Moscow, 1968), 592 pp.

196. Lipshits, A.M., Rashevskii, P.K., *Veniamin Fedorovich Kagan*, (Moscow, 1969), 44 pp.

197. Gamov, G.A., *My World Line. An Informal Autobiography* (New York, 1970), 184 pp.

198. Kudriavtsev, P.S., 'Stanovlenie sovetskoi fiziki [The Establishment of Soviet Physics],' in *Osnovateli sovetskoi fiziki [The Founders of Soviet Physics]*, (Moscow, 1970), pp. 5–82.

199. Putiata, T.V., Fradlin, B.N., *Iaroslav Ivanovich Grdina*, (Moscow, 1970), 112 pp.

200. Gulo, D.D., *Nikolai Alekseevich Umov*, (Moscow, 1971), 320 pp.

201. Itenberg, I. Ia., 'Erenfest i teoriia otnositel'nosti [Ehrenfest and the TR],' in P.S. Erenfest, *Otnositel'nosti. Kvanty. Statistika [Relativity, Quanta and Statistics]*, (Moscow, 1972), pp. 301–307.

202. Mandel'shtam, L.I., *Lektsii po teorii otnositel'nosti v kvantovoi mekhanike [Lectures on the TR in Quantum Mechanics]*, (Moscow, 1972).

203. Delokarov, K. Kh., *Filosofskie problemy teorii otnositel'nosti (na materiale filosofskikh diskusii v sssr v 20–30-e gody) [Philosophical Problems of the TR (On the Basis of Material of the Philosophical Discussions in the USSR in the 1920s and 1930s)]*, (Moscow, 1973).

204. *Vospominaniia ob A.F. Ioffe [Reminiscences of A.F. Ioffe]*, (Leningrad, 1973), 252 pp.

205. *Erenfest-Ioffe. Nauchnaia perepiska [Ehrenfest-Ioffe. Scientific Correspondence]*, (Leningrad, 1973), 310 pp.

206. Frenkel', V. Ia., 'Novye materialy o diskussii Einshteina i Fridmana po reliativistskoi kosmologii [New Materials on the Discussion of Einstein and Friedmann on Relativistic Cosmology],' in *Einshteinovskii sbornik 1973 [1973 Einstein Collection]*, (Moscow, 1974), pp. 5–18.
207. Pertsov, V., *Maiakovskii. Zhizn' i tvorchestvo [Maiakovskii. Life and Works]*, Vol. 3, (Moscow, 1976), p. 315.
208. Frenkel', V.Ia., 'Einshtein i sovetskie fiziki [Einstein and Soviet Physics],' *Vorposy istorii estestvoznaniia i tekhniki*, Vyp. 3 (52), (Moscow, 1976), pp. 25–30.
209. Rumer, Iu. B., 'Neizvestnye fotografii A. Einshteina [Unknown Photographs of Einstein],' *Priroda*, 1977 (9), pp. 108–111.
210. Frenkel', V. Ia., *Paul' Erenfest*, (Moscow, 1977), 192 pp.
211. Shirokov, A.P., 'Razvitie geometricheskikh idei lobachevskogo v kazanskom universitete [The Development of Geometrical Ideas of Lobachevskii in Kazan University],' in *Vsesoiuznaia nauchnaia konferentsiia po neevklidovoi geometrii "150 let geometrii lobachevskogo". Plenarnye dolkady [The "150th Anniversary of the Geometry of Lobachevskii" All-Union Scientific Conference on Non-Euclidean Geometry. Plenary Papers]*, Kazan, June 30–July 2, 1976. (Moscow, 1977), pp 22–31.
212. *Al'bert Einshtein i teoriia gravitatsii [Einstein and the Theory of Gravitation]*, (Moscow, 1979).
213. Kedrov, B.M., 'Ob Einshteine i o vzgliadakh Engel'sa i Lenina [Einstein and the Views of Engels and Lenin],' *Voprosy filosofii*, 1979 (3), pp. 21–31.
214. Stachel, J., 'Einstein and the Rigidly Rotating Disk,' in *General Relativity and Gravitation*, Vol. I, ed. A. Held, (New York, 1980), pp. 1–15.
215. Vizgin, V.P., 'Leninskii analiz sostoianiia fiziki na rubezhe XIX i XX vv. [The Lenin's Analysis of the State of Physics at the Turn of the 19th and 20th Centuries],' in *Leninskoe filosofskoe nasledie i sovremennaia fizika [The Lenin's Philosophical Heritage and Contemporary Physics]*, (Moscow, 1981), pp. 222–262.
216. Vizgin, V.P., *Relativistskaia teoriia tiagoteniia (istoki i formirovanie, 1900–1915 gg.) [Relativistic Theory of Gravity (Sources and Formation, 1900–1915)]*, (Moscow, 1981), 332 pp.
217. Rytov, S.M., 'Iz davnikh vremen [From Distant Times],' in *Vospominaniia o I.E. Tamme [Reminiscences of I.E. Tamm]*, (Moscow, 1981), pp. 185–188.
218. Gorelik, G.E., *Pochemu prostranstvo trekhmerno? [Why Is Space Three Dimensional?]*, (Moscow, 1982).
219. Gorelik, G. E., *Razmernost' prostranstva [The Dimensionality of Space]*, (Moscow, 1983), esp. Chapter Five, 'Kvantovye granitsy klassicheskoi geometrii GTR [Quantum Limits of Classical Geometry of the GTR],' pp. 73–114.
220. Pauli, W., *Teoriia otnositel'nosti [The RT]*, 2nd Edition (Moscow, 1983), 336 pp. (The first German edition was published in 1921; the first Russian edition in 1947).
221. *K.E. Tsiolkovskii v vospominaniiakh sovremennikov [K.E. Tsiolkovskii in the Reminiscences of Comtemporaries]*, (Tula, 1983), 289 pp.
222. Frank, I.M., 'Razvitie predstavlenii o priprode izlucheniia vavilova-cherenkova [The Development of the Concepts of the Nature of Vavilov-Cherenkov Radiation],' *UFN*, 1984, Vol. 143, pp. 111–127.
223. Vizgin, V.P., *Edinye teorii polia v 1-oi treti XX v. [Unified Field Theory in the First Third of the 20th Century]*, (Moscow, 1985).

224. Lezhneva, O.A., 'Maksvell v rossii [Maxwell in Russia],' in *Maksvell i razvitie fiziki XIX–XX vekov [Maxwell and the Development of Physics in the 19th and 20th Centuries]*, (Moscow, 1985).

Note added in proof. Additional materials on the reception of TR in Russia see in "Einsteinovskii sbornik, 1984–1985", Moscow, 1987 (in press).

Academy of Sciences of the U.S.S.R.
Institute of the History of Science and Technology

BRONISŁAW ŚREDNIAWA

THE RECEPTION OF THE THEORY OF RELATIVITY IN POLAND

INTRODUCTION: THE GENERAL SITUATION OF SCIENCE IN POLAND IN THE NINETEENTH AND FIRST HALF OF THE TWENTIETH CENTURY

The situation of science in Poland in the nineteenth and in the first years of the twentieth century was not favorable. From the last years of the eighteenth century the whole territory of Poland had been divided and occupied by Russia, Prussia and Austria. The occupying powers were by no means interested in the development of science in the annexed country. Therefore, many decades of the nineteenth century were occupied by the struggle of the Polish community for preservation and re-Polonization of academic schools. This struggle succeeded only in the Austrian sector. There, at the turn of the twentieth century, four academic schools were active where science was taught and scientific research was performed, among them the two universities: Jagellonian University in Cracow (founded in 1364) and Lvov University (established in 1661). Physics and mathematics were also taught and cultivated in the Technical Academy in Lvov and in the Agricultural Academy at Dublany, near Lvov.

A very important role in the development of Polish science was played by the Academy of Sciences and Letters in Cracow, founded in 1873 as the continuation of the Cracow Scientific Society, established in 1816.[1] When Poland regained independence, this institution was revived in 1919 as the Polish Academy of Sciences and Letters.[2] The Academy, connected closely with the Jagellonian University, was the leading all-Polish scientific institution, active until 1951 when, after the First Congress of Polish Science, it was liquidated and the Polish Academy of Sciences was founded.[3]

In the Russian sector science was repressed by the tsarist authorities. Vilna University, established in 1578 by the Polish king Stefan Batory, was closed in 1831. Later it was reestablished in 1862–1869 as the "Main School" and then turned into Warsaw Russian University, boycotted by the Poles. When in 1915, during the First World War, the Russian army

327

Thomas F. Glick (ed.), The Comparative Reception of Relativity, 327–350.
© 1987 *by D. Reidel Publishing Company.*

evacuated Warsaw, the Polish Warsaw University was refounded. In the same year the Polish Warsaw Technical School was opened.

The closing of the Main School, however, did not end scientific activity in Warsaw. The Polish community founded institutions and societies where unofficial scientific and educational work was continued and editorial activity was cultivated.[4] In the Prussian sector of Poland no Polish academic school was active. The University of Breslau was a German university until 1945 and played no role in the development of Polish science.

When Poland regained independence in 1918 the conditions under which Polish science existed changed radically. Besides the hitherto existing scientific centers in Cracow, Lvov, and Warsaw, Vilna University was refounded and a new university created in Poznan.

1. THE EARLY RECEPTION OF RELATIVITY AT THE JAGELLONIAN UNIVERSITY BY AUGUST WITKOWSKI

Let us first deal with the reception of relativity among Polish physicists. The earliest traces of interest in relativity could be seen in Cracow. The Jagellonian University, active without break in spite of many obstacles and difficulties set before it by the Austrian authorities, became after its full re-Polonization in 1870 the leading all-Polish center of science. There in 1883 the professor of experimental physics Zygmunt Wróblewski and professor of chemistry Karol Olszewski liquified nitrogen and oxygen and, later, other gases also. Wróblewski died in 1888 and in the same year August Witkowski (1854–1913) from Lvov Technical University was appointed as his successor.[5] In Cracow, Witkowski continued Wróblewski's experimental research, investigating mainly thermodynamic properties of air. But Witkowski was also deeply interested in theoretical physics. He included theoretical considerations in his experimental papers and lectured on both experimental and theoretical physics.

Witkowski possessed remarkable physical intuition and the ability to understand new ideas. He was one of the first physicists who recognized the significance of Einstein's fundamental paper, "Zur Elektrodynamik bewegter Körper"[6], in which the principles of the special theory of relativity were formulated. As is well known, this paper, which appeared in 1905, did not evoke any significant repercussion in scientific literature for some years after its publication. In 1906, for example, only

Planck, in a paper entitled "Das Prinzip der Relativität und die Grund-lagen der Mechanik", alluded to Einstein's theory. In the introduction he noted that "the principle of relativity introduced recently by H.A. Lorentz and in more general formulation by A. Einstein, if confirmed, will allow for the considerable simplification of all problems of the electrodynamics of moving bodies."[7]

But, as Leopold Infeld wrote many years later:[8]

Yet I know that there were physicists who read Einstein's paper very carefully in the interim and who saw in it the birth of a new science. My friend Professor Loria told me how his teacher, Professor Witkowski (and a very great teacher he was!), read Einstein's paper and exclaimed to Loria: "A new Copernicus has been born! Read Einstein's paper." Later, when Professor Loria met Max Born at a physics meeting, he told him about Einstein and asked Born if he had read the paper. It turned out that neither Born nor anyone else had heard about Einstein. They went to the library, took from the book-shelves the seventeenth volume of *Annalen der Physik* and started to read Einstein's article. Immediately Max Born recognized its greatness and also the necessity for formal generalizations. Later Born's own work on relativity theory became one of the most important early contributions to this field of science.

In 1908, well after Loria's talk with Born and before the appearance of Born's 1909 paper, only one article was published in which the author refers to Einstein. It was Minkowski's paper, "Die Grundgleichungen für die elektromagnetischen Vorgänge in bewegten Körpern".[9] In this article, Minkowski based his considerations on Lorentz's papers, but remarked that "A. Einstein expressed most distinctly that this postulate [i.e., that of relativity] is not an artificial hypothesis but rather a new one, imposed by phenomena, the concept of time." Then Born's paper of 1909 and his further work on relativity[10] in the next few years, together with Hermann Minkowski's famous lecture, *Raum und Zeit*, delivered at the 80th Congress of the German Society of Naturalists and Physicians in 1908, contributed in a decisive way to the propagation of relativity.

The fact that Witkowski quickly became aware of the significance of the theory of relativity was not accidental. For many years he had been interested in the foundations of physics and especially in the concept of the ether, to which he devoted some of his philosophical lectures. As early as 1887 he presented in a lecture "On Recent Views of the Theory of Light", held at the General Assembly of the Copernicus Society of Polish Naturalists, the contemporary situation of the theory of the ether as the medium through which energy is propagated.[11] At this time he was inclined to accept the theory of the ether, saying: "The principal

BRONISŁAW ŚREDNIAWA

assumption of the theory of the ether consists in accepting that the so-called ether exists; all people agree with this assumption and today it is almost certain." But referring to Maxwell's theory of electromagnetism, Witkowski stressed in his lecture that "it cannot be denied that the further development of electromagnetic theory can bring great profits to the theory of light and electricity as well. But in virtue of what we presently know about this theory, we can presume that it will be a transient one and in the end will accord with the elastic theory, probably contributing to the explanation of electrical phenomena."

Witkowski's doubts about the real existence of the ether were expressed in his lecture entitled "Ether", delivered in 1902.[12] There he discussed the difficulties connected with the concept of a moving ether, drawing conclusions from Lodge's and Michelson's experiments and from the phenomenon of the aberration of light. He concluded that "while the concept of matter is the result of our external, sensorial experience, so the concept of ether is the white board on which our mind draws a colorful picture of mutual relations between material bodies. Such an ether is entirely sufficient for us. In such a sense the ether surely is, was and will be."

In 1909, at a session of the Cracow Academy of Science, Witkowski delivered a lecture "On the Principle of Relativity", in which he presented, in a clear and simple way, the ideas of special relativity.[13] In it, he discussed the problem of relative motion and the Michelson-Morley experiment, mentioned the Lorentz contraction, and then formulated Einstein's axiom of relativity and presented its consequences: the contraction of measuring rods and the lengthening of time intervals. The lecture closed with remarks on the theory of the ether. Witkowski stated that "We needed the ether so long as we strayed through the forest of various loose facts with no guiding thread or clear idea. After regaining our eyesight and understanding their mutual relations, let us restrict ourselves to the conclusion that light phenomena occur to us in such a way as if they consisted in the motion of waves running from one place to another at the known, vast velocity. Let us not ask about anything more, because we shall get no more knowledge altogether. From these entities we should erase the ether definitively and firmly." Yet many years of the habit of using the concept did not allow Witkowski to abandon the ether decisively, since he immediately added: "It will, however, not disappear from science. The textbooks of the optics of bodies at rest will not change in any detail. They will preserve the ether,

I presume, forever, but rather as a concept of didactic nature, as a means for visualizing, not for explaining, the laws of nature."

2. EARLY POLISH PUBLICATIONS AND LECTURES ON RELATIVITY

The text of the above-mentioned lecture was not, however, the earliest Polish publication on relativity. It was Jakob Laub (1881–1962) who published the first Polish article on the subject.[14] Laub began his studies at Jagellonian University and continued them in Vienna and in Göttingen, where he became an ardent adherent of the theory of relativity. Next, he worked in Würzburg. In 1908 he published the paper "Zur Optik bewegter Körper", in which he discussed the principles of special relativity and calculated Fresnel's convection coefficient.[15] Earlier, in July 1907, he had reported the results of this work at the Tenth Congress of Polish Physicians and Naturalists in Lvov, evoking discussion among physicists and mathematicians.[16] Laub published the contents of this work in a comprehensive article in Polish entitled "Contributions to the Electrodynamics of Moving Bodies."[17] Soon Laub became Einstein's collaborator, writing three papers jointly with him.[18]

In Cracow, original work on relativity was initiated by Witkowski's former assistant, the physicist and oculist Kamil Kraft (1873–1945), who published five papers on special relativity in 1911 and 1912.[19] In two of them he derived an identity for expressing the D'Alembertian of arbitrary bivector by four-dimensional divergences and curls and applied this identity to the solution of Maxwell's equations. He obtained among other results a formula equivalent to the so-called Sommerfeld fundamental formula.[20] In the next two papers, Kraft calculated in a simple way the coefficients of the Lorentz transformations between two reference systems with non-parallel axes, simplifying considerably the formulas derived earlier by Minkowski.[21] Then he derived the expressions for the components of electric and magnetic quantities in moving material media. In his last paper on relativity, Kraft formulated and proved the following theorem: If two (rectangular) coordinate systems S (x, y, z, t) and S' (x', y', z', t') are in uniform translational relative motion with constant velocity, then for appropriate choice of the origins of the systems and of the units of time and length the following relation holds

$$\varepsilon \ (x'^2 + y'^2 + z'^2) - t'^2 = \varepsilon \ (x^2 + y^2 + z^2) - t^2 \qquad (1)$$

where ε is a constant. Here Kraft did not assume that the velocity of light is the same in all inertial coordinate systems, and therefore his relation is more general than the relation

$$x'^2 + y'^2 + z'^2 - c^2 t'^2 = x^2 + y^2 + z^2 - c^2 t^2 \qquad (2)$$

(where c is the velocity of light), valid in special relativity. For $\varepsilon = 11c^2$, formula (1) gives the Lorentz transformation and consequently the laws of the special theory of relativity, and for $\varepsilon = 0$, it yields the Galileo transformation and thus Newtonian mechanics. The first of Kraft's papers was cited by Wilson and Lewis[22], the last one by Infeld.[23]

At the same time, Ludwik Silberstein (1872–1948), former *privatdocent* at Lvov University, who was then lecturer of theoretical physics at the University of Rome, published two papers: "The Quaternionic Form of Relativity" and "Second Memoir on Quaternionic Relativity."[24] In the first, he defined quaternions corresponding to four-vectors and called them physical quaternions; derived the Lorentz transformation in quaternionic form; and then, referring to some formulae which he had previously derived[25], calculated the equations of electromagnetic potentials and Lorentz force in quaternionic form. In the second paper he calculated the quaternionic form of force, stress and density of energy. He presented the first paper to the Warsaw Scientific Society in 1911, also publishing it in Polish in the reports of this Society.[26]

In the same period, Silberstein published a textbook, *Electricity and Magnetism*, in Polish.[27] In the second volume of this book he included remarks about relativity and announced that in the third volume a chapter on relativity would appear. But the first part of this volume (the only part published) appeared without the announced exposition of relativity.

In 1911 two more articles about relativity appeared in Polish. Czeslaw Bialobrzeski (1878–1954), a *privatdocent* at Kiev University (later professor of theoretical physics at Warsaw University), published an article entitled "The Principle of Relativity and Some of its Applications," a clear elucidation of the principle of special relativity.[28] Likewise the Polish physicist Henryk Merczyng (1860–1916), professor of electrotechnics and mechanics at the Technical University of St. Petersburg, wrote an article "On the Principle of Relativity in the Physical Meaning of Time and Space," in which he presented the principles of special relativity, illustrating them with numerical examples, showing the differences between the results of Newtonian and relativistic physics.[29]

Marian Smoluchowski, a convinced adherent of relativity, wrote about it in his *Manual for Self-Study* (which was for many years very popular among students and young scientists): "a new branch of physics has emerged, the so-called theory of relativity. It is a domain of basic significance for the whole structure of theoretical physics, since it forms, in a certain sense, a new synthesis of all phenomena depending on motion, entailing a major revision of the old concepts of time and space. . . . In such an excellent way it creates uniformity and harmonic simplicity [in physics] whereas, without it, the chaotic complexity of different mutually-unconnected phenomena would reign."[30]

3. LEOPOLD INFELD'S DISSERTATION

Kraft and Silberstein published their papers in 1911 and 1912. No further original paper on relativity appeared in Polish until 1921. But the theory of relativity was not absent from university courses, although not the sole subject of any separate course. In Cracow, relativity was included in the five-year cycle of lectures delivered by the professor of theoretical physics Wladyslaw Natanson (1864–1937). One of Natanson's students was Leopold Infeld (1898–1968) who wrote in his memoirs[31]:

I heard Einstein's name for the first time in the second-year [course on theoretical mechanics] at the Jagellonian University. At the end of the academic year, Professor Natanson devoted several hours to Einstein's special relativity. I heard for the first time about the Lorentz transformation as formulated by Einstein. These lectures were a revelation to me . . . I remember that Professor Natanson called Einstein "a genius among geniuses". . . . I was not sufficiently prepared to understand fully the structure of the theory of relativity, but I knew that I would return to it.

The lectures described by Infeld were delivered in the academic year 1917–18, and it is very probable that Natanson also lectured on relativity in his earlier courses on theoretical mechanics.

After four years of study Infeld moved to the University of Berlin, where he attended the lectures of Laue and Planck and got acquainted with Einstein in the winter semester 1920–21. During his stay in Berlin Infeld wrote his doctoral dissertation entitled "Light Waves in the Theory of Relativity," which he presented in Cracow to Professor Natanson.[32] Infeld obtained his doctorate in 1921.

His dissertation was composed of two parts. The first was devoted to special relativity. There, Infeld restricted himself to geometrical optics, considering waves of very high frequencies. He showed that in this case

Maxwell's equations can be solved when the eikonal equation is satisfied. This implies that the three-dimensional Poynting vector has the same direction as a light ray in a moving material medium. The second part of Infeld's dissertation was devoted to general relativity. He showed that for high-frequency waves in a static gravitational field, the Poynting vector also exists and is parallel to the light vector (whose world line is a geodesic). Thus Infeld proved that geometrical optics is the zero-order approximation to Maxwell's equations in special, as well as in general, relativity.

It is interesting that even at that time the remnants of the terminology of the ether can be traced, since Infeld wrote in his dissertation: "Propagation of each action, whose background is the aether, can take place only along null geodesic lines in our four-dimensional continuum." Infeld's dissertation was the first Polish paper concerning general relativity.[33]

4. RELATIVITY IN THE ACADEMIC SCHOOLS OF WARSAW, LVOV AND VILNA

The theory of relativity was fully accepted by physicists in other Polish academic centers. At the reestablished Warsaw University the first lecture about general relativity was delivered in 1917 by the professor of experimental physics Stefan Pieńkowski, immediately after his return from Belgium to Warsaw. Some years later, lecture courses devoted to special and general relativity began at Warsaw University, the first delivered in the years 1923–24 and 1924–25 by the mathematician Witold Pogorzelski (1895–1963). Pogorzelski had completed his *habilitation* in theoretical physics at the Jagellonian University in 1921;[34] as part of the *habilitation* procedure he lectured on "The Principle of Relativity and Electromagnetic Phenomena". Lecture courses on relativity were also given by Czeslaw Białobrzeski, who in 1921 became professor of theoretical physics at Warsaw University; and by *privatdocent* Myron Mathisson who, in the academic year 1935–36, lectured on relativity and cosmology.[35] Notes from Pogorzelski's lectures were elaborated and edited by Stanislaw Wahrhaftmann as *Outline of the Theory of Relativity*.[36] This was the only Polish textbook on relativity until 1956 when Bazański's *Classical Field Theory*, written according to Infeld's lectures, appeared.[37]

In the early 1920s we must note one paper on general relativity written by Gustaw Doborzyński, a teacher in the Technical High School of Wawelberg and Rotwand. The paper was entitled "Invariants of

General Relativity and Schwarzschild's Formula."[38] After reviewing Schwarzschild's calculation, the author derived the formula by applying the principle of equivalence in a simpler way than in Schwarzschild's original calculations. In the Free Polish University in Warsaw only one seminar in relativity, directed by Professor Stanislaw Kalinowski in 1937–38, can be noted.

The first lecture course on relativity in Lvov was given by Professor Wojciech Rubinowicz (1889–1974) at the General Faculty of Lvov Technical University in 1921–22 (and repeated over the next years until this faculty was abolished in 1934). Courses on relativity at Lvov University started much later, namely in 1931–32. They were given by the *docent* Leopold Infeld and, the following academic year, by *docent* Szczepan Szczeniowski.

In Vilna, relativity was cultivated by Jan Weyssenhoff (1889–1972).[39] In 1914–1919 Weyssenhoff had been in Zurich, where he had the opportunity to meet Einstein. Many years later he recalled his acquaintance with Einstein in an article entitled "Remarks about Einstein's Life and Activity Based on my own Reminiscences."[40] Their first short meeting, during which Weyssenhoff told Einstein about his doctoral dissertation on quantum theory, took place in 1916. Closer scientific contacts between them followed during Einstein's two-month stay in Zurich in 1919. In the above-mentioned article Weyssenhoff wrote: "After two years Einstein returned to Zurich in order to give a two-month series of lectures about both theories of relativity in January and June 1919. At that time he attended physics seminars, in which I was one of the two most eager speakers and then, particularly during common meetings in a café, I had the opportunity for closer contacts with Einstein."

Once appointed professor at Vilna University, Weyssenhoff began to give courses on relativity, the first in the academic year 1923–24 (and then, from 1927–28, every year until 1934–35). When in the latter year Weyssenhoff was called to the Jagellonian University, the relativity course was continued by his successor in Vilna, Professor Szczeniowski. Weyssenhoff also delivered his course on relativity in Cracow in 1936–37.[41]

5. ATTITUDE OF POLISH MATHEMATICIANS TOWARDS RELATIVITY

The theory of relativity also awoke interest among Polish mathematicians. As early as 1909, Minkowski's lecture *Raum und Zeit* was translated into Polish.[42] In 1913 the translation of Varičak's "Nichteuklidische

Interpretation der Relativitätstheorie" from the *Jahresbericht der Deut-schen Mathematiker Vereinigung* of 1912 was published in Polish.[43] In 1918 the young Warsaw mathematician Stanislaw Sachs delivered two lectures about Einstein's theory of gravitation at the Mathematico-Physical Student Circle of Warsaw University, which concluded with remarks on the possibilities for verification of the general theory. These lectures were subsequently published.[44] There were, however, no origi-nal papers on relativity published by Polish mathematicians until the 1920s.

Among the overwhelming majority of Polish mathemeticians relativ-ity encountered no objections. There was, however, one important exception: the objections raised by the outstanding Cracow mathemati-cian and professor at the Jagellonian University, Stanisław Zaremba (1863–1942), expressed in his paper entitled "The Theory of Relativity against the Facts Stated by Experiment and Observation" (1922) in which he declared that the way the foundations of relativity were formulated was unsatisfactory.[45] The contents of this paper and the discussion which ensued are interesting because they clearly demon-strate the difference between the modes of thought of physicists and of a certain kind of mathematician.

Zaremba's mathematical activity was strongly connected to physics. In his autobiography he wrote: "I have from my youth been attracted by problems arising in mathematical physics. This very fact, and my deep conviction that investigations in the field of mathematical physics attain their full scientific value when they are completely exact, determined the character of my activity."[46] Zaremba's main interests and achieve-ments lay in the theory of partial differential equations of the second order. He also published papers and textbooks on physics. In these he assumed the point of view of a mathematician and discussed the founda-tions of physical theories, attempting to axiomatize them.

The above-mentioned paper reflected this orientation of Zaremba's research and expressed his initially critical attitude towards relativity. Its aim was "to investigate thoroughly whether the theses, extracted by relativists as the logical consequences of the premises of relativity, really follow from these premises." The first part of the paper was devoted to attempts to axiomatize the foundations of both special and general relativity and, in particular, to a discussion of the properties of space-time. In the second part of the paper, Zaremba examined the connec-tion between relativistic concepts and experiments, concluding that because of the non-existence of rigid bodies in special relativity the principal concepts of this theory cannot be confirmed experimentally. Also in the frame of general relativity, the shift of Mercury's perihelion cannot be interpreted in a unique way because, in this theory, the distance cannot be defined univocally. But Zaremba stressed that his aim was by no means to reject the theory of relativity. He wrote that "according to the foregoing state of affairs one can neither confirm nor deny the facts stated by any experiment or observation. To make such verifications possible, the set of premises should be gradually supplemented."

Zaremba's paper provoked a sharp controversy in 1923 with Tadeusz Banachiewicz, professor of astronomy at the Jagellonian University. In the course of the polemic, Banchiewicz published an article wherein he stressed that the concept of the rigid body is unnecessary in relativity, and Zaremba's definition of a rigid body was only an approximation, inadmissible when one considers phenomena explained by the theory of relativity.[47] Moreover, some of Zaremba's axioms concerning measurements were arbitrary and inappropriate to physical situations in which measurements are performed. A shortened version of Zaremba's paper, published in *Scientia* under the title "An Attempt to Explain the Standpoint of the Theory of Relativity", evoked in similar fashion the criticism to the *Journal de Physique*'s reviewer, J. Rossignol, who wrote that Zaremba, "in order to develop his propositions . . . bases himself only on vague sentences and eschews all accuracy."[48]

But Zaremba himself changed his critical stance towards relativity and became its adherent.[49] His 1924 paper, entitled "On the Mobility of Bodies Underlying Mr. Lorentz' Contraction in the Direction of Velocity", contained no criticism of relativity.[50] Here the author proves valid (under special relativity) the theorem that the body undergoing the Lorentz contraction and moving with respect to the ether performs a uniform motion. This theorem can be formulated in modern terminology to state that a body undergoing a Lorentz contraction performs uniform, rectilinear motion with respect to any inertial system. (Note that in spite of the passage of almost twenty years from the time of the formulation of special relativity, the unnecessary term "ether" was often still used, indicative of the difficulty in shedding the habits of many years.)

Also worthy of mention is a brochure by the mathematician Alojzy Stodółkiewicz, probably a provincial high school teacher, who published his papers in *Mathematical News*.[51] In this brochure, entitled "Remarks about the Theory of Relativity", he criticized the foundations of both theories of relativity, adopting Kant's views of the concept of time, maintaining that geometry was built without connections with experiment and that the laws of physics depend on the conditions in which mankind lives on Earth and therefore are not trustworthy even in the case of classical physics, and much more so in the case of relativity. The contents of this pamphlet cannot be taken seriously.

Wladyslaw Ślebodziński (1884–1972), a specialist in differential geometry from Poznan, also maintained an interest in relativity in the 1920s. In 1924 he published two papers, one on the differential geometry of Riemannian spaces entitled "Contribution à la théorie des courbes et des congruences d'une espace riemannienne",[52] and the second on relativity, "Recherches géometriques sur le champ statique de gravitation",[53] where applying the result of the just mentioned geometrical paper he obtained, besides Schwarzschild's and Treffz' solutions of Einstein's equations for symmetric static gravitational fields, two new solutions for these equations.

Finally, some words about astronomers. In 1921, Banachiewicz gave a lecture in the Polish Academy of Sciences and Letters, where he remarked that the effect of the deflection of light rays in a gravitational field may influence in a somewhat different way the observations of eclipses depending on the distance from the earth of the celestial bodies emitting light.[54]

Three articles about relativity appeared in successive volumes of the Yearbook of the Cracow Observatory – one by the Poznan astronomer Józef Witkowski on relativity in astronomy, discussing the verification of general relativity [55], and two translations: H.N. Russell on "The Modifications of our Concepts about Nature" and E.v.d. Pahlen on "Infinitude of the Universe and the Theory of Relativity."

6. INFELD AND WEYSSENHOFF ON THE FOUNDATIONS OF RELATIVITY, 1927–1935

We have already noted that Infeld and Weyssenhoff addressed the foundations of relativity by responding, directly or indirectly, to Zaremba's critique of 1922. A direct reply to Zaremba's objections was contained in Infeld's two-part paper, "On Measurements in Classical Physics and in the Theory of Relativity."[56] Here he discusses the problem of measurements of time and length in Newtonian and relativistic physics, paying special attention to the definition of the rigid body, taking into account conceptions of Einstein, Hilbert and Eddington. He concluded that the definition of rigid bodies is dependent on the laws of physics, and beyond them rigid bodies are equally impossible to define in classical and in relativistic physics. Classical physics and the theory of relativity both contain implicit definitions of rigid bodies and ascribe different properties to the same really existing objects. Here lies the principal difference in constructing physical and mathematical theories, because in mathematics definitions are formulated independently of theorems, and this is what Zaremba demanded with respect to relativity.

Weyssenhoff's paper, "Commentaries on the Theory of Relativity, I", written in the same year as Infeld's, contained an implicit answer to Zaremba's critique of relativity, although the latters's name was not mentioned.[57] In this paper, Weyssenhoff defines his methodological standpoint precisely. He starts with the remark that there are two ways to establish the foundations of physical theories. One is axiomatization, following the pattern of axiomatization in mathematics. The other is the concrete visual representation of the foundations of physics down to the smallest detail, by means of 'ideal experiments' designed to measure each newly introduced physical quantity. This was the approach accepted by Weyssenhoff and applied to the discussion of the properties of space-time, the fundamental concept of relativity.

Weyssenhoff returned to relativity in a series of papers written in the mid-1930s. In the two-part paper, "Illustrations of the Theory of Relativity," he discusses the components of the metric tensor and their interaction in special and general relativity.[58] In the last paper of this series, "Metric Field and Gravitational Field," he specified the difference between the metric field as the region of space-time and the gravitational field as a metric field in the presence of matter.[59] Weyssenhoff, incidentally, spent the summer months of 1935 at the Institute of Advanced Study in Princeton, where he again met Einstein.

These papers of Infeld and Weyssenhoff opened a new period in the history of relativity in Poland, one of systematic scientific research in this branch of physics. Infeld was working in Cambridge and Lvov on the relativistic theory of elementary particles and, later, in Princeton with Einstein and in Toronto on the problem of motion in general relativity. After his return to Warsaw in 1950, he created an important

center of relativity research there.[60] Weyssenhoff worked in his later
years with collaborators in Cracow on the theory of relativistic spin
particles.[61] During the 1930s, Mathisson (in Warsaw, Cracow and Cambridge) studied the motion of spin particles in gravitational and electromagnetic fields.[62]

7. PROPAGATION OF RELATIVITY AMONG POLISH INTELLECTUALS

Before World War I, interest in relativity among Polish intellectuals was
rather moderate. People interested in the development of science could
find information on relativity and the problems connected to it in the
articles mentioned above or, for instance, in translations of articles by
Poincaré, Abraham, Salpeter and others, published between 1905 and
1914 in the monthly popular science magazine *Wszechświat* [The Uni-
verse].

Broader interest in relativity on the part of intellectuals burst out
suddenly in 1920 after publication of the results of the 1919 eclipse
observations. As early as February 8 and 9, 1920, an article entitled
"Einstein and Newton", by Maria Sulkowska, appeared in the Cracow
newspaper *Czas* [Time]; it gave a report of the meeting of the Royal
Society on November 6, 1919, devoted to the results of the Sobral and
Principe eclipse expeditions confirming the predictions of general relativity. The principles of special and general relativity are also recapitulated correctly in this article.

In the years 1920–1923 the gradual acceptance and adoption of
relativistic ideas followed among circles of intellectuals interested in the
development of natural sciences. In university towns discussions took
place, articles were written in newspapers, popularizing brochures and
books were published, disputes and polemics broke out. The discussion
in Lvov was particularly intense – not suprisingly, because Lvov was a
city where intensive intellectual life flourished. Stanislaw Ulam gives the
following testimony to this fact in his memoirs: "From today's perspective Lwów may seem to have been a provincial city, but it is not so.
Frequently lectures by scientists were held for the general public, in
which such topics as new discoveries in astronomy, the new physics and
the theory of relativity were covered. These appealed to lawyers,
businessmen and other laymen. . . . Around 1919–1920, so much was
written in newspapers and magazines about the theory of relativity that I
decided to find out what it was all about."[63]

In Lvov three ardent and energetic adherents of relativity were especially active at time (1920–22): the physicist Stanisław Loria, then professor of theoretical physics (and, in later years, of experimental physics) at Lvov University; the technician Maksymilian Tytus Huber (1872–1950), the well-known professor of technical mechanics at Lvov Technical University; and the philosopher Zygmunt Zawirski (1882–1948).

The public discussion of relativity started when the philosopher Julian Zachariewicz published an article on "The Theory of Relativity and Albert Einstein" in the Lvov newspaper *Słowo Polskie* [The Polish Messenger] on October 10, 1920. This aggressive article revealed not only misunderstanding of the theory of relativity but also the mathematical ignorance of the author who wrote, attacking Einstein: "Nobody has succeeded in presenting his theory, in recapitulating it, in showing what it relies upon, because in Einstein's books there is no theory. . . . What proofs does Einstein give for all that? He gives none. There are endless terms and mathematical columns designed to prove or supplement the text; we see, indeed, that this mathematics, these figures, replace the text." Huber responded to Zachariewicz's attack in a series of five articles entitled "Albert Einstein and his Theory," published in the same newspaper between November 14 and 19, 1920. He blamed Zachariewicz for ignorance, aversion to understanding new ideas and anxiety about introducing the public to science, and he recapitulated the principles of special and general relativity in brief.

At the end of November and the first days of December 1920, Loria, who had met Einstein in Vienna in 1912 at the Assembly of German Physicists, delivered a series of public lectures entitled "Relativity and Gravitation. A. Einstein's Theory," organized by the Polish Polytechnic Society in Lvov. Both Loria's lectures and the discussion were summarized in *Słowo Polskie* in December 1920. Encouraged by the interest of the audience, Loria published these lectures in a book of the same title.[64] Its first edition was an almost literal repetition of the lectures, supplemented by a description of experiments mentioned in them. In the second edition, however, Loria abandoned the form of the lectures and elaborated the topic in a more extensive way – in fact, a new book of five chapters constituting a useful introduction to the systematic study of relativity.

In the second half of December 1920 a series of four articles appeared in *Słowo Polskie* under the common title "In Defense of the Absolute," written by the engineer Waclaw Wolski. Wolski assumed the concepts of absolute space and time, independent of measurement. From this stance, he rejected relativity and Einstein's law of addition of velocities in particular. Wolski was answered in January 1921 by Zawirski, who published in the same journal an article entitled, "The Arguments in Defense of the Absolute," wherein he stated that concepts which are independent of measurements do not pertain to physics, but to metaphysics. Some days later he continued the discussion in a two-part article entitled "Time and Space in the Representation of Great Philosophers," where among other topics he discussed relativistic concepts of time and space. These articles were also published as a brochure in 1921.

In January 1921 Huber delivered a series of lectures entitled "Time, Space, Matter and the Universe," organized by the Polish Copernican Society of Naturalists and published both in the journal of this society and as a brochure.[65] In February of this year, Huber gave a new cycle of lectures, "On the Theory of Relativity," more than the preceding one, organized by the Literary Scientific Society of Lvov. The text of these lectures was published in *Słowo Polskie* during the first half of that month.

In March 1921, Loria delivered an inaugural lecture at Lvov University entitled "Ether and Matter". In this lecture (also published later as a brochure), he discussed the development of the concept of ether and the reasons why it became unnecessary in relativity.[66] In the same month, Zawirski published a long series of fifteen articles in *Słowo Polskie* (and also in a separate brochure) under the title "The Physical Theory of Relativity and Philosophical Relativism". After some introductory epistemological remarks, he presented Avenarius's psychophysical parallelism, the monistic conception of reality, the views of Kant, Petzoldt, Einstein, Bergson, Schlick and others. The series of articles ended with the conclusion: "We see the extraordinarily great importance of general relativity precisely in the fact that, by breaking with the standpoint of critical realism, which until recently was supported by a physics which believed in empty space and atoms, it speaks extraordinarily in favor of phenomenalism and therefore of the idealistic view of the world."

Huber, in his last popular lecture, delivered on June 20, 1925 with the title "The Role of the Theory of Relativity in the Evolution of Fundamental Concepts of Physics," maintained that Newtonian mechanics (according to the opinion common at that time) will continue to be used in everyday practice and relativity will retain its [special] epistemological meaning.[67]

Interest in relativity was also strong in other Polish towns, though there are no traces of polemics over it in the Warsaw, Cracow or Vilna press of these years. Relativity was propagated in these towns by the local sections of the Polish Physical Society which organized series of popular lectures, usually in the early spring.[68] In Warsaw, Pogorzelski delivered a lecture on "Dynamics of the Electron and the Theory of Relativity" as part of a cycle on the structure of matter, in March 1920. In February and March 1922, a series of lectures devoted to relativity was organized. The speakers and their topics were: S. Kalinowski, "On Gravitation;" Pieńkowski, "Optical Phenomena in Moving Bodies (two lectures); Pogorzelski, "The Theory of Electrons;" and Białobrzeski, "The Theory of Relativity" (two lectures). The average attendance at these lectures was 300 listeners. On the basis of his presentation, Bialobrzeski wrote a book entitled *Popular-Scientific Lectures on Relativity*, in which he presented the topic and its philosophical aspects in a notably clear way.[69] Jan Weyssenhoff gave a series of four lectures on relativity in Cracow in March 1920. The following year, he became a professor at the University of Vilna where, in February and March 1922, he gave another cycle of four lectures on relativity.

We have already mentioned some of the numerous books, brochures and translations that appeared in the early 1920s. The engineer Stefan Berman wrote a popular volume entitled *Introduction to Einstein's Theory of Relativity*.[70] This good, critical exposition of relativity finished with the remark: "The importance of relativity lies principally in the fact that it gives in a consistent way, without referring to hypotheses constructed *ad hoc*, a truthful and uniform picture of hitherto observed physical phenomena. At the same time, it introduces a series of new laws, astonishing in their audacity. Born on the basis of experiment, it announces a series of new experiments absorbing the attention of the whole scientific world."

The last brochure of these years was written by the physicist and writer Bruno Winawer (1883–1944) of Warsaw and was entitled "Still about Einstein. Relativity from the Air."[71] He presented in a brief and joking way the principles of relativity, adding: "We have not yet touched upon one problem, because we don't understand it. Why does this mathematically beautiful theory from time to time provoke protests, objections or furious attacks by quite incompetent people? . . . These are questions which do not pertain to physics but to the psychology of the crowd."

As an example of such an attack, we mention the brochure of Paulin Chomicz entitled *Einstein's Theory of Relativity in the Light of the Absolute Philosophy of Hoene-Wroński.*[72] This author, adopting Hoene-Wroński's mechanics, based on the concept of active force and the force of inertia, admits the possibility of the infinite velocity of light, of the existence of material ether which forms the background of light and matter, and even remarks that ether can be observed in the tails of comets. Nobody answered such lucubrations.

In these years of intense interest in relativity the following books were translated into Polish: Beer's *Relativitätstheorie und ihre historische Grundlage*;[73] Einstein's *Über spezielle und allgemeine Relativitäts-theorie,*[74] *Geometrie und Erfahrung,*[75] and *Äther und Relativitäts-theorie*;[76] Freundlich's *Die Grundlagen der Einsteinschen Relativitäts-theorië*;[77] and Moszkowski's *Einstein, Einblicke in seine Gedankenwelt.*[78]

Although the public polemics about relativity died out gradually in the mid-1920s, intellectuals remained interested in the theory. As examples of popularizing lectures and articles published over the next few years, see three interesting essays by Ślebodziński and Feliks Burdecki in the periodical *Mathesis Polska* in 1926–1928. Interest in relativity was also evident in the theater, although somewhat later. In 1934 the dramaturgist and actor Antoni Cwojdziński (1896–1972), trained as a physicist, wrote a witty comedy called *Einstein's Theory*, in which he satirized the hasty and senseless drawing of absurd conclusions from scientific theories such as relativity.[79] This comedy, which won the Reynal prize, enjoyed a run of 525 performances during the years 1934–1936.

8. PERCEPTION OF RELATIVITY AMONG PHILOSOPHERS

Until the twenties, Polish philosophers did not manifest any considerable interest in natural philosophy, leaving its problems to physicists. In the first two decades of the century mainly Natanson and Smoluchowski were active in natural philosophy.[80] Until 1920 no article on relativity appeared in philosophical journals, no brochure was published, no lectures on relativity at the meetings of the Philosophical Society were noted. Interest in relativity among Polish philosophers was stimulated in 1920, when the public discussion of the theory became general. As we have seen in the preceding section, philosophers also took part in the discussion, some of them as adherents, others as opponents, of relativity. The period of philosophical interest in relativity, which began at this time, extended to the last years of the decade.

Research on philosophical aspects of relativity was undertaken in the 1920s by Zawirski. In a paper, "Philosophical Reflections on the Theory of Relativity" (1920), he discussed the properties of space and time and stated that if the theory of relativity breaks deep-rooted intuitive convictions in a compact and consequent theory which agrees with experiment, then one must give up those convictions. "No theory," he wrote, "can prove that the absolute time of absolute movement doesn't exist; it only proves that these concepts are useless for science."[81]

In 1924, Zawirski published two relevant papers. In the first, entitled "Axiomatic Method and Science," he discussed the problem of the applications of axiomatic method to physics and, in particular, to special and general relativity and to the problem of measurement in relativity.[82] In conclusion, he stated that "physical knowledge does not present absolute reality to us; at most, it only points to it. The mathematical form of the laws of nature may be applied, according to the axiomatic method, to reality, which is independent of consciousness, provided we could interpret them suitably. This defines the subject of physics only when we can ascribe to them some sensual elements, or elements which are connected to them. In this way, sense-data are not only starting points but represent something which provides the essence to the mathematical form of the laws of nature." The second paper of that year was his *Habilitation* lecture entitled "The Connection of the Principle of Causality with the Principle of Relativity."[83] There Zawirski, after referring to the views of Newton, Mach and Einstein, tried to connect causality with continuity. He expressed doubts as to whether the principle of causality is true in inhomogenous gravitational fields. He maintained also that causality in relative motion (for instance, the appearance of centrifugal force) follows from the fact that the mass of the system considered is much smaller than the mass of the Universe.

In later years, Zawirski was interested in the concept of time. To this topic he devoted a paper, "Evolution of the Concept of Time," in 1935 and, in 1936, a monograph bearing the same title.[84] At the same time, Henryk Mehlberg published the papers "Physical and Extraphysical Time" and "Essai sur la théorie causale du temps."[85] In all these papers, the authors refer to the theory of relativity.

During many years Leon Chwistek (1884–1944), professor of logic at Lvov University, philosopher, mathematician, logician, painter and historian of art, was deeply interested in relativity.[86] He began its study in 1909 at Göttingen and then returned to it around 1934. He held numerous discussions with physicists and mathematicians. His attitude towards relativity was critical, inasmuch as he could not agree with the relativistic concept of space-time. He expounded his point of view on space-time in a paper, "Some Remarks on the Fundamental Laws of Propagation of Light" and in his book, *The Limits of Science*.[87] He maintained that two kinds of space-time exist, namely material space-time and the space-time of light, used by astronomers. He proposed to assume in the space-time of light a time other than that of relativity. The views of Chwistek were not highly· regarded either by physicists or philosophers.

Articles popularizing philosophical aspects of relativity were published in *Przegląd Powszechny* [Catholic Review] in 1926–29 by Rev. Feliks Hortyński and in the periodical *Droga* [The Way] by Rev. Augustyn Kubisiak, both adherents of relativity.[88] We may also note the polemical article of Józef Tyszkiewicz in *Catholic Review* entitled "Einsteinism and Christian Philosophy" where the author protested against intemperate drawing of philosophical conclusions from relativity and using them in the struggle with religious views. Anachronistic and inconsistent with the generally accepted views on relativity was the statement of Rev. Kazimierz Wais, author of a textbook for theological seminars entitled *Detailed Cosmology*.[89] There he advocated the theory of the ether and observed that "the theory of relativity has hitherto had many opponents."

In *Philosophical Quarterly*, five reviews of books by Einstein, Bergson, Hennings, Wentzel and Weyl, published in Germany and France, appeared. They were written by Zawirski and Marceli Metallman. Between 1922 and 1926 several lectures on relativity were delivered at sessions of the Philosophical Society in Warsaw, Cracow and Poznan by Zawirski, Hortyński, Kozłowski and the Cracow mathematician Witold Wilkosz.

At the end of the twenties activity to popularize relativity decreased considerably, in the press, as well as at the sessions of the Philosophical Society. The reception of relativity among philosophers and intellectuals had been accomplished.

NOTES

1 Akademia Umiejetnosci w Krakowie.

2 Polska Akademia Umiejetności.

3 Polska Akademia Nauk.

4 K. Maślankiewicz, ed., *Zarys dziejów nauk przyrodniczych w Polsce* (Warsaw, Wiedza Powszechna, 1983), pp. 64–86.

5 K. Zakrzewski, *Wiadomości Matematyczne* [Mathematical News], 17 (1913); 211; and B. Średniawa, *History of Theoretical Physics at Jagellonian University in the 19th and in the First Half of the 20th Century*, Zeszyty Naukowe UJ, Prace fizyczne/ Universitas Jagellonica, Acta Scientiarum Litterarumque, Schedae Physicae, No. 24 (Warsaw-Cracow, Polish Scientific Publishers, 1985), §2,6 and 3,9.

6 *Annalen der Physik*, **17** (1905), 821; **18** (1906), 639.

7 M. Planck, *Verh. der deutschen Physik, Ges.* 8 (1906), 136.

8 L. Infeld, *Albert Einstein: His Work and its Influence on our World* (New York, Scribner's, 1950), p. 44. See Born's important article, "Die träge Masse und das Relativitätsprinzip,' *Annalen der Physik*, **28** (1909), 571.

9 H. Minkowski, *Nachr. von der Königl. Ges. d. Wiss. zu Göttingen phys. math. KL.*, **53** (1908).

10 See *Scientific Papers of Max Born* (Edinburgh, Oliver Boyl, 1953).

11 A. Witkowski, in *Kosmos*, **12** (1881), **71** (in Polish).

12 Witkowski, *Eter* (Cracow, Czasu, 1903).

13 Witkowski, *O zasadzie względności* (Cracow, Akademia Umiejetności, 1909).

14 L. Pyenson, 'Einstein's Early Scientific Collaborators,' *Historical Studies in the Physical Sciences*, **7** (1976), 83–123; and 'The Incomplete Transmission of a European Image: Physics at Greater Buenos Aires and Montreal, 1890–1920,' *Proceedings of the American Philosophical Society*, **122** (1978), 92–114; and B. Średniawa, *Theoretical Physics at Jagellonian University* (n. 5, above), §6,2.

15 J. Laub, *Annalen der Physik.* 23 (1907), 738, and 25 (1908), 175.

16 Laub, *Optyka cial ruchomych, Sprawozdania z posiedzeń naukowych X Zjazdulekavzy i przyrodnikow polskich* (Lvov, Altenberg, 1907–08).

17 Laub, *Prace Matem.-Fiz.*, **19** (1908), 63.

18 Einstein and Laub, *Annalen der Physik*, **26** (1908), 532, 541; **28** (1909), 445.

19 Średniawa, *Theoretical Physics at Jagellonian University*, §4,5; K. Kraft, *Bulletin International de l'Academie de Sciences et de Lettres de Cracovie, Classe de Sciences math. et ant., Serie A, Sciences math.* (1911), 537, 564; 596; (1912), 384, 952.

20 A. Sommerfeld, *Annalen der Physik*, **32** (1910), 749.

21 H. Minkowski, *Nach. d. Kön. Ges. d. Wiss. zu Göttingen* (Leipzig, Teubner, 1910).

22 E.B. Wilson and G.N. Lewis, in *Proceedings of the American Society of Arts and Sciences*, **28** (1912), 289; see especially p. 454.

23 See note 56, below.

24 L. Silberstein, *Philosophical Magazine*, **23** (1912), 790; and *ibid.*, **25** (1913), 135.

25 Silberstein, *Annalen der Physik*, **22** (1907), 759.

26 Silberstein, *Comptes Rendues de Séances de la Société de Sciences de Varsovie*, **4** (1911), 506 (in Polish).

[27] Silberstein, *Elektryczność i magnetyzm*, vol. I (Warsaw, E. Wende, 1908; vol. II, 1909; vol. III, pt. I, 1913). A Polish translation of Silberstein's *Theory of Relativity* (London, Macmillan, 1914) was likewise announced – in M. Smoluchowski's *Manual for Self-Study* – but never appeared in print; Smoluchowski, *Poradnik dla Samouków*, vol. II (Warsaw, A. Heflich, St. Michalski, 1917), p. 243, where he also praises Silberstein's *Electricity and Magnetism*.

[28] Cz. Bialobrzeski, *Wektor*, **1** (1911), 1 (in Polish).

[29] H. Merczyng, *Wszechświat* [The Universe], **31** (1911), 657, 678, 690 (in Polish).

[30] M. Smoluchowski, *Poradnik dla Samouków* (n. 27, above), p. 352. On Infeld, see E. Infeld, I. Białynicki-Birula, A. Trautman, *Leopold Infeld. His Life and Scientific Work* (Warsaw, Polish Scientific Publishers, 1978).

[31] L. Infeld, *Moje wspomnienie o Einsteinie* (Warsaw, Iskry, 1956), p. 7.

[32] Infeld, *Prace Matem. Fiz.*, **32** (1922), 33 (in Polish).

[33] In addition, one can mention two seminar lectures on relativity held in 1920 at the sessions of the Polish Physical Society in Cracow: Weylich's, 'Critical Remarks on the Theory of Relativity, and Rev. F. Hortyński's 'On the Theory of Relativity;' *Sprawozdania i Prace Polskiego Tow. Fiz.*, **1** (1920–21), 16.

[34] A. Piekara, *Postępy Fizyki*, **35** (1984), 177; and Średniawa, *Theoretical Physics at Jagellonian University*, see §6,2.

[35] *Ibid.*, see §7,3, App. 6, and *Postępy Fizyki*, **33** (1982), 373.

[36] St. Wahrhaftmann, *Zarys teorii względności. Opracowane według wykladów Dr. W. Pogorzelskiego* (Warsaw, Kolo Mat.-Fiz. Stud. U.W., 1923).

[37] St. Bazański, *Klasyczna teoria pola, według wykladów Prof. L. Infelda* (Warsaw-Łódź, PWN, 1956).

[38] G. Doborzyński, *Wiad. Matem.*, **29** (1925), 1 (in Polish).

[39] J. Rayski and B. Średniawa, *Acta Physica Polonica*, B. 3, no. 5, III (1972).

[40] J. Weyssenhoff, *Postępy Fizyki*, **6** (1951), 481 (in Polish).

[41] Information concerning these relativity lectures and courses was taken from the catalogs of the appropriate academic institutions.

[42] *Wiadomości Mat.*, **13** (1909), 23.

[43] *Ibid.*, **17** (1913), 149.

[44] *Ibid.*, **22** (1918), 179.

[45] S. Zaremba, *Teoria względności wobec faktów stwierdzonych doświadczeniem i spostrazeżeniem* (Cracow, Wyd. Min. Wyzn. Rel. i Ośw Publ., 1922), *Journal de Mathematique Pure et Appliquée*, 1 (1922), 105.

[46] Archives of Jagellonian University, WF II 122.

[47] T. Banachiewicz, *Rocznik Astronomiczny Obserwatorium Krakowskiego* [Annual Yearbook of the Cracow Observatory], II, 1 (1922).

[48] Zaremba, in *Scientia*, **31** (1922), 341. The later papers of Infeld and implicitly those of Weyssenhoff contain responses to Zaremba's critique (see notes 56 and 57, below); J. Rossignol, *Journal de Physique*, 3, no. 9 (1923).

[49] Professor J. Weyssenhoff, personal communication.

[50] S. Zaremba, *Bull. Société Math. de France*, **52** (1924), 596.

[51] A.J. Stodołkiewicz, *Uwagi o teorii względności* (Warsaw, Gebethner i Wolf, 1922).

[52] W. Ślebodziński, *Prace Matem.-Fiz.*, **34** (1925–26), 65.

[53] Ślebodziński, *ibid.*, **91** (1925–26).

[54] T. Banachiewicz, *Spraw, z czynności i pos. PAU*, vol. XXVI (1921), 11.

[55] J. Witkowski, *ibid.* (1923), 113.

[56] L. Infeld, *Spraw. i Prace Polskiego Tow. Fiz.*, **3** (1927), 5 and 117 (in Polish with French summaries).

[57] J. Weyssenhoff, *ibid.*, **3** (1927), 295.

[58] Weyssenhoff, *Zeitschrift für Physik*, **95** (1935), 391 and 107 (1937), 64.

[59] Weyssenhoff, *Bulletin Acad. Pol. de Science et de Lettres. Classe de Sc. Math. et Nat.*, Serie A (1937), 253.

[60] E. Infeld *et al.*, *Leopold Infeld* (n. 39, above).

[61] B. Średniawa, 'Relativistic Equations of Motion of Spin Particles,' in P.G. Bergmann and V. de Sabbata, eds., *Cosmology and Gravitation, Spin, Torsion and Supergravity* (New York and London, Plenum, 1980), p. 434.

[62] See E. Infeld *et al.*, *Leopold Infeld*, and B. Średniawa, *Postępy Fizyki*, **33** (1982), 374 (in Polish).

[63] S. Ulam, *Adventures of a Mathematician* (New York, Scribner's, 1983), p. 18.

[64] S. Loria, *Względnośc i grawitacja, Teoria A. Einsteina* (Lvov, H. Altenberg, 1921; 2nd ed., 1922).

[65] M. Huber, *Kosmos*, **46** (1921), **19** (in Polish); and *Czas, przestrzeń i materia i kosmos* (Lvov, Zwiazkowa Druk, 1921).

[66] S. Loria, *Eter i materia* (Lvov, H. Altenberg, 1921).

[67] Huber, *Rola teorii względności w ewolucji fundamentalnych pojęć fizyki* (Cracow, Druk U.J., 1925; Archiwum Tow. Nauk. we Lwowie).

[68] *Sprawozdania i Prace Pol. Tow. Fiz.*, **1** (1920–21), 14, 17; **2** (1921–22), 54, 59.

[69] Cz. Białobrzeski, *Wyklady popularno-naukowe o teorii względności* (Warsaw, Trzaska, Evert i Michalski, 1923).

[70] S. Berman, *Wstęp do teorii względności A. Einsteina* (Warsaw, K. Trepte, 1923).

[71] B. Winawer, *Jeszcze o Einsteinie. Teoria względności z lotu ptaka* (Warsaw, Polska Skladnica Pomocy Szkolnych, 1924).

[72] P. Chomicz, *Teoria względności Einsteina w świetle filozofii absolutnej Hoene-Wrońskiego* (Warsaw, Prace Tow. Mesjanistycznego ks. Kuncewicza i Hoffmanna, 1922).

[73] F. Beer, *Teoria względności i jej podstawa historyczna* (Vienna, 1921).

[74] A. Einstein, *O szczególnej i ogólnej teorii względności*, M.T. Huber, trans. (Warsaw, Książnica-Atlas, 1921; 2nd ed., 1922). The second edition also included a translation of Einstein's article Dialog über die Einwände gegen die Relativitätstheorie, *Naturwissenschaften* (1918).

[75] Einstein, *Geometria a doświadczenie*, Prof. Gotfried, trans. (Vienna, Renaissance, 1922).

[76] Einstein, *Eter a teoria względności*, L. Freudenheim, trans. (Vienna-Lvov, renaissance, 1922).

[77] E. Freundlich, *Zasady Einsteina teorii grawitacyjnej* (Warsaw-Lvov-Vienna, Renaissance, 1923).

[78] A. Moszkowski, *Einstein, rzut oka na jego myśli*, T. Dropiowski (Lodz, Fischer, 1924).

[79] Teoria Einsteina, in A. Cwojdziński, *Komedie Naukowe* [Scientific Comedies] (Warsaw, Państw. Inst. Wyd. (Warsaw, 1968).

[80] I. Dąmbska, *Zagadnienia Naukoznawcze*, **1** (1979), 57; and Średniawa, *Theoretical Physics at Jagellonian University*. §4, 8.

[81] Z. Zawirski, in *Przeglad Filozoficzny* [Philosophical Review], **23** (1920), 343.

[82] *Ibid.*, **2** (1924), 129.

[83] *Ibid.*, **12** (1935), 397.

[84] *Ibid.*, **12** (1936), 48, 99; and Zawirski, *L'Evolution de la notion du temps* (Cracow, Academie Pol. des Sciences et de Lettres, 1936).

[85] H. Mehlberg, *Przegląd Filozoficzny*, **37** (1934), 978; and *Studia Philosophica*, **1** (1935), 119 (in French).

[86] K. Estreicher, *Leon Chwistek, Biografia Artysty (1884-1944)* (Cracow, PWN, 1971).

[87] L. Chwistek, *Arch. Tow, Nauk we Lwowie*, Sec. III, Vol. IX, fasc. 4 (1937) (in Polish, with French summary); and *idem.*, *Granice Nauki* (Lvov-Warsaw, Książnica-Atlas, 1935).

[88] A detailed list of articles and lectures about relativity up to 1939 is to be found in the volumes of the periodical *Ruch Filozoficzny* [Philosophical Movement], *passim.*

[89] K. Wais, *Kosmologia Szczególowa*, 2 vols. (Gniezno, 1931–32), in II, 19.

Jagellonian University

TSUTOMU KANEKO

EINSTEIN'S IMPACT ON JAPANESE INTELLECTUALS

The Socio-Cultural Aspects of the "Homological Phenomena"

1.

In response to an invitation from a Japanese publisher, Einstein decided to visit Japan in order to satisfy his yearning for the distant Orient and to escape from the oppressive anti-Semitism prevalent in Germany since the assassination of Foreign Minister Rathenau.[1] Departing for Japan on the N.Y.K. liner *Kitanomaru* which sailed from Marseille on October 8, 1922, he arrived in Kobe on November 17 and stayed forty-three days in this Far East land, sparking off an "Einstein boom" in Taisho Japan (1912–1926). Sailing from Moji on the N.Y.K. liner *Harunamaru* which departed on December 29, Einstein arrived on February 1 in the following year at Port Said, where he went ashore and headed for Palestine. Including the time spent aboard ship on the outward and homeward voyages, his visit to Japan lasted a full four months.

In a postcard to Mr. and Mrs. Max Born in Berlin written several days before he left Japan, Einstein describes his impressions of Japan as follows.[2] "Dear Borns – Splendid sunshine at Christmas. A happy, beautiful country, with a delicate, sensitive people." (23 December 1922). In a letter to Niels Bohr in Copenhagen written aboard ship near Singapore on the way home, he writes "The trip is splendid. I am charmed by Japan and the Japanese and am sure that you would be too." (11 January 1923).[3]

Prior to visiting Japan, Einstein had toured the Netherlands, Czechoslovakia, Austria, the United States and Britain from 1920 into 1921, and in the spring of 1922 he had also visited France. These visits, however, had a major political significance in that he was acting as a cultural envoy or lending support to the Zionist movement, and though culturally speaking he belonged to the European sphere, where the scars of World War I were still fresh, he was dogged by the Jewish issue. Let me emphasize the fact that at the same time as evoking an extraordinary response towards the Japanese scenery and people, as shown in the above remarks, this trip to Japan, where prejudice against Jews was

Thomas F. Glick (ed.), The Comparative Reception of Relativity, 351–379.
© *1987 by D. Reidel Publishing Company.*

virtually non-existent, was Einstein's first and last Asian experience, and heightened his awareness as a cosmopolitan.[4]

For an invitation to a single scientist, the Einstein boom in Japan was of an unprecedented scale far exceeding the post-World War II boom centering on Yukawa Hideki, Japan's first Nobel Prize winner who was living in America at the time. An account [5] sent to the home office by the German Ambassador to Japan at that time, Wilhelm Solf, illustrates this fact.

"His trip to Japan resembled the parade of a general returning from a triumphant campaign. During the visits of the Prince of Wales and General Joffre, the Imperial Court and the military put on a display and there were planned preparations and a semiofficial response on the part of the media. However, there was none of this in the welcome accorded Einstein – on the contrary, the whole Japanese populace from the highest dignitaries down to rickshaw coolies participated spontaneously and without any preparations or compulsion! When he arrived in Tokyo such multitudes thronged the station that the police could only look on with folded arms at the waves of humanity which made one fear for one's life. (Omitted) Since it is hardly conceivable that the Japanese people are interested in the theory of relativity, which is beyond the understanding of the layman, most Germans here have reached the conclusion that he was being feted out of a consideration to accord a German equal treatment with that accorded Britain (Prince of Wales), America (Denby) and France (Joffre). (!) (Remainder omitted)"

As a German in Japan, Solf is somewhat modest in attributing the Einstein boom to a "consideration for equal treatment". Needless to say, civilization and enlightenment in the Meiji period were carried out under the initiative of the Meiji government, but perhaps because of the France-connected reaction to the shogunate, the Constitution, the military, the mode of production and so on had become increasingly slanted towards German culture. German idealism lay at the center of the national universities, and medicine meant German medicine. In a sense the whole of Japan was captivated by German things. After World War II American culture took the place of German culture. Perhaps a certain Japanese mentality lies behind the magnitude of this swing.

Most Japanese viewed Einstein as the flower of German science, and treated him as a German scientist. In Germany he was attacked as a non-German Jew, but overseas his position as a representative German scientist was emphasized to exaggeration. Einstein himself was forced to put up with this double-faced Janus-like position about which he could do nothing. At a luncheon hosted by the mayor of Osaka, Einstein's statement that he accepted the elaborate welcome of the German national anthem and German flag not as an individual but in the name of science was probably his own response to this ambivalent situation.

Apparently the manner in which the Japanese made a cult out of German science made a great impression on Einstein, as his travel diary (unpublished) reveals. He also felt this admiration when he visited the Seiryoden at the Kyoto Imperial Palace.

"In a spacious room with a view over the garden from the Emperor's throne, the portraits of about 40 Chinese politicians hang as a tribute to the cultural fruit that Japan has received from China. This admiration of foreign scientists is also evident today amongst the Japanese people. The almost emotional manner in which the many Japanese who have studied in Germany extol German scholars. As if a shrine should exist to commemorate the bacteriologist Robert Koch. There is no sense of cynicism or even of skepticism in their earnest praise. This is a characteristic of the Japanese, this pureness of soul which is not found in any other people. We must love and respect this country."[6] (December 10)

Lacking in cynicism, Einstein's admiration for Japanese intellectuals, whom he characterized as having "no sense of cynicism or even of skepticism", applies today even sixty years later to, for example, the self-affirmative words of a representative Japanese intellectual in a dialogue[7] with an Austrian-style economist: "In Japan there is an expression 'to efface oneself', and the Japanese have a self-effacing tendency to be humble and to learn foreign things in earnest". Of course the Japanese adulation of German culture is merely one clue to learning the background behind the Einstein boom in Japan.

Above all, there is no doubt that the eagerness (even if cloaked in enterpreneurial intentions) of the Kaizosha president, Yamamoto Sanehiko, and his colleagues, whose invitation had sparked off the boom, and the enthusiasm and physical stamina of Einstein staged a gala which packed a "full year's worth of experience"[8] into the crammed 43-day schedule. Probably the foresight of Yamamoto and his colleagues in planning competitive cooperation with central and local newspapers such as the Osaka Asahi, Osaka Mainichi, Shin-Aichi and the Fukuoka Nichinichi Shimbun at the time of Einstein's provincial lectures was also one factor behind this success. Through Einstein, Kaizosha certainly succeeded in a kind of amalgamation of leading "medium communication" newspapers that had widely supported Taisho democracy. On the business side, *Kaizo's* December 1922 special issue was sold out and reprinted, and Kaizosha's "Einstein Zenshu" (four volume-set), the first complete works of Einstein to be published in the world, sold an amazing 4,000 copies during and after the boom.

A social class demanding extensive scientific knowledge was already in existence at the time of the Einstein boom.

After World War I Japan rapidly became industrialized, and large numbers of engineers, skilled laborers, teachers and students appeared, and these converged on the cities and added to the intellectual class. This was a period in which Japan was swiftly adopting foreign culture, what with the issue of weekly magazines, the popularity of movies, cafés, bars, dances, photos and Western clothing, as well as visits to Japan by celebrated musicians and dancers and invitations to foreign scientists. It was at this time that the first popular scientific periodicals appeared – *Kagaku Chishiki* (Scientific Knowledge) (founded in 1921) and *Kagaku Gaho* (Science Illustrated) (founded in 1923). Despite Terada Torahiko's misgivings that "popularizing science is a good thing, but it should not be vulgarized", these periodicals put even greater emphasis on increased visual effect through improvements in photoengraving technology. Thus the heightened interest in science really flared up in one go as the result of Einstein's visit to Japan.

It was often said that only seven people in the world understood the theory of relativity at that time. Nevertheless, the reason that many people gathered to hear about it was because the general intellectual class which supported the wide-spread Taisho democracy movement had expectations of it not as an isolated physics theory, but as an idea opening up new horizons.

As soon as World War I ended, U.S. President Woodrow Wilson urged that "the Allied victory is a victory for democracy", and in revolutionary Russia V.I. Lenin cried that "From now on the world should be ruled by the proletariat". On his return to Japan from Europe, Professor Yoshino Sakuzo, in consideration of conflict with the constitutional system, cautiously preached not *"minshu-shugi"* (democracy) but *"minpon-shugi"*, a politics which gave top priority to the interests of the people. Though mindful of the eyes of authority, Taisho democracy swayed the whole nation amidst the upsurge in the suffragist movement and labor movement and also had a keen effect on magazine journalism.

Attracting particular attention amidst all this was the new general magazine *Kaizo* (*Reconstruction*), which was founded in 1919. The magazine *Taiyo* (*Sun*), which had been in publication since the Meiji era, was losing its former influence, and *Chuo Koron* was continuing an unrivalled run through the activity of its noted editor, Takita Choin. The process by which *Kaizo* introduced itself and through its more radical editorial policy carved out a new class of readers which had not been in the existing framework testifies vividly to the extent of the base of Taisho democracy.[9]

In particular, the year in which Einstein visited Japan, 1922, was the year in which such noted persons as the elder military statesman Yamagata Aritomo, the politician Okuma Shigenobu, the writer Mori Ogai and the scientist Takamine Jokichi had passed away, so that it was a time of change in the leadership and government of modern Japan. At the same time, as demonstrated by the establishment of the Japan Communist Party, the National Federation of Students Self-Government Associations and the Japan Federation of Employers' Associations, 1922 was also a year of clear differentiation and confrontation between the ruling class and the ruled.[10]

The biggest project carried out in its early stages by Kaizosha, which had built up its assests through the major bestseller *Shisen o koete* (Crossing the deathline) (by Kagawa Toyohiko) the year after the company was founded, was to invite to the Far East the three world-renowned personages of Bertrand Russell (arrived in Moji on July 16, 1921 and departed from Yokohama on July 30), Margaret Sanger (arrived in Yokohama on March 10, 1922 and departed from Shimonoseki on April 5) and Albert Einstein in the space of a mere eighteen months. All three were revolutionary thinkers, with Russell scathingly demolishing the supremacy of chauvinism in international politics, Sanger advocating earthly birth control and Einstein breaking away from the absoluteness of the space-time framework of perception. Russell's visit to Japan acted as the trigger for the invitation to Einstein, and Russell and Sanger were both placed under the strict surveillance of Japanese authorities during their visits, foreshadowing the dark eclipse of Taisho democracy.

The connections amongst the people who had a hand in inviting Einstein bring out in tacit relief Japan's intellectual circle of the time. It all began when Yamamoto Sanehiko, the president of Kaizosha, heard about Einstein from Nishida Kitaro, the great philosopher at Kyoto

University. He immediately visited Professor Ishiwara Atsushi (Jun) (Japan's first theoretical physicist to carry out joint research with Einstein in Zurich) at Tohoku University to hear about Einstein's work in detail. As soon as he returned he said to his staff members Akita Tadayoshi and Yokozeki Aizo "I've discovered a great scientist. Let's invite him to Japan." Yokozeki later recalled that "When I asked him whom he was referring to, he replied Einstein. I thought he meant Eduard Bernstein (German social democrat)." A month later Yokozeki visited Professor Kuwaki Ayao (the first Japanese theoretical physicist to meet Einstein in Bern) and again learned of the greatness of Einstein's achievements. Yokozeki then returned to Tokyo and consulted Professor Nagaoka Hantaro of the University of Tokyo, the doyen of the Physical Society of Japan, but this notoriously irascible old man merely grunted in reply, so apparently he became somewhat worried. This was at the end of 1920.

After this Yokozeki went to negotiate with Russell, who despite illness was in China for a year at the invitation of the University of Peking, and made him promise to call in at Japan on his way home. Arriving in Kobe in July 1921 with Dora Black, his mistress and secretary who later became his second wife, Russell received an enthusiastic welcome from a sea of red flags. Yamamoto shot his pet question at this uncompromising philosopher. "Who are the three greatest people alive today?" Russell replied "First Einstein, then Lenin. There is nobody else." Reportedly this strengthened Kaizosha's resolve to invite Einstein.

Yamamoto, who had managed to obtain the approval of Nagaoka, who originally had frowned on this private project, requested Kuwaki, Ishiwara and others to take up the matter with Einstein by letter. He also telegrammed the reporter Murobuse Takanobu, who was currently in London, to hurry to Berlin and negotiate directly with Einstein. With the cooperation of the Japanese embassy in Germany, Murobuse contacted Einstein three times from the end of August into early September, and obtained his unofficial consent. Then in the early spring of the following year, 1922, Akita travelled to Germany and sought the cooperation of the philosopher Tanabe Hajime, an assistant professor at Kyoto University who was studying in Germany, and entered direct negotiations, and after several rounds of talks a formal written consent arrived.

Broadly speaking, therefore, the persuasion and assistance of the Kyoto school was responsible for both the initial and final stages of this invitation, but it originated in the non-central parts of academism represented by Ishiwara, who was to be driven out of Tohoku University over a love affair, Kuwaki, who had become engrossed in the history of science at Kyushu University, and the fearless journalists who had begun to gain confidence through developing a new class of readers, and it materialized in the form of having pressed the Imperial Academy for confirmation. Here lies the uniqueness of Taisho democracy.

During his 43-day sojourn in Japan Einstein was accorded an enthusiastic welcome by both the government and the people. Giving a total of seven general lectures (Admission fee of three yen for adults and two yen for students. This was equivalent to the cost of ten ordinary lunches, so it was a considerable sum.) – two in Tokyo and one each in Sendai, Nagoya, Osaka, Kyoto and Fukuoka – Einstein spoke with fervor for as long as five hours on occasion, addressing a full house wherever he went.[11] To give an idea of this wild enthusiasm, several dozen people who had entrance tickets for the lecture at Kanda in Tokyo were unable to enter because of the great congestion, so Kaizosha paid their return fare to Sendai, the site of the next lecture. Einstein also had some contact, though slight, with ten of the sixteen universities in Japan at that time, attending welcome parties given by students at five universities, and his magnetic personality directly left a great impression on Japanese youth.

2.

It is my contention that at the point in time when Einstein visited Japan in 1922 Japanese society and culture clashed actively with his personality and ideas in various fields and at many levels. In order to demonstrate this, I intend to highlight his influence by picking out *homological phenomena* common to the worlds of culture and ideas.

The science historian Thomas Kuhn refers to the beliefs, values and techniques common to the members of the scientific community as a paradigm.[12] Focusing here on patterns or styles of sensibility and thought, I have given the name "homological phenomena" to isomorphic patterns or styles that exist in the semantic space of different fields of science, culture and thought in terms of topological mapping or superposing.

What then constitutes the special patterns of sensibility and thought characteristic of the theory of relativity, which revolutionized Newtonian physics and created a new paradigm? Broadly speaking, I believe that it has two features. First, the theory of relativity dissents radically from man's commonsense feeling that he is the center of the universe.

Previously it had been thought that by using watches it would be an extremely simple matter to determine that it was the same time. According to the theory of relativity, however, if twins were to hold two perfectly-synchronized watches and one travels in space while the other remains on earth, immediately the pace of the watch hands and the speed at which the twins age would come to differ. The weight of a person running at a speed of 260,000 kilometers a second would double, and so attempts by fat people to lose weight by running are a complete contradiction according to Einstein's theory. Thus the theory of relativity set a theoretical limit to man's five senses. Upon reflection, the fact that man discovered Euclidean geometry, was able to establish Newtonian dynamics and thought up the Kantian concepts of absolute time and absolute space was also related solely to man's view of himself as the "center" of the universe. The theory of relativity made it evident that this viewpoint is not guaranteed without reflection, and that in no way does it constitute the origin of a privileged coordinate. We must also realize that the abstract space-time of the relativity theory makes a clean sweep of the conventional anthropocentric model, deposing man from his central position so that he becomes no more than an infinitesimal manifestation in the universe. The grounds for man's claim to a special position in the universe had already begun to shake under Copernicus' attack, but now they crumbled completely in the face of Einstein's space-time.[13]

The second feature of the theory of relativity is the fact that, differing from mere relativism and pluralism, it aims at realizing an absolutism through thoroughgoing relativisation. According to the theory of relativity, it should be requested that any law of nature have the same form to any observer who may deduce this law, irrespective of his motion relative to any other observer, and irrespective of the particular language that may be used to express this law.

As Hermann Minkowski says, contrary to its name the theory of relativity ultimately depicts an absolute world, an idea-like world. Individual events that occupy actual space and follow time are the shadows of this idea-like world. Even so, unless *things* exist this absolute world (space-time) will also cease to exist, so it is not a Platonic idea. In the expressive world denoted by the theory of relativity, it would be possible to express *objective* laws about nature using a set of continuous parameters of space and time as *subjective* linguistic elements to make all of the natural laws be preserved in form. However, by this alone the selection of *subjective* linguistic elements would become completely arbitrary, so it is necessary to provide a "language transformation rule" which directs this transformation amongst subjective linguistic elements. This rule was established in the special relativity theory in the form of the "Lorentz transformations".

In sum, the pattern of sensibility and thought in the "relativity theory viewpoint" can be seen in the fact that it shuts out the intervention of ordinary feelings and commonsense and the fact that through the denial of a priority viewpoint (frame of reference) it attempts to grasp an absolute, overall reality. In cases in which such an isomorphic state of mind or similar mentality can also be found in the semantic space of another culture sphere, I contend that a homological phenomenon exists between them. When it is of the same type as the first, let us refer to it as Homological Phenomenon I, and when it is of the same type as the second let us call it Homological Phenomenon II. Furthermore, if their chronological relationship is clear, all the necessary conditions are present to judge that the influence has been registered. In addition, when these homological phenomena are totally applied to explain various values seen amongst thoughts and manners, it can be said that a spirit of the times (Zeitgeist) – a mentality common to the era, an "ethos" or grand style – has arisen.[14]

These homological phenomena must also be found not only in the semantic space of the theoretical languages of science, the arts and thought, but also in the semantic space of the everyday language that supports these. It is not until these are totally integrated that the social and cultural impact of Einstein and the theory of relativity can be portrayed in its entirety. In the words of Michael Polanyi,[15] in time our attention must shift from a proximal attitude which views the near-at-hand, partial and individual to a distal attitude which views far-off, overall and unified things. Thus from the specific individualness that is the *basis*, the moulded totality that is the *picture* can be portrayed.

3.

Japanese intellectuals manifested various reactions towards the socio-cultural event of Einstein's visit to Japan. As one lead to gauging these reactions, let us look at a questionnaire conducted at the time by a representative literary magazine.[16]

Replies to the questionnaire were received from 37 intellectuals in

various circles. Judging from the issue number of this magazine, probably the questionnaire was sent out during December before Einstein
had left Japan. Ten of the respondents worked in the field of science,
such as scientists and doctors, seventeen were in the literary line, such as
writers, while the remaining ten included newspaper reporters amongst
their number. Owing to the nature of the magazine, none of the
respondents belonged to the new middle class consisting of such people
as politicians, ordinary company workers, government officials, military
personnel and primary or junior high school teachers. Rather they were
of the cultural elite that had risen from this new middle class, but even
so they represent quite an average distribution of Taisho intellectuals at
the time.

A single reading of the results of this survey suffices to show that there
were diverse levels to these reactions. Some respondents reacted positively, while others reacted negatively. Even amongst the positive
reactions there was a diversity ranging from a reaction focusing on the
human aspect of Einstein's personality, a reaction extolling the great
achievements of his scientific research, a political reaction sympathizing
with his anti-war views, and an ideological reaction to his revolutionary
theory. The negative reactions included an elitist reaction rejecting the
reaction of the masses, a critical reaction opposing the capital staging
the boom and a mystic reaction pointing out the limits of science. Out of
the total of thirty-seven respondents, twenty-one viewed the effect of
Einstein's visit to Japan positively (six in the scientific line, eleven in the
literary line and four others), while twelve viewed the visit negatively
(three in the scientific line, five in the literary line and four others), and
four people were undecided (one in the scientific field, one in the
literary line and two others). As shown in the figure, I have divided the
catalytic agent Einstein into his "human nature" and his "scientific
nature" (theory of relativity), and the passive stage of Japan into a dual
structure of "Japanese culture and society" and the "Japanese science"
which is connoted by this. Based on the replies to this questionnaire, the
table shows whether Einstein had any effect on each of these. Though
only to be expected, seven-tenths of the literary men and "others"
(newspaper reporters and so on) focused on Einstein's human nature,
and the number of those who recognized that Einstein had some kind of
influence on Japanese culture and society outnumbered those who said
that he had no influence by 10 to 5, or two to one. The majority of those
in the scientific field focused on how Einstein's scientific nature would

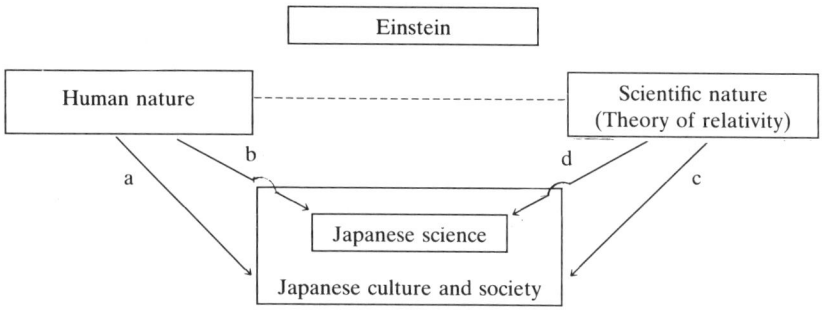

		a	b	c	d	Total	Undecided
Science field	Positive	–	1	4	1	6	1
(10)	Negative	1	–	2	–	3	
Literary field	Positive	7	–	3	1	11	1
(17)	Negative	3	–	2	–	5	
Others	Positive	3	1	–	–	4	2
(10)	Negative	2	–	2	–	4	
		16	2	13	2	33	4

affect Japanese culture and society, and here too those viewing his visit positively outnumbered those with a negative attitude by two to one.

As well as those people who came in contact even momentarily with Einstein's personality, there were also many who had never met him in person but who felt friendly towards him as the result of the daily newspaper articles and Okamoto Ippei's cartoons, as indicated by the following:

"It is perfectly natural that such a theory was born from such a person. Lately when I look at the clear starry sky at night I find myself thinking of the Professor's character and mental attitude, and it gladdens my heart." "He has enabled me to take a fresh look at nature and life, and has given fluidity to my fixed view of life."

It can easily be perceived how Einstein's personality had an incalculable effects on how public viewed science and scientists. The following is a typical example of this.

Miyake Yasuko was married to a scientist and studied under the writer Natsume Soseki. After the death of her husband she established herself as a writer and was active as a critic, novelist and lecturer, as well as a social crusader. She writes: "I had the impression that scientists are

narrow-minded, restless and like a sneak thief waiting for an empty
chair, so the great and noble light in his eyes was an incredible wonder
to me. His gentle attitude of total absorption in his research almost
moved me to tears. The theory of relativity is not something that
everybody can understand. However, the Professor's visit to Japan has
had ample significance merely in the fact that it has enabled many
people to come in contact with the "personage" of a "scholar", which is
impossible in Japan."

This is indeed a case in which the scientist Einstein wrought a 180°
change in the image of scientists as "sneak thiefs".

Naturally there were also writers who, though recognizing the scientific necessity, took the negative attitude that inviting Einstein to Japan
would serve no purpose but to gladden the hearts of capitalists. From
the cynical attitude that "Instead of advocating Science! Science!, we
should first practice Euclidean geometry over and over again." to the
negative view that "To us it is merely the waft of a spring breeze.", what
these have in common is the criticism that "they are after money," "are
following blindly after a publisher" and "an attitude as if handing over a
cheap article at a night stall". This was also a backlash against the
publicity onslaught launched by Kaizosha. There were also some who
said that "These days even science is nothing more than a kind of
superstition that makes one lose one's "balance of mind".

For better or for worse, the pendulum of views amongst these contemporaries who in the midst of the turmoil of the Einstein "shock"
were trying to assess the significance of Einstein's visit in order to reply
to the questionnaire tends to be amplified greatly. It also goes without
saying that the extent of his influence, which will become apparent later
on, is not yet appreciated.

Next I would like to move on from the impressions reflected in this
questionnaire to the issue of how the theory of relativity was transmitted
to each field of society.

4.

Scientific information is transformed in the process of being transmitted,
and the manner of this transformation depends on the recipient. According to R.G.A. Dolby,[17] there are three phases in this transmission.
Firstly the recipient realizes the newness of the information, and either
becomes interested in this or feels resistance, and then either he

adopts this or rejects it – i.e. awareness – interest/resistance – adoption/ rejection. I would like to point out that in the third phase there is also a temporal familiarity – domestication – which is neither of these. This is thought to be the usual pattern in which the daily language system reacts to such theoretical linguistic terms as "energy" and "entropy". Thus the information comes to anchor in the ocean of daily language even though it is incomprehensible. In many cases this involves misunderstanding.

Whether they were aware of it or not, the name relativity principle misled and amused many Japanese, who have a tendency to jump to conclusions. The Japanese word for relativity is "Sotaisei", but many Japanese misread this as "Aitai-sei", as the Chinese character used to write the "So" in "Sotai-sei" can also be pronounced "Ai". Amongst young people an "Aitai-sei stroll" meant a lovers' stroll, and "carrying out the Aitai-sei theory" meant being in love. The term "Aitai-jini" was used instead of the word "Shinju" (lovers' suicide), whose use had been prohibited by the shogunate. Moreover, at that time there was a flood of articles on "sei" (sex), and spectacular love affairs which had no regard for class or age were creating a sensation, so it is understandable that the theory of relativity (Sotai-sei) was immediately mistaken for sex between lovers (Aitai-sei). Original Japanese ditties known as the "Einstein Aitai-sei Bushi" enlivened the gay quarters, and all of these were love songs.

Some businessmen who saw that the subject of Einstein's lectures was "Aitai-sei" even regarded scientists as immodest. Consequently some scientists suggested that the Sotai-*sei* principle be known scientifically as the Sotai principle, omitting the "sei" part which caused the confusion with sex. This is an instance in which public misunderstanding drove scientists to unify their terminology.

This interest in the theory of relativity was by no means confined to the general public, but also extended to the imperial court, political circles and the business world. The "Osaka Mainichi Shimbun" of the time records that after a budget meeting of the Kato Cabinet a debate flared up amongst Cabinet members as to whether or not people could understand the theory of relativity.[18] The distinction between "understanding" and "not understanding" depends on whether or not one can successfully incorporate the concept into the semantic space of one's own linguistic system. So the depth of one's "comprehensibility" usually differs in its degree.

In the popular response to the theory of relativity it was common to pass through the two phases of awareness and interest and then to run into a wall of incomprehensibility resulting in a state of wordly vulgarization, but the responses of intellectuals were more complex and pluralistic.

One response was the deliberate silence of the central figures of the Taisho literary world, which had already taken form. If silence or disregard is also considered as a phase two reaction, this constitutes a kind of air pocket, and is in striking contrast to the active reactions in the rest of society. Since the I-novel, a neorealistic work which brought reality to personal life, was already enjoying great popularity, as long as this "I" had no interest in science the Einstein boom would become merely a completely unrelated flash in the pan. In the Shirakaba school of writing, which had a kind of modernism, about the only person to touch on this subject was Arishima Takeo. Arishima must have attended the reception for Einstein at the Imperial Hotel on December 1 and heard his violin recital. In a short piece entitled "Chishiki kaikyu to iu mono" (On the intellectual class), which was written the following spring several months before he committed double suicide with his mistress, Arishima cites the theory of relativity and expresses the extremely interesting view that it is undeniable that nowadays the people's lifestyle is reflected even in the investigation of truth and fact. "Nowadays there is no longer a fixed criterion in all laws owing to the confusion in life, and it is necessary to treat all advantages and disadvantages, good and evil and beauty or ugliness without any standards. Did not the penetrating minds of those scholars analytically feel this situation?"

Tsuchii Bansui, whom the literary world thought would write no more anthologies, wrote a long epic poem entitled "An dem Grossem Einstein"[19], which was a vigorous paean to Einstein in the fashion of a heroic poem, which was his forte. Up in a corner of Tohoku, Miyazawa Kenji, while obviously being exposed to this Einstein ideology, built up a poetic universe in a form that had no connection with the literary world. Ishiwara Atsushi, who had parted with the Araragi school, began to make moves towards a Shin Tanka (New Tanka) which incorporated ideological aspects such as new science. The attempt by Yokomitsu Riichi, the founder of the Shinkankaku school of writers, to take in the scheme of relativity through the Great Kanto Earthquake was a very rare exception.

It is no mere coincidence that Ishiwara Atsushi's Shin Tanka movement took an identical course to the upsurge in his science criticism, which boldly tackled the fundamental issues of the theory of relativity and the quantum theory. As a scientist and science thinker, both directly and indirectly Ishiwara had himself participated in and witnessed the era in which the concepts of time, space and matter, which are the basic framework of modern science, had been criticized and completely revolutionized by such giants as Einstein, M. Planck, N. Bohr and A. Sommerfeld. The poet Ishiwara had lived through an important period of transformation in scientific thought – the destruction of the dynamic view of nature that had prevailed up until the 19th century and the formation of a new relativistic, quantum theory view of nature. From the crucible of this vivid experience, homologically he focused critically on the unchanging Tanka framework of 31 syllables (5, 7, 5, 7, 7) and advocated its abolition, and no doubt this was owing to the trend of the times (Homological Phenomenon II).[20]

It is uncertain how Yokomitsu Riichi, who had already lost his father at the age of 24 and had travelled to Korea, came in contact with the Einstein boom. Homologically, however, he was awakened *by the external world* to the "inadequacy of man's five senses" from the way in which such speed variables as automobiles and aeroplanes flitted around the burnt-out wilderness left by the Great Kanto Earthquake, which occurred the year after Einstein's visit to Japan, and from the way in which the radio broadcast deformed sound (Homological Phenomenon I). Thus through a disaster that occurred in an extremely concentrated and extraordinary period of time, Yokomitsu portrays a stratified, multitemporal world in some short stories by means of a new sense that he worked out intellectually. This is the birth of the so-called "Shinkankaku school".

Already here the naturalism that depended on man's five senses and the new realism that emphasized privileged, subjective individuality had been completely washed away, and the position of the highly-skilled Shinkankaku school, which could shift its viewpoint at will ommatidially and multitemporally, loomed clearly.[21] This position is indeed isomorphic and homological with the idea of the construction of an absolute reality through the medium of thoroughgoing relativism established anew through the theory of relativity (Homological Phenomenon II). Yokomitsu's first full-length novel "Shanhai" is said to incorporate all of the methods of the Shinkankaku school. He writes "Shanghai appeared, becoming a seaport *as a moving body of the external world including nature*".

Neither the word "Einstein" nor the idea of the "theory of relativity" appears in the work of the four-dimensional poet Miyazawa Kenji, who has earned a high reputation particularly of late. However, attention has focused on the fact that most ideas such as the "fourth dimension" or "four dimensions" appear in the valuable "Nomin Geijutsu Gairon Koryo" (General Summary of Rural Art) (1926), which is the sole source revealing Miyazawa's views on art. He writes that "I would like to discuss it based on an agreement amongst the corroboration of modern science, the experiments of the seekers after truth and our intuition."

The period when he demonstrated explosive creative energy in his representative work "Haru to shura" (Spring and Asura) (1924), which consists of 69 mental sketches, was for Miyazawa the twenty-two months during which his younger sister died and Einstein visited Japan. In the preface to this anthology he declares "All of these propositions/Are asserted within a fourth-dimensional extension/As mental images and the nature of time itself". Scholars have debated the significance of the fact that in a copy preserved for the author's correction he added the two lines "Each of the following sketches/Follows a four dimensional structure", but later erased them.

Opinions have been divided over the "fourth-dimension" and "four dimensional structure" which form Miyazawa's keywords, splitting into the two opposing views of Minkowski=Einstein-like four-dimensional world and that of the Bergson-like fourth-dimensional aspect.[22] I have suggested the following view on this point.[23]

Miyazawa aspired to unify the Absolute with the Buddha nature which is the law itself of existence of the Minkowski=Einstein-like four-dimensional space-time (world). In the actual task of expressing this as a poet, however, the only course open to him was to compose the mental sketches that use the enduring Bergson-like self as a lever. When Miyazawa says "All of these propositions/Are asserted within a fourth-dimensional extension/As mental images and the nature of time itself", the phrase "these propositions" refers to each poem in "Haru to shura", and each of these are mental sketches which have a characteristic coloring and arrangement – i.e. each create a characteristic image space in the time series, that is in the fourth-dimensional extension. Furthermore, as in the added sentence that he crossed out in red, undoubtedly Miyazawa prided himself on the fact that through relativizing the characteristic image space of each of these poems, the whole is transformed into an absolute and permanent work as a four-dimensional image continuum and as "immortal fourth-dimensional art" (Homological Phenomenon II). No doubt the reason that he later crossed this out was that his confidence on this matter temporarily wavered. I believe that the issue of choosing between Einstein and Bergson was solved and merged together for Miyazawa by the complementary world-method relationship between the *world to be expressed and the method of expressing it*.

5.

As mentioned earlier, when Kaizosha's Yamamoto Sanehiko asked Bertrand Russell who are the three greatest people in the world, Russell replied "First Einstein, then Lenin. There is nobody else." These words reveal Russell's deep sympathy with the selfless and courageous action of Einstein who was aiming at solidarity with the human race. This was one factor behind the fervor of the Einstein boom. Here the theory of relativity is transformed into a weapon of social thought.

"Visit by Einstein, the scientific world's revolutionary son" – This headline[24] in a local newspaper announcing a presentation by Einstein conveys the fervent atmosphere of the time and leaves a vivid impression. After asking "Why has he come?", it continues as follows. "His achievement in destroying the absolute view of space-time held by all scientists and philosophers in the past and creating in its place a relative and democratic world warrants being ranked along with the achievement of Lenin in destroying the autocratic and absolute czarism and creating instead a democratic Soviet Russia."

Worthy of note is the fact that this not only classes Einstein along with Lenin, but also that it regards the relative, extraordinary, physical world as "homological" with the democratic, ordinary, political world. This is inconceivable to us nowadays, but at that time it was an intellectual "atmosphere" that was readily accepted. This intentional homological phenomenon of politicizing relativism is also obvious from the headline and contents of another newspaper editorial.[25]

Here Einstein's theory of "relativity" is firstly assumed as a new idea running through civilized life, liberalism and natural society, and is set against nationalism, which is a "past, stereotyped privilege " and an "unnatural, absolutist form". This views the theory of relativity in connection with social revolution in the sense that it places "time on the side of truth" and brings about a "new world in future human society".

In addition, the writer of this editorial regards both society and the nation as non-"absolute", and in order to relativize these and have them coexist he aims at integrating them by again using the more comprehensive concept of "the relativity of nature" (Homological Phenomenon II).

An editorial entitled "A relative view of labor and management"[26] is another good example of a homological response. This was about night work by child workers, which was to be prohibited. This editorial reasons that the positions of both labor and capital have only a relative value, that this is merely placing oneself at the center of a static system, and that is necessary to transcend this viewpoint and take a fresh look from the long-term system (motion system, future system) of the nation and society. Along with its humanitarian standpoint, this advocacy of the creation of a world that transcends this relative standpoint is noteworthy as a Homological Phenomenon II example that shows the views of a journalist at that time.

On the night of December 11 the newly-inaugurated Zenkoku Gakusei Rengo (National Federation of Students' Self-Government Associations) held an extraordinary general meeting and unanimously adopted the following resolution, which it sent to Einstein in Kyoto by telegram.[27] "We pay our great respect to your attitude during the World War of all-out resistance to capitalist nations and the war they brought about."

The radical group known as the"Nippon Puroretaria Domei"(Japanese Proletarian Alliance), which centered on Tane Maki-sha (The Sowers) – which had been greatly influenced by the Clarté movement in France – the proletariate and the lower classes, also sent a more or less identical telegram to Einstein, along with some questions.

1. What are your views on the – – – – – imperialist government of Japan?
2. What do you hope for from Japanese youth?

The dashes here probably stand for "aggressive", but at that time this expression was not allowed to be used in newspapers. The main person behind these questions was Komaki Omi, the central figure of Tane Maki-sha. He welcomed the pacifist Einstein as a rebel comrade, and attempted to protest against official intervention and the increasing militarism by making use of Einstein's fame and popularity.

The long-awaited reply to these questions finally arrived from Einstein.[28] It was written aboard ship on the homeward voyage, and was dated January 22, 1923. In this letter Einstein cites as problems facing Japan the fall in labor wages as the result of overpopulation, and the danger of militarization. Though differing in nuance, this was virtually identical to what Russell had said when he visited Japan in the previous year. Einstein also wrote that he hoped that young people in Japan would move away from ultranationalism and tread a course towards an internationalism that was "linked to international cooperation and organization".

Kagawa Toyohiko, the author of *Shisen o koete* (Crossing the deathline) and the mentor of the Kansai district Yuaikai (Friendship Association), was also involved in the farmers' association movement and consumers' association movement, and was well-known for his practical Christian-like social activities. Kagawa went to meet Einstein when he arrived in Kobe, and also met him again when Einstein visited Kobe from Kyoto to give a lecture. He writes "Last year when I met Bertrand Russell I received the impression that he was an unapproachable person. By contrast, Einstein is as mild as a dove. He said that Rathenau was an opportunist and that Marx transcended his times."[29] Meanwhile, Einstein in his travel diary merely states "Travelled to Kobe. Had lunch at a fishing village (Japanese restaurant) in Kobe with Yamamoto and an *important young socialist politician.*"[30] Unfortunately there was no in-depth discussion of religion.[31]

Special mention must be made here of the fact that there was a very heated discussion between Einstein and a certain Japanese social thinker over their differing evaluation of the intelligentsia. The person involved was Yamamoto Senji, commonly known as Yamasen, who was a biologist and lecturer in the Faculty of Medicine at Kyoto University and popular amongst the people as a social activist. At the same time he was influenced by Margaret Sanger, and his branching out into political criticism as a pioneer advocate of sex education also occurred around the time of Einstein's visit to Japan in 1922. He evolved a legitimate proletarian movement, and was active as a leftist Diet member representing Kyoto until being stabbed to death by a rightist in 1929.

Yamasen visited Einstein at the Miyako Hotel in Kyoto in order to receive a recommendation for an anti-war book[32] he had translated. After receiving the recommendation, "When we self-interestedly tried to bid farewell, Professor Einstein detained us and requested us to stay for a while, and asked us various questions." Thus a discussion developed between Yamasen and Einstein on the role of intellectuals, and this dialogue deserves special mention amongst the events that occurred during Einstein's stay in Japan. Yamasen describes the gist of this discussion.[33]

This dialogue acutely reflects the differences in the viewpoints of Einstein, who held out hope towards the role of the intellectual classes in the peace movement (this is why he accepted the work of the League of Nations Intellectual Cooperation Committee and Yamasen, who rejected this outright as a dream and ceaselessly pinned his hopes on the workers' movement rather than the words and actions of scholars who discuss "concepts that are meaningless in content and uselessly grandiose in outward form only". Rather than being a mere difference between the political situations of Germany and Japan, this was also a difference between Einstein, who had a global viewpoint, and Yamasen, who was challenging Japanese reality.

Einstein maintained that "The best method for intellectuals to promote international peace and brotherly love amongst mankind is through their scientific contributions and artistic achievements. Creative work raises man high above individual and self-centered national goals." Einstein had a consistently high opinion of the role of intellectuals in overcoming nationalism. Meanwhile Yamasen wrote that "For better or for worse, the Japanese intellectual class is overrated by the Professor.", and disparaged the Japanese intellectual's emphasis on formality as follows. "The reason is that the necessary conditions for belonging to the Japanese intellectual class nowadays are an awareness of oneself as an intellectual and a career and dignity and self-respect appropriate to this position. In other words, self-conceit plus a diploma plus an attitude of stoicism. Naturally, as I said before, this is a question of form alone, and the content is a separate matter."

A young intellectual who was present at this dialogue between Yamasen and Einstein has written that Einstein threw up his hands and argued furiously in German that "Peace will not come to the world unless you attack the militarism and chauvinism of the Japanese government."[34]

Unfortunately, Einstein's travel diary makes no mention at all of this memorable meeting.

6.

So far our discussion of the impact on Japan of Einstein's thought has centered on the socio-cultural aspects. No space remains for a description of the philosophical and scientific aspects, so I would like to close this paper merely by citing the names of those who were connected with the outline.[35]

The book *Tetsugaku Izen* (Before Philosophy), which was completed by the philosophy historian Ide Takashi in the summer holidays of 1921, sold over 10 000 copies in the six months or so immediately prior to the Einstein boom. The second essay in this book, "Tachiba to sekai" (Viewpoints and the world), regards democracy as an ideological attitude that evaluates and respects various ideological viewpoints in their proper position, and claims that "philosophizing" (philosophieren) is the act of standing in the "viewpoint of viewpoints" that further unifies these various viewpoints. This bookish-sounding word "philosophizing" immediately became fashionable. For people who were familiar with the idea of the "viewpoint of viewpoints", this probably acted as a psychological preparation for understanding the gist of the idea of "from the relative to the absolute" that is expounded in the theory of relativity, which is ideologically homological to this (Homological Phenomenon II). It opens up a vague path to accepting the theory of relativity.

There were also philosophers who grappled more directly with the issue of the philosophical significance of the theory of relativity. Saegusa Hiroto, who commenced from a study of Kant and Hegel and evolved a theory of technology, and Tosaka Jun, who together with Saegusa founded a society for studying materialism in 1932 and who became a pioneer of scientific philosophy and died in prison during the war, were amongst those who sought the subject matter for their first published papers in the theory of relativity.[36]

There can be no gainsaying the fact that at least up until prior to the war Nishida Kitaro and Tanabe Hajime were the two greatest authorities in Japanese philosophical circles. As described earlier, these two had a major influence in inviting Einstein to Japan, and Tanabe's deep involvement with the theory of relativity in the early formation of his thought continued right through until late in life.[37] It is evident from his correspondence and other writing that at that time Nishida evinced extraordinary interest in the theory of relativity. This interest bears fruit as the third paper[38] in his collection of papers entitled *Hataraku Mono kara Miru Mono e* (From acting to seeing). This collection of papers is a compilation of nine papers written over the five years between 1923 and 1927. Nishida's thought since *Zen no Kenkyu* (The study of virtue) takes an enormous leap forward in this work, which is said to have established the so-called "Nishida philosophy" that fused Oriental Zen thought with Western logic. In this third paper he evaluates the appearance of the theory of relativity as a victory that "skilfully captures the principles that constitute physical knowledge, and penetrates to the truth of physical phenomena." Nishida then claims that the four-dimensional space-time introduced by this theory is regarded as the objective world of the truly active self, and that this world is divided into two depending on whether or not it has a temporal content. If it is empty time it is regarded as the physical world, and if it is time which has content, it is regarded as the spiritual world. Without doubt the theory of relativity acted as a turning point for Nishida's philosophy, which was still in the process of formation.

There was also a dispute amongst representative Japanese economists of the time about the issues that constitute economic theory. Fukuda Tokuzo, who took an empirical, psychologistic standpoint, attacked the transcendental concept of "economic cultural values" of Soda Kiichiro, who took an axiomatic, logicalistic standpoint. The weapon used in this attack was the theory of relativity. Just as ether had been banished from the physical world, he attempted to remove the concept of *a priori* cultural values from the economic world. Ishiwara Jun, a physicist who was also versed in philosophy, joined in this dispute which continued for several years.[39] Fukuda attended a lecture by Einstein during his visit to Japan, and also talked with him in person, and contributed to a clarification of the points at issue by quoting from Hans Reichenbach.

In Japanese scientific circles the theory of relativity had already begun to gain acceptance quite some time before Einstein's visit to Japan. Einstein's ideas had been widely introduced by the physicist Kuwaki Ayao, who on 17th March 1909 became the first Japanese to meet Einstein, who was working in the patent office in Bern, and by Ishiwara Jun, who in the same year became the first Japanese to publish a paper on the theory of relativity, as well as by Tanabe Hajime and others. This level of enlightenment was quite high, comparing favorably with that in the West.

This was also a period of great change from private science to industry-based science. Up until then scientific research had been carried out in the tiny laboratories of professors or in the back rooms of inventors' homes. However, the period around 1920 saw the arrival of an era of organized science centering on corporate research institutions, which were backed by abundant funds, and on university departments of science and engineering and various national research and test institutions, which were expanded in response to the establishment of the corporate institutions.[40]

Original Japanese scientific research had already appeared in each field in the early twentieth century, and steady progress was continuing towards increased independence. However, this sporadic research was either carried out by Japanese studying in Europe, or even if carried out in Japan in many cases it was isolated research. In order to have a broad and profound influence on each field of science it was necessary to hasten the preparation of an adequate research setup. This was a task facing Japanese scientific and technological circles from the 1910s into the 1920s, or throughout the whole Taisho period. Einstein's visit to Japan fitted in with this mood of promoting science and technology. There must have been a great gap in the degree of familiarity with science on the part of the general public in America, where Thomas Edison, who had become a national hero through 19th century science and his inventions, was still alive, and Japan, which lacked such a personage. A certain Tohoku University assistant professor who attended one of Einstein's university lectures wrote to a local newspaper that he felt such fondness towards Einstein that he felt like calling him "Father". This term also symbolizes the relationship between the Japanese scientific world and that of Europe, at the top of which stood Einstein. This topological relation is even carried through in the welcoming address given at the Imperial Academy.[41]

"(Omitted) How fortunate we are today to be able to welcome here to Japan in this manner a great and famous man of the world who has created a major revolution in the scientific world. Carried away by our admiration of your noble character, we members of the Imperial Academy are greatly honored by your presence, and it is truly an overwhelming joy to have this opportunity to meet with you and to drain this cup of pleasure to the dregs. (Remainder omitted)"

Naturally Oriental modesty is reflected in the words "admire the noble character" of "a great and famous man of the world", but at the same time these words reveal the international position of Japanese scientific circles, which were not yet fully mature.

This was also apparent in the series of lectures given at the University of Tokyo. Extending over a total of five days, Einstein's lectures here were specialized lectures on the theory of relativity, but their content was virtually identical to the lectures he had given at Princeton University more than a year ago, a record of which had already been published.[42] Truly regrettable for the Japanese physics world is the fact that instead of drawing out in this lecture the full faculties of their distinguished guest, Japanese scientists meekly let slip this once-in-a-lifetime opportunity by taking the attitude of a master-apprentice relationship – despite the fact that some discord was caused by the appearance of Doi Fuzumi[43], who opposed the theory of relativity. In this respect the lecture at Kyoto University was a success in that, at the request of Nishida Kitaro, it was a personal and vivid description of how Einstein developed his theory of relativity.

In a letter to a friend, Terada Torahiko, a physicist well-known for his essays, laments the fact that the older generation were monopolizing Einstein, but the fact that no opportunity was provided for Terada and Ishiwara and the younger generation to hold a free discussion with Einstein was also related to the traditional Japanese attitude towards learning. The Confucian precedence of the elder over the younger was still strict, and learning was regarded mainly as knowledge leading to success, and the general intellectual climate lacked consideration for learning as a process.

In closing, let me emphasize that Einstein's visit to Japan had a major influence on the young generation including Yukawa Hideki and Tomonaga Shinichiro, who were still at junior high school, and acted as a stimulus producing many physicists.[44]

Acknowledgements: I am very much indebted to Dr. Otto Nathan of the Einstein Estate, Professor Charles C. Gillispie of Princeton University and Professor John Stachel who provided me with the opportunity to investigate the unpublished materials concerning Einstein's visit to Japan in the Einstein Archives. My deep gratitude also goes to Mrs. Judy Wakabayashi who helped me in the translation.

NOTES

[1] *Albert Einstein in Berlin 1913–1933*, Teil I, Berlin, 1979, p. 230, Dokument Nr. 153. In this letter (20th December, 1922) to the German ambassador to Japan, Wilhelm Solf, Einstein frankly describes his motive for visiting Japan.

[2] *The Born-Einstein Letters* (tr. by Irene Born), London, 1971, p. 74.

[3] A.P. French (ed.), *Einstein, A Centenary Volume*, Cambridge, 1979, p. 274.

[4] Most biographical literature on Einstein views his visit to Japan as unimportant and treats it as mere recreation. The voluminous work by Ronald W. Clark: *Einstein, The Life and Times*, New York, 1971, even spitefully describes it as a disillusioning experience. By checking unpublished materials concerning Einstein at Princeton University's Archives, I

have filled in this blank and corrected this misunderstanding in the following paper in particular: Tsutomu Kaneko, 'Einstein's View of Japanese Culture', *Historia Scientiarum*, No. 27 (1984), which was compiled from a reading of Einstein's travel diary for October 6th, 1922 to March 12th, 1923, as well as in the book *Einstein Shock* (Vols. I, II, Tokyo, 1981, in Japanese). Based on these, I have drawn up this paper in order to highlight the controversial points.

[5] Dokument Nr. 153, *op cit.* Solf's dispatch is dated 3rd January, 1923.

[6] German original: Im Saal, wo vom Hof aus die Krönungssessel sichtbar sind, Bildnisse von etwa 40 – chinesischen – Staatsmännern als Anerkennung für die kulturelle Befruchtung, welche Japan durch China empfangen hat. Diese Verehrung für fremde Lehrer lebt heute noch unter den Japanern. Rührende Anerkennung vieler Japaner, die in Deutschland studiert haben, an ihre deutschen Lehren. Es soll sogar zur Erinnerung an den Bakteriologen Koch ein Tempel existieren. Ernste Hochschätzung ohne eine Spur Zynik oder auch nur Skepsis für Japaner charakteristisch. Reine Seelen wie sonst nirgends unter Menschen. Man muss dies Land lieben und verehren.

[7] These are the words of the representative humanities scholar of modern Japan, Kuwabara Takeo, in a dialogue with F.A. von Hayek: 'Keizai Hatten to Nihon no Bunka' (Economic development and Japanese culture) (in Hayek and Imanishi Kinji *Shizen, Jinrui, Bunmei* (Nature, man and civilization), Tokyo, 1979.

[8] Yamamoto Sanehiko, 'Sannin no Hinkaku' (Three distinguished visitors), April 1933 issue of *Kaizo*.

[9] Described in detail in Seki Tadaka et al. (ed.), *Zasshi 'Kaizo' no 40-nen* (The 40 years of the 'Kaizo' magazine), Tokyo, 1978, and Yokozeki Aizo, *Watashi no Zakkicho* (My notebook), Tokyo, 1965.

[10] Shinobu Seizaburo, *Taisho Demokurashii-shi* (A history of Taisho democracy), Vol. III, Tokyo, 1959, p. 695.

[11] According to the agreement exchanged between ┌ ᵎstein and Kaizosha (dated 15th January 1922), Einstein's honorarium was 2000 British pounds. The travel expenses of both Einstein and his wife and the Professor's living expenses during their stay (approximately 700 British pounds) were borne by Einstein. (Kaizosha paid for the living expenses of Mrs. Einstein). The original plan was for Einstein to give six general lectures and six scientific lectures (Tokyo), but eventually he gave eight general lectures. In a letter to Yamamoto, Einstein protested that "This is a bad custom in the name of charity", but in the long run he failed to have his way. In this respect, it is inevitable that Einstein later referred to Yamamoto as a "cunning publisher".

[12] Thomas S. Kuhn, *The Structure of Scientific Revolutions*, Chicago, 1962, 1970.

[13] Mendel Sachs, *Ideas of the Theory of Relativity*, Jerusalem, 1974 was extremely useful in considering this pattern of relativistic thought.

[14] E.G. Boring, 'The Dual Role of the Zeitgeist in Scientific Creativity', in P. Frank (ed.), *The Validation of Scientific Theories*, 1961. In considering someone's influence, if the people who are thought to have been affected have written about this in their diaries or works, this acts as formal written evidence of this influence. Without doubt this is important in historical research, but depending on this alone involves the risk of failing to grasp the undercurrents of the influence. Thus at the same time as picking out such material, I intend to portray the impact of Einstein's visit to Japan by highlighting the homological phenomena.

[15] Michael Polanyi, 'The Logic of Tacit Knowing', in *Knowing and Being*, Chicago, 1969.

[16] Published in the February 1923 issue of *Shinshosetsu* (New novels). On the theme of 'Opinions on the Dr. Marusawa alchemy affair and Dr. Einstein's visit to Japan in these times of emphasis on science'.

[17] R.G.A. Dolby, 'The Transmission of Science', in *History of Science*, XV, 1977.

[18] An account of this was reproduced in the English language newspaper *Japan Weekly Chronicle*, and became so widely known abroad that it was also recorded in the bulky biography by Ronald Clark.

"The report described a discussion "of quite unusual nature" by the Cabinet Council: "One of the Ministers asked whether ordinary people would understand Professor Einstein's lectures on the theory of relativity," it began.

Mr. Kamada, Minister of Education, rather rashly said of course they would. Dr. Okano, Minister of Justice, contradicted Mr. Kamada, saying that they would never understand. Mr. Arai, Minister of Agriculture and Commerce, was rather sorry for Mr. Kamada, so he said that they would perhaps understand vaguely. The headstrong Minister of Justice insisted that there could be no midway between understanding and not understanding. If they understood, they understood clearly. If they did not understand, they did not understand at all. A chill fell on the company. Mr. Baba, the tactful director of the Legislation Bureau, said that they could understand if they made efforts. Their efforts would be useless, persisted the Minister of Justice. He had himself ordered a book on the theory of relativity when the theory was first introduced into Japan last year and tried to study it. On the first page he found higher mathematics, and he had to shut the book for the present." (Remainder omitted)

[19] Announced in a three-line headline at the top of page three of the *Kahoku Shimpo* of 3rd December 1922 (the day Einstein was to arrive in Sendai). This was translated into German by Georg Würfel and presented to Einstein while he was still in Moji.

[20] Ishiwara Jun, 'Tanka no Shin-keishiki ni tsuite' (On the new form of Tanka), in the November 1923 special edition of *Shukan Asahi*.

[21] Yura Kimiyoshi, 'Hae no Kamera-Ai' (The camera-eye of a fly), in Yura Tetsuji (ed.) *Yokomitsu Riichi no Bungaku to Shogai* (The literature and life of Yokomitsu Riichi), Tokyo, 1977.

[22] Tanikawa Tetsuzo, 'Daiyonjigen no Geijutsu' (The Fourth-dimensional Art), in *Miyazawa Kenji no Sekai* (The world of Miyazawa Kenji), Tokyo, 1970. Ono Takaaki, *Miyazawa Kenji no Shisaku to Shinko* (The thought and belief of Miyazawa Kenji), Tokyo, 1979.

[23] Kaneko Tsutomu, 'Miyazawa Kenji no Yonjigen Genso' (The four-dimensional fantasies of Miyazawa Kenji), in *Einstein Shock* II.

[24] *Shin-Aichi*, 2nd December 1922.

[25] *Tokyo Nichinichi Shimbun*, 3rd December 1922. The writer of this editorial was Masumoto Uhei, a marine engineer who was the Japanese government delegate to the first meeting of the International Labor Organization (ILO) in Washington at the end of October 1919.

[26] *Shin-Aichi*, 14th December 1922.

[27] *Yomiuri Shimbun*, 13th December 1922.

[28] 'Einstein Hakase Toben' (Replies by Professor Einstein) in February 1923 issue of *Kaizo*. The whereabouts of the German original are unknown. The following is the complete English translation from the Japanese.

Dear friends,

I was unable to answer your letter earlier because I had lost both it and the address. Mr. Yamamoto has again furnished me with the address, so I would like to reply to your questions, but am unable to recall them in detail.

Firstly I must point out that my observations about Japanese social and political conditions are so very limited that even I cannot rely on my own judgment. Concerning the first point [In the list of questions this was about the aggressive Japanese imperialism], I observed two matters which at a glance seem to be incompatible. There is neither conspicuous poverty nor a lack of money, but nevertheless piecework carried out at home is dreadfully ill-paid for the most part. As far as my observations showed, I believe that this riddle can be interpreted as the result of the fact that the people have few desires, their way of life is suitable, and in addition they are particularly moderate in their consumption of alcohol. Be that as it may, in any case I believe that this country will become increasingly industrialized, and owing to the political situation it will become necessary to organize the working classes. If this organization is to be of value to the whole nation, it must not turn into a malicious movement which is carrying out opposition merely for the sake of oppositon, such as happened with us in Europe for a long time. You must realize in particular that the main factor behind the lowering in wages for home work lies in the overpopulation of this country, and therefore it cannot be done away with by purely political methods. On the other hand, the struggle against militarism seems to me to be a purely political issue. In my opinion this constitutes a real danger to this nation. This is because, owing to its geographical position, Japan is fortunate enough to require little military protection. The Washington Conference created the first opportunity enabling us to hold out some hope on this matter. I am convinced that in the future efforts by the people will be linked to international cooperation as well as organization, and will never combine with military planning. I hope that Japan will draw a conclusion from this for its own sake and for the sake of all countries in the world.

Special regards,
A. Einstein

[29] Kagawa Toyohiko, 'Shimpen Zakki' (Personal notes), in January 1923 issue of *Kumo no Tsue*, (published in the 24th volume of the collection of his works, *Kagawa Toyohiko Zenshu*).

[30] The original German in Einstein's diary of his visit to Japan: Reise nach Kobe. Mittagessen mit Yamamoto und dem bedeutenden jungen Sozialpolitiker im Fischerdorf von Kobe.

[31] However, instead of Kagawa, the Kirisuto-kyo Seinen-kai (Association of Christian Youth) of Tokyo sent four questions on religion to Einstein. His reply was dated 22nd January 1923, the same day as his reply to Komaki and his colleagues, and was written aboard ship on the return voyage. The German original of this reply can be found in the February 1922 issue of *Kaicho*.

[32] Translation of Georg Friedrich Nicolai, *The Biology of War*, Zurich, 1916. Nicolai was a private lecturer at Berlin University at the time, and at the start of World War I he was persecuted because he and Einstein had issued an anti-war declaration (Anruf an die Europäer) known as the Nicolai-Einstein declaration.

[33] Yamamoto Senji, 'Hikyo-naru Jisho Chishiki Kaikyu' (Cowardly self-styled intellectuals), in April 1923 edition of *Kaizo*. The following is taken from this.

[*Discussion on the Intellectual Class*]

Einstein: "What is the attitude of Japanese intellectuals towards the peace movement?"
Yamamoto: "On the whole individual intellectuals have an understanding of and are in sympathy with the peace movement, and would like to welcome demands for thorough-going arms reduction. Even though each individual entertains such hopes and ideas, however, each puts his own safety above all else and is merely hiding away inside a shell, and lacks the courage and enthusiasm to risk his life in defending his beliefs. Consequently we cannot expect any major contribution towards realizing world peace from Japanese intellectuals, as they have no unity. In my opinion, if chauvinists in Japan nowadays were to sound off loudly in an attempt to wage war against another country, the only ones who would attempt in a rage to oppose this stupid jingoism and who could be equipped in the future with the actual ability to smother this in advance would be the workers' organizations. In order to achieve our ends, therefore, rather than depending on cowardly intellectuals who take no action, I believe that in the present case it is urgently appropriate to attempt to supply the necessary preliminary knowledge to workers' organizations, which have unshakable confidence and abundant ability to implement this knowledge." (It seems that Mr. Inagaki, who is an expert diplomat, translated my answer here in a somewhat ambiguous manner.)
Einstein: "I feel that the influence of workers' organizations on actual politics is not as great as you imagine. I remain convinced that education and propaganda work amongst intellectuals is necessary and of the first consideration. However, I am completely ignorant of the different situation in Japan, so I would like to avoid any definite statement on this matter."
(Omitted)
Einstein: "I know nothing about Japanese intellectuals, but even during the short time that I have been in this country the impression that I have received from the people whom I have met on the streets is that they are extremely gentle and warmhearted, and that they have an ardent love of peace. Inspiring this people with pacifism would be no difficult task."
Yamamoto: "Of course the people of any nation are gentle and warmhearted and love peace. In Japan today, however, there is no appropriate method of expressing the political views of the people, so their peace-loving attitude has no direct effect at all. Even if the workers are opposed to war there is nothing they can do about it, as there is no Social Democrat Party such as exists in Germany."

[34] Yasuda Tokutaro, *Omoidasu Hitobito* (People I recall), Tokyo, 1976, p. 32. Yasuda was a medical student at Kyoto University at that time, and was a cousin of Yamamoto Senji. He later carried out research into the history of manners and customs and the history of science.

[35] For further details see the author's *Einstein Shock* Vol. II, Chapters 6 and 7.

[36] Saegusa Hiroto, 'Sotai-sei Riron no fukumeru Tetsugakuteki Mondai' (Philosophical issues in the theory of relativity), in *Shiso*, No. 10 (July 1922). Tosaku Jun, 'Butsuri-teki Kukan no Jitsugen' (Materialization of physical space), in *Tetsugaku Kenkyu*, **10**, (7) (November 1925).

[37] Tanabe Hajime, 'Sotai-sei no Mondai' (Problems of relativity) in *Tetsugaku Zasshi*, No. 302 (1912). (In *Tanabe Hajime Zenshu* Vol. 14). Tanabe Hajime, *Sotai-sei Riron no Benshoho* (Dialectics of the theory of relativity), Tokyo, 1955. (In *Tanabe Hajime Zenshu* Vol. 12).

[38] Nishida Kitaro, "Butsuri Gensho no Haikei ni Aru Mono" (What lies in the back-

ground of physical phenomena), in *Shiso* No. 27 (January 1924). (In *Nishida Kitaro Zenshu*, Vol. 4).

[39] Fukuda Tokuzo, 'Yukizumareru Sekai to Sono Tenkai' (The world at an impasse, and its development), in *Jitsugyo no Sekai*, December 1920. Soda Kiichiro, *Bunka Kachi to Kyokugen Gainen* (Cultural values and ultimate concepts), Tokyo, 1922, 1952. Ishiwara Jun, 'Fukuda Hakase no iwayuru Appuriori no Hitei to iu Koto ni Tsuite' (On the negation of the alleged *a priori* of Dr. Fukuda) in *Jitsugyo no Sekai*, January 1921.

[40] For example, the optical industry was established centering on the military (Nagaoka became adviser to Nippon Kogaku), and over thirty national research institutions had been established by 1922, including the University of Tokyo's Aviation Research Institute (1921) and Tohoku University's Metal Materials Research Institute (1922). In addition, the total number of private research institutions throughout the nation as of the end of 1923 stood at 162.

[41] In addition to president Hozumi Nobushige, about 40 members of the Imperial Academy attended the welcoming reception held at the Koishikawa Botanical Gardens belonging to the University of Tokyo on 20th November, including Nagaoka Hantaro, Inoue Tetsujiro, Kitazato Shibasaburo, Fukuda Tokuzo and Minister of Justice Okano. The welcoming address was drafted by Nagaoka, who was on the welcoming committee, and Hozumi, Fujisawa Rikitaro, Inoue, Mikami Sanji and others joined together in changing the wording on two occasions. Along with a German translation, it was then placed in a box ordered specially from a school of fine arts and presented to Einstein.

[42] The special lectures were held every day except Sunday between Saturday, 25th November and 1st December 1922 from 2 p.m. for about one and a half hours, followed by one hour of questions and answers in the lecture theater of the Physics Department of the Tokyo Imperial University. Most of the audience were faculty members, researchers or students from the University of Tokyo, Kyoto University, Tohoku University, Waseda and Keio Universities, the 1st and 4th Senior High Schools, and nearly all prominent physicists and mathematicians of note put in an appearance. The attendance register lists a total of 135 people.

[43] Doi Fuzumi was a pupil of Nagoka Hantaro, and attacked the principle of the constant speed of light and the principle of a fixed electronic charge. These two English papers were condemned by Nagaoka as "a disgrace to the nation", and came under a general attack in the midst of scientific academism. These papers are included in Doi Fuzumi, *Einstein Sotai-sei Riron no Hitei* (Rebuttal of the Einstein theory of relativity), Tokyo, 1922.

[44] For example, see Yukawa Hideki, *Tabibito* (Travellers), Tokyo, 1958, p. 143.

Postscript: Following normal practice in historical works, the Japanese names in this paper are written in the Japanese order of family name first.

University of Osaka Prefecture

THOMAS F. GLICK

CULTURAL ISSUES IN THE RECEPTION OF
RELATIVITY

1.1. COMPARATIVE RECEPTION STUDIES

The objective of an enterprise of comparative scholarship, such as the present volume, is to set out parallel cases of the reception of a single set of ideas, in order to get an overall picture of the balance between general and culture-specific processes. The unit of analysis is the nation and there is no presumption that any one case is more normative than any other. In textbook treatments of reception, the experience of an innovation or new idea in its society of origin may be taken as a normative phenomenon against which its reception elsewhere must be weighed. In the case of paradigmatic change, the bias is even greater: national scientific communities which made no contribution to the paradigm, although receiving and adopting it, are reckoned as uninteresting or derivative. What is the interest, therefore, of the reception of relativity in a country that made no original contribution to it? There are several lines of response to this kind of question. One is to posit, as Mannheim did, that the differential receptivity of cultural or social groups to specific ideas reveals something significant about social and cultural processes. A second is to suppose that such differentiation illuminates science as a social and cultural phenomenon. A third is to challenge received views of what is normative in science by expanding the number of cases examined and broadening the basis of study to include scientific communities thought "peripheral" to mainstream science. A fourth is to change the focus of study from the internal development of ideas to social and cultural contexts of their development and diffusion. A fifth is to cast a social definition of science as ideology; to do so is to finesse the question of what is normative experience, to rid the research program of ethnocentric bias, and to even out the distinction between invention and innovation, discovery and reception. As Tzvetan Todorov has observed in another context, "the reception of . . . statements is more revealing for the history of ideologies than their production."[1] Of course. The proof is in the kind of enterprise this volume undertakes, because such comparisons and contrasts as the national case studies present ulti-

381

Thomas F. Glick (ed.), The Comparative Reception of Relativity, 381–400.
© *1987 by D. Reidel Publishing Company.*

mately reveal ideological distinctions (or, rather, a range of social, cultural and ideological distinctions of which the latter are most significant). Put another way, Goldberg's "national style" involves different ways of doing science, different ways of thinking about science, the embeddedness of science in different value systems, and the attaching to science of different ideological markers. Even if ideology may not inhere in a given scientific idea as relativity, it appears as part of the baggage that ideas carry across cultural boundaries.

1.2. COMPARATIVISM OR ESSENTIALISM

Is the acculturation of an idea simply a constraint upon some cognitive core, or is the cognitive structure itself transformed as it crosses a cultural frontier? Biezunki's *glissements semantiques* are a universal snare of reception studies. Different groups may use the same terminology but attach vastly different meanings to it. Do the contributors to this volume assume a common unit of analysis ("relativity") or would they admit (as David Hull observes with respect to the diffusion of Darwinism) that conceptual systems change – over time, he says, and, we might here add, over space as well.[2] The reception of relativity has much in common with the prior reception of Darwinism (the similarities and differences were noted at the time), as well as with the contemporary reception of Freudian psychology.[3] But it is also true that the reception of relativity, when compared to Darwinism or Freudianism, was, in both cognitive and social senses, a much more sharply bounded phenomenon. Due to the complexity of the ideas the boundary between scientific and non-scientific comment would appear more easily drawn than was the case in the other two receptions where that line was blurred. It is easier to talk of a "French Freud" or "American Darwin" to indicate sharp transformations of the cultural system as German and English ideas, respectively, were acculturated elsewhere. It is less easy to speak of a Russian or Spanish Einstein. The "national style" in relativity was something less than the cultural reinvention that took place in the reception of Freud, but still something more than what is suggested by unquestioning assumption of an essential conceptual core.

A scientist could accept virtually all the tenets of special relativity and still not be a relativist (Poincaré). Or, we can point to the Spanish engineers Herrera, Pérez del Pulgar and Burgaleta who reckoned themselves relativists while rejecting one of the Einstein's principle tenets, the constancy of the velocity of light. In the comparative approach, therefore, we must assume the probability of different semantic structures, socially or culturally based, in different societies and take these structures at face value. The "national style" must involve a process of fusion between a set of ideas and the value system which constitutes any specific culture.

2.1. THE SOCIAL AND CULTURAL STRUCTURES OF SCIENCE

Clearly the "national style" of reception, or any one society's experience, is tightly linked to the professional structure of its scientific community. In the case of relativity national receptions were shaped by the strength and philosophical disposition of the three receiving disciplines – physics, mathematics and astronomy – and the balance of roles played by each discipline in the reception. Elsewhere, I have presented a model which seeks to appraise and define the ability of disciplinary groups to deal with new ideas when they diffuse across cultural boundaries and characterize "active" and "passive" modes of disciplinary reaction.[4] In an "active" reception members of a discipline have direct access to the new ideas (through personal contacts and ability to remain abreast of current research in the language of the new idea); they have both the maturity and sufficient personnel to debate the new ideas critically and institutional structures sufficient to propagate them authoritatively. At its most "active", a disciplinary group will contribute to the further development of the new paradigm through its own research. Among disciplinary groups not as highly institutionalized, reception may be limited to a "passive" role of serving as a conduit for the diffusion of the new paradigm. Passive disciplinary groups typically have no direct contact with mainstream science, are dependent upon translations in order to follow developments in it, and display both temporal and cognitive lags in assimilating new concepts. Such "active" and "passive" traits can be related to the broader social context through the structure of the educational system, the grade of acceptance of the scientific ethos by the elite, and so forth.

2.2. DISCIPLINARY RECEPTIONS

The interrelationship of differently weighted disciplines sets the tone of scientific reception. Thus, in Spain, only mathematics had sufficient weight as a discipline to interpret relativity for the scientific community at large and the public (the participation of individual physicists notwithstanding). But the inability of Spanish mathematicians to contribute to the further development of the paradigm still marks their reception as "passive". The balance among the three disciplines was similar to Italy where, however, mathematicians contributed actively to relativity's development. Within national communities of physicists the balance between theoreticians and experimentalists was obviously important. In this regard two noteworthy patterns of disciplinary substitution present themselves: first, the substitution of mechanics or mathematical physics (considered mathematical subdisciplines) for theoretical physics – a salient feature of French, Italian and Spanish receptions); second, the substitution of engineering for physics wherever the latter was underdeveloped.

Paty notes that in France there was no theoretical physics in the true sense until the 1930s; before that time mathematicians (those who cultivated mathematical physics) dominated the theoretical discussion of relativity and, since they viewed it in formal terms, this proved an obstacle to its introduction as physical concept. Judith Goodstein shows a similar situation in Italy.[5]

Most of our authors did not follow the engineering debate. Reeves discusses the positive views of Lori (an electrical engineer, like many who supported Einstein), and Sredniawa provides examples of both favorable and hostile Polish engineering opinion. Vizgin and Gorelik allude to the hostility of engineers (and applied physicists) to the counterintuitive conclusions of relativity. Only in Spain was the substitution of engineering for physics apparent, although that was the standard situation in Latin America. In his trips to Spain, Brazil, Argentina and Uruguay, Einstein spoke mainly to engineers who comprised the rank-and-file of those with interest in his ideas. It is interesting to note that Argentina did not have in the 1920s a mathematics community comparable in sophistication to that of Spain. Questions posed to Einstein at the Mathematical Society of Madrid in 1923 struck him as interesting and reasonable, while those addressed to him by Argentinian physicists two years later he found difficult to meet with a straight face.[6] Here we have two societies socially and culturally similar, whose scientific communities were quite differently structured.

Disciplinary traditions are of obvious importance. The tradition of non-Euclidean geometry in Russia, the revolt against purism in geometry in Italy and Spain, provided a congenial atmosphere for relativity, one that was congruent with the traditional concerns of the discipline, amongst mathematicians there.

Scientific ideas cross national borders with relative ease; disciplinary boundaries are far harder to breach. As Giuliano Pancaldi observed with respect to the reception of Darwinism, if scientists must reorganize the way they do research as the price of accepting a new theory, there will be resistance. It is harder, in this view, to free oneself from an old disciplinary tradition than it is to shed old theories.[7] Crelinsten notes, with regard to the reception of relativity by American astronomers, the reception of a revolutionary paradigm has the effect of stimulating a process of self-assessment within the discipline.[8] In Spain, this led both to self-confident statements (by J.M. Plans) that Spanish mathematics was no longer isolated from the European mainstream, as well as defensive confessions (by E. Terradas, in a letter to Levi-Civita) that Spain was simply infertile soil for mathematics.[9]

2.3. PHILOSOPHICAL CLEAVAGES

Philosophical cleavages in national science communities were highly significant, with positivism a touchstone. In Spain, Einstein's Catholic followers were at least situational anti-positivists although they didn't deal with the issue in any explicit way. In Brazil, relativity precipitated a show-down in the Academy of Science between anti-relativity positivists and pro-relativity idealists (mainly mathematicians).[10] In Germany, Planck made support of relativity part of his anti-positivist position.[11] In France, entrenched mechanistic positivism in physics preordained a hostile reception. Biezunski argues that French physicists drew on various philosophical traditions including Comtean positivism (which worked against relativity) and Cartesian mechanism (which could work either for or against it). In the United States, however, Goldberg shows that special relativity was "positivized" to the extent that it was typically presented as experimentally based.

3.1. COMMONALITY OF SCIENTIFIC AND NON-SCIENTIFIC CIRCLES: LANGUAGE

Students of Darwinism and psychoanalysis accept as given a high level of interchange between the cognitive structures involved and their social contexts. Both participants in the development and reception of relativity and latter-day historians of the phenomenon insists that "popularizations" of relativity were lacking in physical meaning. The Monsieurs Homais of the world had their reasonably informed say on Darwin and Freud, but were barred from any real appreciation of relativity.[12]

In this context, it is interesting that some contributors to this volume stress the blurring of the science/non-science boundary. The general point is that which has been made most eloquently by Darwin scholars: that scientists participate in other solidarities within their societies and transpose into those solidarities their scientific concerns.[13] Reeves and Kaneko see language linking scientific and non-scientific spheres of discourse. By the very nature of relativity's common language, relativists themselves could not help but participate in various aspects of the "popular" discussion. Indeed their very participation ensured that the "semantic slides" really slid across the science/non-science border. Biezunski gives press coverage of soirées where the process actually was observed in progress.[14] Illuminating is his notion that when scientists do not agree on the "language", the general public is free to appropriate what it will. There is a logic here: the public is attracted by controversy, but the ideas/images they pick up during such a period may extend well into the period following the formation of scientific consensus. For example, in the first three decades of this century, when Darwinism was "in eclipse" and all kinds of competing evolutionary theories were under discussion,[15] the American public (and perhaps others) received a Lamarckian message which persists to this day. The same was true of the diffusion of the "new" psychology in the 1920s and 30s when competing Freudian, Jungian and Adlerian models of personality and development made available a profusion of concepts and images for public appropriation, frequently in combinations which the leaders of these schools found contradictory.

3.2. INCOMPREHENSIBILITY

The theme of relativity's incomprehensibility has been developed at length by Biezunski.[16] His view is that the cognitive issue is inextricably linked to Einstein's persona: Einstein did not fulfill any of the commonly-held expectations of what a scientist was supposed to be like. That made *him* incomprehensible which, in turn, provided a convenient popular explanation for the incomprehensibility of his ideas. The fact that he expressed complex ideas using everyday terms like time, space, and so on, further added to the confusion, as many commentators have noted.

Ultimately popular notions of the incomprehensibility of Einstein's theories simply reflected, I believe, what was said of them in the scientific community. On scientific incomprehension I find Jean Eisenstaedt's analysis particularly convincing: Einstein's theory was not empirically based yet experience confirmed it.[17] The disjuncture between abstract formulation and observational proofs accounted both for the sense of awe on the part of those scientists whom Einstein convinced and disgust – incomprehension – of those who were opposed. The "contras" were disgusted because the method by which the theory was generated was unthinkable to traditional experimentalists. We are still far from understanding the processes whereby the general public picked up clichés already established in the scientific community and marshalled them to justify what were, at base, ideological stances favoring or opposing Einstein. According to Paty, most French scientists had not heard of Einstein before 1919. In countries where the same was true (all but Germany?) scientists and the general public would have formed their opinions contemporaneously, in which case one would have to assume a complex interplay between scientific and public opinion in the formation of the dispositions of each towards Einstein and relativity (see 5.1).

But there may be a simpler, more universal explanation. According to Pierre Bourdieu, "that the public is irretrievably doomed to incomprehension [is] . . . so profoundly embedded in the social definition of the intellectual's vocation that it tend[s] to be taken for granted."[18] Incomprehension is what intellectuals expect of the public. Following Bourdieu's logic we could argue that lines of force which manifested themselves in the "intellectual" field of a given country during the relativity debate were those demarcating various groups of Einstein's supporters and detractors among scientifically educated and lay members of the intelligentsia. Such groups manifest both competitive relations and those of functional complementarity as well. Here there is complementarity among, for example, anti-relativity scientists, *gens du monde* displaced as arbiters of high culture, and elements of the "public" beyond the intellectual field proper (see section 7, below).

3.3. EXTRA BAGGAGE

In the process of interchange and appropriation, ideas inevitably pick up connotations and interpretations that have no logical or necessary connection with the base idea: for example, Darwinism and progress or the rhetoric of economic competition. In the same wise, cultural and philosophical relativism antedated relativity but, after 1919, were assumed to have been influenced by it. In some, many, domains of discussion, these connotations virtually displaced the base ideas. This semantic association functioned on many levels and bore approximately the same relationship to relativity as the idea of progress did to Darwinism. In typical appropriations, Einstein's theory was assumed to have provided "scientific proof" for a previously held position. Thus Ortega y Gasset was accused of favoring relativity because it appeared to legitimize his own concept of perspectivism, previously formulated. Opponents of philosophical relativism, such as traditional Catholic clerics assumed Einstein to have supported that position.

Relativism, on the whole , was baggage for the non-scientist. But even for scientists, relativity bore ineluctable connotations that had nothing to do with Einstein's theories per se but which become attached to them in the process of reception. As an idea in diffusion, relativity entered specific receiving cultures associated with the prestige of leading scientists. Clearly to diffuse as an already prestigious idea rather than in a more neutral form, enhances a favorable reception. In a Spanish textbook, for example, relativity was identified (in 1916) as the "Einstein-Planck" theory.[19] That relativity was labeled a product of German science surely eased its reception in Russia and Poland, hindered it in France. For a whole slew of experimentalists, applied scientists and engineers, relativity bore the stigma of mathematicism, and of exaggerated abstraction or formalism which justified its rejection *a priori*.

Paul Johnson has observed that "the impact of relativity was especially powerful because it virtually coincided with the public reception of Freudianism"; both doctrines appeared to the public as bearers of the same message: that "the world was not what it seemed".[20] In the light, the association by American commentators of Einstein's ideas with dream experience (Missner) acquires a specific cultural context. That relativity should, in this way, acquire psychoanalytic baggage attests to the defraction of multiple ideas through the same cultural field contemporaneously.

Many traits associated with Einstein's persona (see 5.3) diffused inseparably from his ideas particularly on the popular level: he was not the image of the standard scientist; his ideas did not conform to scientific standards, etc. His science was Jewish just like Einstein; it was revolutionary and so was Einstein.

Relativity then came to be associated with differing images of science held by different cultural, social or professional groups. In all the non-European countries visited by Einstein in the 1920s (Japan, Argentina, Uruguay, Brazil) and also in Spain, he and his ideas were symbols of modernization. In commenting on the insufficiency of mathematics instruction in the Spanish Naval Academy, a technical institution, a critic of the school noted as proof of his complaint that relativity was not imparted there.[21] This was a standard message: you cannot modernize without achieving mastery in science; the symbol of mastery of science is relativity.

To some, relativity revealed basic attitudes towards science. To some science was not allowed to generate doubt (Biezunski's French positivists); or, to turn the prescription around, science had to generate certainty (traditional Catholics). For those inclined towards relativistic perspectives, on the other hand, science does not discover absolute truths (Reeves), an antipositivist view of science which leads to very different philosophical and political conclusions.

4. INTERRELATEDNESS OF NATIONAL RECEPTIONS

To what extent were national experiences colored by the experiences of other societies? Marshall Missner insists that the American response to Einstein in 1921 was sui generis because it was during this trip that his persona became known to the public. It is then that the main elements of the Einstein myth, including the incomprehensibility of relativity, were established.[22] We would then have to assume that subsequent public discussion of Einstein and relativity in the European press was highly colored by the American experience. This supposition would have to be verified by searching the daily press of different countries. In Spain, to cite a counter-example, Einstein's American tour of 1921 was not covered, but his French visit the following year was. Those members of the Spanish intelligentsia who followed the latter episode, as a result, were supplied with a full range of clichés to confront the Einstein phenomenon in Spain in 1923. The Argentinian leg of his 1925 tour was in a like manner set up by extensive coverage in the Argentinian press of the Spanish tour.

The interrelated responses of various scientific groups is clear: German theoretical physics (and Ehrenfest particularly) in Russia; Italian mathematics (Levi-Civita particularly) in Spain; German physics (Jakob Laub particularly) in Argentina.[23]

5.1. EINSTEIN'S PERSONA: WHAT CAUSED HIS FAME?

The issue is related to incomprehensibility, insofar as incomprehensibility was a key element in his fame. The most categorical hypothesis is that adduced recently by Lewis Elton: Einstein's fame arose from controversies surrounding the verification of his theory and from the myth of its incomprehensibility, both of which originated in the Anglo-Saxon world.[24] For Missner, the American press was *the* instrument that made Einstein a celebrity. These two conclusions are complementary because, Missner continues, relativity was famous first and only subsequently did Einstein himself become a celebrity. What actually made him famous was, in Missner's view, the contingent association of the many elements we have been discussing. The contingency created the illusion of a great press campaign guided by an invisible hand.[25]

There are real problems of historical and analytic value here. Did incomprehension originate among scientists or in the press? I think it could be shown that most of the clichés about relativity originated with anti-relativity scientists before 1919 and diffused quickly after November of that year throughout many domains of discourse. In the Anglo-American world or elsewhere? Probably what was said in the English-language press was the great fount of what was said later on everywhere. But early on a variety of different styles emerged. The French journalistic style, in which (Biezunski shows) snobbism was one of the keynotes, weighed heavily in Spain among the non-scientific intelligentsia. The tonality of the Spanish newspaper account of Einstein's visit and relativity was that of the French literary bourgeoisie, with local inflections. What Spanish mathematicians said was tied into the Italian discussion and some of the popular engineering discussion was attuned to the popular discussion in England.

5.2. HIS TRAVELS.

Einstein's much publicized trips (to the United States and Italy in 1921, France in 1922, Japan and Spain in 1922–23, Brazil and Argentina in

1925) raise a number of issues. What did the lionizing crowds who pressed in on him throughout his travels get from such an exercise? Various professional and political groups sought to associate Einstein's prestige with their own cause or objectives. With regard to the professions, Einstein's presence reinforced the prestige of dominant groups (physicians in Spain and Argentina – note his visits with former and future Nobel laureates Santiago Ramón y Cajal and Bernardo Houssay, respectively[26] – mathematicians in Italy and Spain; engineers in the Iberian nations generally).

The interaction between Einstein and self-serving admirers caused a certain backlash. Thus Einstein is reported in Japan to be jobbing "a cheap article at a night stall" (Kaneko), while his staged performances in Spain were likened to a music-hall act. Sometimes Einstein is presented as an establishment pawn: he is carted by officials from place to place, where he opens his bag of three speeches for another uncomprehending audience.[27]

It should be remarked in passing that Einstein's travels generated a tremendous avalanche of local commentary, mainly in newspapers, constituting a great and generally unmined source of information on Einstein as a personality and public figure and the cultural impact of his ideas. American historians of ideas are unused to the newspaper as a major source of data because newspapers have not played the focal role in American culture as they have in Europe. Although Einstein insisted, when travelling, on not giving newspaper interviews, more than one enterprising reporter managed to break down his resistance and record his thoughts. Much of the information gleaned (in Spain and Latin America for instance) consists of trivial comments spun out for the benefit of local readers (the genius' opinion of the local folk dance, architecture, politics, and so forth) and reflected Einstein's knack for telling people what they wanted to hear without thereby revealing too much about himself. Still, when combined with his travel logs, a picture begins to emerge of how he viewed foreign cultures, together with a running comment on relativity, its scientific acceptance, and what the public could or could not derive from it (see Appendix).

What stands out about his travels, viewed in comparative perspective, is the universality of public response to him. The positive mass response occurred everywhere, whether there was a Jewish community or not (the Argentinian experience was very similar to the Spanish, in spite of Einstein's involvement with the Argentinian Zionist movement), or whether the culture was western or non-western, as in Japan. The Japan trip, unlike the others, was a planned media event in that it was sponsored by a publishing house, but the texture of the visit and his interaction with the public was very similar to his reception elsewhere. Comparative history is frequently only contrastive; it is easier to identify the anomalies in a structure than it is to explain the similarities in the face of strikingly different social and cultural contexts. Inevitably, then, the comparative method leads us away from Einstein; it was not he who accounted for the similarity in public response, but something in the nature of those cultures and the way science was viewed in them.

5.3. DIFFERING IMAGES

A number of authors (Biezunski, Missner) stress the theme of Einstein the destroyer as one of the most prevalent images. This is, of course, a negative image, associated with the supposed destruction by Einstein of the Newtonian worldview or the traditional Western cosmology. The destroyer image persists to this day. The Argentinian generals of the 1980's according to Amos Elon, "associated psychoanalysis with Freud, and Freud with Einstein (who was said to have undermined the Christian sense of time and space), and Einstein with Marx and Communist subversion."[28] In the 1920s, antirelativists, particularly those originating in traditional Catholic circles, included relativity in stock lists of modern horrors which they condemned, including modern art, modernist philosophy and socialism.[29] The generals simply present a stripped down list, the more sinister because of its scarcely concealed anti-Semitism.

What strikes me as unusual is that our authors do not mention the opposite image – Einstein the magus, the master (not the destroyer) of time and space – which was just as widely diffused an image. On an excursion to Toledo in 1923, Einstein asked Ortega y Gasset how an abstract idea like relativity could be of interest to the masses. Ortega thought the mass appeal had to do with the conjunction of a new cosmological theory with the post-World War I loss of faith in European society: "In such a circumstance there appeared your work, in which laws are promulgated for the stars, which obey them. The human masses have always perceived astronomical phenomena as religious. In them, science is conjoined with mythology and the scientific genius who masters them acquires a magical halo. You, Sr. Einstein, are the new magus, the confidant of the stars."[30] This image, which underlies much of the popular and "pop" commentary on Einstein and relativity, was similar in content and identical in conception with the much disseminated American myth concerning Thomas Edison, "the *wizard* of Menlo Park". To invent the electric light, the myth held, Edison had to refabricate the myths of Nature. No one else had been able to accomplish this feat but Edison, because he was a wizard, could perceive secrets of nature hidden to others. The public viewed Edisons's "magic" with awe, tempered with the inference of evil that inhered in wizardry.[31] The same was true of Einstein. The awe that scientists expressed over Einstein's magical formulation fed this aspect of the myth and in turn was reinforced by it. The magical creator was also a destroyer and the two images fused seamlessly in the post-1945 myth regarding Einstein and atomic energy.[32]

6. EINSTEIN AND POLITICS

Einstein had a well-delineated image owing to his notoreity in the anti-war movement. That was unusual for a scientist. Others had political images but they were associated with those of their country. Einstein's was unusual, striking, and international. This image was extended in the 1920s by his espousal of pacifism and his adoption of Zionism.

6.1. Varieties of Political Appropriation

Political appropriation is perhaps the most direct way to examine the symbolic values that different cultural and social groups were ready to vest in Einstein. During the course of Einstein's week in Barcelona in 1923, Catalan nationalists made every attempt to associate the physicist with their cause. Such an enterprise was consistent with Einstein's espousal of the right of small, oppressed cultural groups, such as European Jews, to self-determination. Catalan nationalism was, on the whole, a politically conservative movement associated with the industrial bourgeoisie. (Einstein was successful, notwithstanding, in convincing a Catalanist socialist leader, Rafael Campalans, to drop the word "nationalist" from his program because of the term's reactionary connotations.) That rightist dictators like Mussolini and Primo de Rivera could also make use of Einstein to justify the "relativization" of their opponents' political legitimacy attests both the immensity of his prestige and the universality of anti-positivist philosophies which relativism/ relativity was seen to support.

Vizgin and Gorelik show, in the case of the Soviet Union, that there was a split among Marxist physicists over how to read relativity ideologi-

cally. Those who rejected relativity in the name of dialectical materialism do not seem so different from others elsewhere who rejected the
new concepts in the name of positivism. Other Marxists, however, were
able to establish a dialectical relation between the relative and absolute
and find in this a confirmation of Marxist philosophy.

Like Darwin, therefore, Einstein's perceived message was susceptible
to appropriation by groups of such differing political orientation. This is
related to the paradoxical use of simple, familiar words like time, space,
relative and absolute to (ostensibly) describe complex concepts. The
everyday meanings of these terms were so broad that, taken together,
they constituted an "open text" which persons of starkly contrasting
ideologies could make over to suit their political preconceptions.[33]

6.2. Xenophobia

Einstein and relativity aroused a xenophobic response in United States.
France and Spain, all for different reasons. In the United States,
incomprehensibility translated socially into an anti-democratic impulse:
in the land of "popular science", relativity was viewed as the imposition
of European elitism (Missner). In France, anti-German hostility in the
immediate post-war period linked up to pre-war antipathy to German
scientific "speculation" (Biezunski). In Spain, anti-relativists com-
plained that any new idea to which a foreign name was attached became
accepted uncritically and Spanish critics couldn't get a hearing; it was
even argued that *anti*-relativists had to cite foreign authorities to get a
hearing.[34] The context of Spanish xenophobia was not political but
cultural, an expression of that society's defensiveness regarding its perceived historical inability to do science, and, at the same time, the need
to redress the stigma of "Spanish decadence" and the "Black Legend".

6.3. Fear of Revolution

If Einstein's ideas were "revolutionary" it followed that revolutionaries
would be inclined to favor (Biezunski), conservatives to oppose (Missner), them on purely political grounds. In Barcelona, Einstein was
invited to lecture a group of anarchist unionists who admired him for his
pacifism. There he was heard by journalists (mistakenly, as it turned
out) to identify himself as a revolutionary, "only in science," not
politics. Einstein spent the next week in Madrid denying he ever made

such a statement and, in his lecture on general relativity there, added a few lines pointing out that he was not a revolutionary in science, only the "translator" of Newton and Galileo into modern physical language.[35] Still, anarchists perceived him as a revolutionary – a testament to the success a certain kind of popularization had in reaching the uneducated – and anarchist journalists made ample use of relativism to justify their political program.[36] Vizgin and Gorelik assert that the relativity boom of the early 1920s corresponded to the general atmosphere of the revolutionary rebuilding of Soviet society. Where there was fear of revolution, as among conservative sectors in most of the countries considered (in particular, the United States) relativity and its pacifist creator could easily be associated with the negative images of the revolutionary left.

7. THE FIELD OF RECEPTION

Bourdieu is sometimes unclear as to the distinction between an intellectual field (composed of intellectuals only) which excludes the public, and the ostensibly broader cultural field which encompasses the entire culture. And yet Bourdieu envisages a functional relationship between public and intellectual field; once that relationship is admitted, there is no reason to exclude the public from the field. I prefer, therefore, to posit a broader field which is functionally defined for specific ideas: in the case of relativity/Einstein the field includes anyone who discusses it/him: scientists, other intellectuals, professionals, and even the masses. Such a broad field can be justified so long as we satisfy the requirement that functional relationships be demonstrated among the field's components.

In societies where science had to buck an adverse cultural tide, interest by the man-in-the-street could be significant. First, the sale of large runs of popularizing books offering "relativity at the reach of all" provides economic legitimation. Mass interest also provides political legitimation in the form of a visible phenomenon that politicians and administrators can easily read. In this way, a significant value change that affects the public climate for science positively is defracted through the field of reception.

The reception of relativity also mobilized the professions. Such mobilization is especially significant in modernizing countries where the professions mobilize against traditional political and cultural elites. As

Bourdieu observes, each science is tied to one of the major professions or to industry – physics and chemistry to the latter, biology to medicine, mathematics to engineering and architecture, and so forth. Through the professions, therefore, individual scientific disciplines receive various kinds of legitimation and support. That is, when a powerful professional group (medical doctors and engineers being the obvious examples) mobilizes on behalf of itself it may pull a pure scientific discipline after it.

The above generalizations sketch out functional relations between different social and cultural groups in a field of reception. Critical lines of force demarcating the constituent groups may divide scientists (experimental versus theoretical physicists, physicists versus mathematicians), between professional groups (physicists versus engineers), between educational groups (scientists and engineers versus lay intellectuals), or political-cum-philosophical ones (liberal materialists versus traditional vitalists), or social ones (middle class intelligentsia versus proletarian unionists). The latter, although poorly educated, may participate in the educational system through union or church organized outreach programs, such as the Spanish anarchists' night schools, where the diffusion of modern scientific ideas was one of the core components informing the program's rationale.

APPENDIX

Some Interviews with Einstein

1. Spain
1.1. Ricardo Baeza, 'Delante del professor Einstein,' *El Sol* (Madrid), July 33, 1921 [from London]. Einstein's philosophical evolution, relativity (the theory that makes the fewest suppositions should be preferred).
1.2. Interview by Andrés Révész, on train from Barcelona to Madrid, *ABC* (Madrid), March 2, 1923. Spanish science, daily routine, smoking, literature (Don Quixote), socialism.
1.3. A. Fabra Rivas, 'Una visita a Einstein,' *El Sol*, March 27, 1930 [in Germany]. Spanish science and reminiscences of his trip to Spain.

2. Brazil
2.1. Interview with Einstein on the *Cap Polonio* (ship) published in *O Paiz* (Rio), March 22, 1925. Einstein says relativity's greatest merit was not to have enlarged the horizons of science but to have restricted them. The infinite cosmos was an erroneous idea. He observes that relativity's absolute components have been ignored by the

public. The velocity of light is an absolute standard. Comment on the Sobral eclipse observations.

2.2. Interview in *O Jornal* (Rio), March 22, 1925. Comments on Rio de Janiero, Portugal, Haldane and the British aristocracy, intellectual solidarity, European culture and science, Rathenau, South American music and dances, Japan.

2.3. A long interview with George Santos, elements of which appear in *O Imparcial* (Rio de Janeiro), May 8, 1925, and *Illustração Brasileira*, June 1925. Einstein talks of literature, classical and popular music, cinema, the international mentality and soccer.

3. Argentina

3.1. Interview in Montevideo, March 24, 1925, published in *La Nación* (Buenos Aires), March 25: America is the land of the future; situation of the Jews; Palestine and Hebrew University; internationality of science; European scientific tradition.

3.2. *La Epoca* (Buenos Aires), March 25, 1925: his lectures: people have exaggerated the revolutionary nature of relativity; state of science (astronomy has to be made over and chemistry retouched; Lorentz and others have already shaken up physics).

3.3. Interview in *La Nación*, March 26, 1925, about the nature of his upcoming lectures: nature of ether as presented by Abraham and Föppl (mechanical properties); not true that relativity is inaccessible to the intuition (notion of relative space is more intuitive than that of absolute space).

3.4. Laro Fernández Arias, 'Einstein y su teoría según él mismo,' *Cara y Caretas* (Buenos Aires), April 4, 1925. At home of Bruno Wassermann. Difficulty of popularizing relativity; science is like music (you have to know the notation); the advantage of exact science is that it is like an international language; world politics.

NOTES

[1] See Todorov, *The Conquest of America* (New York, Harper & Row, 1985), p. 54, where he draws attention to the "receivability" of texts in specific socio-cultural settings.

[2] David L. Hull, 'Darwinism as a Historical Entity: A Historiographic Proposal,' in David Kohn, ed., *The Darwinian Heritage* (Princeton, Princeton University Press, 1985). pp. 773–812.

[3] On Darwinism, see Thomas F. Glick, *The Comparative Reception of Darwinism* (Austin, University of Texas Press, 1974), containing case studies for England, the United States, Germany, France, Russia, the Netherlands, Spain, Mexico, and the Islamic World; and articles by P. Bowler (Britain and the United States), P. Corsi and P. Weindling (Germany, France, and Italy), and F. Scudo and M. Acanfora (Russia) in Kohn, *The Darwinian Heritage*. On psychoanalysis, see the comparative *aperçus* in Jack J. Spector, *The Aesthetics of Freud* (New York, Praeger, 1972), pp. 187–209; and O. Mannoni, *Freud* (New York, Vintage, 1974), pp. 166–193. On the receptions of Darwin and Freud compared, see David Shakow and David Rapaport, *The Influence of Freud on American Psychology* (New York, International Universities Press, 1964) ch. 2.

[4] Thomas F. Glick, 'La transferencia de las revoluciones científicas a través de las fronteras culturales', *Ciencia y Desarrollo* (Mexico), Jan.–Feb. 1987, pp. 77–89.

[5] Judith R. Goodstein, 'The Italian Mathematicians of Relativity,' *Centaurus*, **26** (1983), 241–261.

[6] Thomas F. Glick, *Einstein in Spain* (Princeton, Princeton University Press, 1988), chapter 4; *La Razón* (Buenos Aires), April 17, 1925; South American Travel log, Einstein Papers, Princeton, April 16, 1925.

[7] Giuliano Pancaldi, *Darwin in Italia* (Bologna, Il Mulino, 1983), p. 13.

[8] Jeffrey Crelinsten, "William Wallace Campbell and the 'Einstein problem': An Observational Astronomer Confronts the Theory of Relativity," *Historical Studies in the Physical Sciences*, **14** (1983), 88.

[9] J.M. Plans, 'Las matemáticas en España en los últimos cincuenta años,' *Ibérica*, **25** (1926), 172–174, on p. 174: "Mathematics has made a great leap in Spain; the lack of phase with respect to other countries has almost been overcome." Terradas to Levi-Civita, March 22, 1922, Levi-Civita Papers, Accademia dei Lincei, copy deposited in the California Institute of Technology Archives: "What a pity that mathematics is so little developed among us!"

[10] Simon Schwartzman, *Formação da comunidade científica no Brasil* (São Paulo, Editora Nacional, 1979), pp. 109–115.

[11] J.L. Heilbron, *The Dilemmas of an Upright Man: Max Planck as Spokesman for German Science* (Berkeley, University of California Press, 1986), pp. 138–139.

[12] See Glick, *Einstein in Spain*, chapter 7. M. Homais, the pharmacist in *Madame Bovary*, appeared in the popular discussion of relativity in Spain as a symbol of the popularization of science among provincial pharmacists, physicians and intellectuals generally.

[13] Two authors who effectively interweave social and cognitive issues in the history of evolutionary biology are Robert M. Young, *Darwin's Metaphor: Nature's Place in Victorian Culture* (Cambridge, Cambridge University Press, 1985), and Adrian Desmond, *Archetypes and Ancestors: Paleontology in Victorian London, 1850–1875* (Chicago, University of Chicago Press, 1982).

[14] Michel Biezunski, *La diffusion de la théorie de la relativité en France*, unpub. doctoral diss., University of Paris, 1981, p. 207, where a journalist describes an informal discussion between Einstein and some French scientists. The scientists play the simultaneous role of *gens du monde* and pose very different kinds of questions.

[15] See Peter Bowler, *The Eclipse of Darwinism* (Baltimore, Johns Hopkins University Press, 1983).

[16] In *Diffusion de la relativité en France*, chapters 2 and 3.

[17] Jean Eisenstaedt, 'La relativité générale à l'étiage: 1925–1955,' *Archive for History of Exact Sciences* **35** (1986), 115–185.

[18] Pierre Bourdieu, "Intellectual Fields and Creative Project," in Michael D. Young, ed., *Knowledge and Control* (London, Collier-Macmillan, 1971), pp. 161–188, on p. 165.

[19] J. Mañas y Bonví, *Optica aplicada* (Barcelona, 1915), Suplemento a la página 427, p. i: 'Hypothesis of Einstein and Planck: The ether does not exist.'

[20] Paul Johnson, *Modern Times: The World from the Twenties to the Eighties* (New York, Harper and Row, 1983), pp. 5, 11.

[21] The mathematics program of the Naval Academy in 1922 was deemed inadequate to the mastery of "the new and already popularized doctrine of Einstein, through inability to form a clear idea of the four dimensions of space-time," A. Azarola, "El estudio de las Matemáticas en la Escuela Naval Militar," *Revista General de la Marina*, **86** (1920), 441–453.

[22] Marshall Missner, 'Why Einstein Became Famous in America,' *Social Studies of Science*, **15** (1985), 267–291.

[23] On Laub in Argentina, see Lewis Pyenson, *The Young Einstein* (Bristol/Boston, Adam Hilger, 1985), pp. 231–232.

[24] Lewis Elton, 'Einstein, General Relativity, and the German Press, 1919–1920,' *Isis*, **77** (1986), 95–103, on p. 103.

[25] In 'Why Einstein Became Famous in America' (n. 22, above) and an earlier unpublished version of the same article.

[26] On Cajal, Glick, *Einstein in Spain*, chapter 3. On Houssay, who acted as Einstein's host on a visit to a private medical laboratory near Buenos Aires, *La Prensa*, April 7. On his travels, Einstein sought out the best (or best-known) scientists, regardless of discipline.

[27] Images from his Spain tour; Tomás Gómez de Nicolás, 'La relatividad de los valores: Alegrémonos de no ser sabios,' *El Imparcial* (Madrid), March 10, 1923.

[28] Amos Elon, 'Letter from Argentina,' *The New Yorker*, July 21, 1986, p. 75, possibly reflecting a statement attributed to an ideologue of the military regime quoted by Anthony Lewis, in a review of Jacobo Timmerman, *Prisoner Without a Name*, New York Times Book Review, May 10, 1981: "Argentina has three main enemies: Karl Marx, because he tried to destroy the Christian concept of society; Sigmund Freud, because he tried to destroy the Christian concept of the family; and Albert Einstein because he tried to destroy the Christian concept of time and space."

[29] Horacio Bentábol, an antirelativist Spanish engineer, lists as examples of modern philosophical disorientation: relativity, non-Euclidean geometry, the Russian Revolution, Cubism, and religious disarray; *Observaciones contradictorias a la Teoría de la Relatividad del profesor Alberto Einstein* (Madrid, 1925), p.86.

[30] José Ortega y Gasset, 'Con Einstein en Toledo,' in *El tema del nuestro tiempo*, 18th ed. (Madrid, Revista de Occidente, 1976), pp. 195–202, on pp. 196–197.

[31] Wynn Wachhorst, *Thomas Alva Edison: An American Myth* (Cambridge, Mass., MIT Press, 1981), pp. 23, 25, 30.

[32] The post-1945 myth of Einstein as genius and destroyer is discussed by Alan J. Friedman and Carol C. Donley, *Einstein as Myth and Muse* (Cambridge, Cambridge University Press, 1985), pp. 171–180. Friedman and Donley do not perceive the magus element in the Einstein myth, even when it occurs in their own evidence (e.g. in the fascinating statements by two psychotics, pp. 190–192, who admired Einstein for his ability to conjure the powers of nature).

[33] On the *Origin of Species* as an "open text", see Gillian Beer, 'Darwin's Reading and the Fictions of Development,' in David Kohn, ed., *The Darwinian Heritage* (Princeton, Princeton University Press, 1985), pp. 543–588, on p. 574.

[34] Bentábol, *Observaciones contradictorias* [n. 29, above], p. 23, explaining why he felt compelled to cite Henri Bouasse in support of his own antirelativity conclusions. The same point was made by José Escofet, 'Crónicas catalanas: Einstein y los matemáticos,' *Las Provincias* (Valencia), March 18, 1923. Escofet's argument, typical of the displaced *gens du monde* (he was editor of *La Vanguardia*, a daily representing the interests of the Catalan bourgeoisie), was that neither supporters nor opponents of Einstein were believed by the Spanish public unless they cited foreign sources. This view was erroneous; the Spanish relativists were believed because of their immense academic prestige.

[35] On Einstein and the Catalan nationalists and anarchosyndicalists, see Glick, *Einstein in Spain*, chapter 3. His Madrid lectures are reproduced in *ibid*, Appendix III.

[36] Francisco Pellicer, 'Revolución científica y revolución económica,' *Redención* (Alcoy), March 22, 1923. The argument here is that physical relativity is the harbinger of a moral relativity which hold that political and economic institutions are relative, not absolute.

Boston University

INDEX

BOSTON STUDIES IN THE PHILOSOPHY OF SCIENCE

Editors:

ROBERT S. COHEN and MARX W. WARTOFSKY

(Boston University)

1. Marx W. Wartofsky (ed.), *Proceedings of the Boston Colloquium for the Philosophy of Science 1961–1962.* 1963.
2. Robert S. Cohen and Marx W. Wartofsky (eds.), *In Honor of Philipp Frank.* 1965.
3. Robert S. Cohen and Marx W. Wartofsky (eds.), *Proceedings of the Boston Colloquium for the Philosophy of Science 1964–1966. In Memory of Norwood Russell Hanson.* 1967.
4. Robert S. Cohen and Marx W. Wartofsky (eds.), *Proceedings of the Boston Colloquium for the Philosophy of Science 1966–1968.* 1969.
5. Robert S. Cohen and Marx W. Wartofsky (eds.), *Proceedings of the Boston Colloquium for the Philosophy of Science 1966–1968.* 1969.
6. Robert S. Cohen and Raymond J. Seeger (eds.), *Ernst Mach: Physicist and Philosopher.* 1970.
7. Milic Capek, *Bergson and Modern Physics.* 1971.
8. Roger C. Buck and Robert S. Cohen (eds.), *PSA 1970. In Memory of Rudolf Carnap.* 1971.
9. A. A. Zinov'ev, *Foundations of the Logical Theory of Scientific Knowledge (Complex Logic).* (Revised and enlarged English edition with an appendix by G. A. Smirnov, E. A. Sidorenka, A. M. Fedina, and L. A. Bobrova). 1973.
10. Ladislav Tondl, *Scientific Procedures.* 1973.
11. R. J. Seeger and Robert S. Cohen (eds.), *Philosophical Foundations of Science.* 1974.
12. Adolf Grünbaum, *Philosophical Problems of Space and Time.* (Second, enlarged edition). 1973.
13. Robert S. Cohen and Marx W. Wartofsky (eds.), *Logical and Epistemological Studies in Contemporary Physics.* 1973.
14. Robert S. Cohen and Marx W. Wartofsky (eds.), *Methodological and Historical Essays in the Natural and Social Sciences. Proceedings of the Boston Colloquium for the Philosophy of Science 1969–1972.* 1974.
15. Robert S. Cohen, J. J. Stachel, and Marx W. Wartofsky (eds.), *For Dirk Struik. Scientific, Historical and Political Essays in Honor of Dirk Struik.* 1974.
16. Norman Geschwind, *Selected Papers on Language and the Brain.* 1974.
17. B. G. Kuznetsov, *Reason and Being: Studies in Classical Rationalism and Non-Classical Science.* 1987.
18. Peter Mittelstaedt, *Philosophical Problems of Modern Physics.* 1976.
19. Henry Mehlberg, *Time, Causality, and the Quantum Theory* (2 vols.). 1980.
20. Kenneth F. Schaffner and Robert S. Cohen (eds.), *Proceedings of the 1972 Biennial Meeting, Philosophy of Science Association.* 1974.
21. R. S. Cohen and J. J. Stachel (eds.), *Selected Papers of Léon Rosenfeld.* 1978.
22. Milic Čapek (ed.), *The Concepts of Space and Time. Their Structure and Their Development.* 1976.

23. Marjorie Grene, *The Understanding of Nature. Essays in the Philosophy of Biology.* 1974.
24. Don Ihde, *Technics and Praxis. A Philosophy of Technology.* 1978.
25. Jaakko Hintikka and Unto Remes. *The Method of Analysis. Its Geometrical Origin and Its General Significance.* 1974.
26. John Emery Murdoch and Edith Dudley Sylla, *The Cultural Context of Medieval Learning.* 1975.
27. Marjorie Grene and Everett Mendelsohn (eds.), *Topics in the Philosophy of Biology.* 1976.
28. Joseph Agassi, *Science in Flux.* 1975.
29. Jerzy J. Wiatr (ed.), *Polish Essays in the Methodology of the Social Sciences.* 1979.
30. Peter Janich, *Protophysics of Time.* 1985.
31. Robert S. Cohen and Marx W. Wartofsky (eds.), *Language, Logic, and Method.* 1983.
32. R. S. Cohen, C. A. Hooker, A. C. Michalos, and J. W. van Evra (eds.), *PSA 1974: Proceedings of the 1974 Biennial Meeting of the Philosophy of Science Association.* 1976.
33. Gerald Holton and William Blanpied (eds.), *Science and Its Public: The Changing Relationship.* 1976.
34. Mirko D. Grmek (ed.), *On Scientific Discovery.* 1980.
35. Stefan Amsterdamski, *Between Experience and Metaphysics. Philosophical Problems of the Evolution of Science.* 1975.
36. Mihailo Marković and Gajo Petrović (eds.), *Praxis. Yugoslav Essays in the Philosophy and Methodology of the Social Sciences.* 1979.
37. Hermann von Helmholtz, *Epistemological Writings. The Paul Hertz/Moritz Schlick Centenary Edition of 1921 with Notes and Commentary by the Editors.* (Newly translated by Malcolm F. Lowe. Edited, with an Introduction and Bibliography, by Robert S. Cohen and Yehuda Elkana). 1977.
38. R. M. Martin, *Pragmatics, Truth, and Language.* 1979.
39. R. S. Cohen, P. K. Feyerabend, and M. W. Wartofsky (eds.), *Essays in Memory of Imre Lakatos.* 1976.
40. B. M. Kedrov and V. Sadovsky. *Current Soviet Studies in the Philosophy of Science.*
41. M. Raphael, *Theorie des Geistigen Schaffens auf Marxistischer Grundlage.*
42. Humberto R. Maturana and Francisco J. Varela, *Autopoiesis and Cognition. The Realization of the Living.* 1980.
43. A. Kasher (ed.), *Language in Focus: Foundations, Methods and Systems. Essays Dedicated to Yehoshua Bar-Hillel.* 1976.
44. Trân Duc Thao, *Investigations into the Origin of Language and Consciousness.* (Translated by Daniel J. Herman and Robert L. Armstrong; edited by Carolyn R. Fawcett and Robert S. Cohen). 1984.
45. A. Ishmimoto (ed.), *Japanese Studies in the History and Philosophy of Science.*
46. Peter L. Kapitza, *Experiment, Theory, Practice.* 1980.
47. Maria L. Dalla Chiara (ed.), *Italian Studies in the Philosophy of Science.* 1980.
48. Marx W. Wartofsky, *Models: Representation and the Scientific Understanding.* 1979.
49. Trân Duc Thao, *Phenomenology and Dialectical Materialism.* 1985.
50. Yehuda Fried and Joseph Agassi, *Paranoia: A Study in Diagnosis.* 1976.
51. Kurt H. Wolff, *Surrender and Catch: Experience and Inquiry Today.* 1976.
52. Karel Kosík, *Dialectics of the Concrete.* 1976.

53. Nelson Goodman, *The Structure of Appearance*. (Third edition). 1977.
54. Herbert A. Simon, *Models of Discovery and Other Topics in the Methods of Science*. 1977.
55. Morris Lazerowitz, *The Language of Philosophy. Freud and Wittgenstein*. 1977.
56. Thomas Nickles (ed.), *Scientific Discovery, Logic, and Rationality*. 1980.
57. Joseph Margolis, *Persons and Minds. The Prospects of Nonreductive Materialism*. 1977.
58. G. Radnitzky and G. Andersson (eds.), *Progress and Rationality in Science*, 1978.
59. Gerard Radnitzky and Gunnar Andersson (eds.), *The Structure and Development of Science*. 1979.
60. Thomas Nickles (ed.), *Scientific Discovery: Case Studies*. 1980.
61. Maurice A. Finocchiaro, *Galileo and the Art of Reasoning*. 1980.
62. William A. Wallace, *Prelude to Galileo*. 1981.
63. Friedrich Rapp, *Analytical Philosophy of Technology*. 1981.
64. Robert S. Cohen and Marx W. Wartofsky (eds.), *Hegel and the Sciences*. 1984.
65. Joseph Agassi, *Science and Society*. 1981.
66. Ladislav Tondl, *Problems of Semantics*. 1981.
67. Joseph Agassi and Robert S. Cohen (eds.), *Scientific Philosophy Today*. 1982.
68. Wuadysuaw Krajewski (ed.), *Polish Essays in the Philosophy of the Natural Sciences*. 1982.
69. James H. Fetzer, *Scientific Knowledge*. 1981.
70. Stephen Grossberg, *Studies of Mind and Brain*. 1982.
71. Robert S. Cohen and Marx W. Wartofsky (eds.), *Epistemology, Methodology, and the Social Sciences*. 1983.
72. Karel Berka, *Measurement*. 1983.
73. G. L. Pandit, *The Structure and Growth of Scientific Knowledge*. 1983.
74. A. A. Zinov'ev, *Logical Physics*. 1983.
75. Gilles-Gaston Granger, *Formal Thought and the Sciences of Man*. 1983.
76. R. S. Cohen and L. Laudan (eds.), *Physics, Philosophy and Psychoanalysis*. 1983.
77. G. Böhme et al., *Finalization in Science*, ed. by W. Schäfer. 1983.
78. D. Shapere, *Reason and the Search for Knowledge*. 1983.
79. G. Andersson, *Rationality in Science and Politics*. 1984.
80. P. T. Durbin and F. Rapp, *Philosophy and Technology*. 1984.
81. M. Marković, *Dialectical Theory of Meaning*. 1984.
82. R. S. Cohen and M. W. Wartofsky, *Physical Sciences and History of Physics*. 1984.
83. E. Meyerson, *The Relativistic Deduction*. 1985.
84. R. S. Cohen and M. W. Wartofsky, *Methodology, Metaphysics and the History of Sciences*. 1984.
85. György Tamás, *The Logic of Categories*. 1985.
86. Sergio L. de C. Fernandes, *Foundations of Objective Knowledge*. 1985.
87. Robert S. Cohen and Thomas Schnelle (eds.), *Cognition and Fact*. 1985.
88. Gideon Freudenthal, *Atom and Individual in the Age of Newton*. 1985.
89. A. Donagan, A. N. Perovich, Jr., and M. V. Wedin (eds.), *Human Nature and Natural Knowledge*. 1985.
90. C. Mitcham and A. Huning (eds.), *Philosophy and Technology II*. 1986.
91. M. Grene and D. Nails (eds.), *Spinoza and the Sciences*. 1986.
92. S. P. Turner, *The Search for a Methodology of Social Science*. 1986.
93. I. C. Jarvie, *Thinking About Society: Theory and Practice*. 1986.
94. Edna Ullmann-Margalit (ed.), *The Kaleidoscope of Science*. 1986.

95. Edna Ullmann-Margalit (ed.), *The Prism of Science.* 1986.
96. G. Markus, *Language and Production.* 1986.
97. F. Amrine, F. J. Zucker, and H. Wheeler (eds.), *Goethe and the Sciences: A Reappraisal.* 1987.
98. Joseph C. Pitt and Marcella Pera (eds.), *Rational Changes in Science.*
99. O. Costa de Beauregard, *Time, the Physical Magnitude.* 1987.
100. Abner Shimony and Debra Nails (eds.), *Naturalistic Epistemology: A Symposium of Two Decades.* 1987.
101. Nathan Rotenstreich, *Time and Meaning in History.* 1987.
102. David B. Zilberman (ed.), *The Birth of Meaning in Hindu Thought.* 1987.
103. Thomas F. Glick (ed.), *The Comparative Reception of Relativity.* 1987.
104. Zellig Harris *et al., The Form of Information in Science.* 1987
105. Frederick Burwick, *Approaches to Organic Form: Permutations in Science and Culture.* 1987.

53. Nelson Goodman, *The Structure of Appearance*. (Third edition). 1977.
54. Herbert A. Simon, *Models of Discovery and Other Topics in the Methods of Science*. 1977.
55. Morris Lazerowitz, *The Language of Philosophy. Freud and Wittgenstein*. 1977.
56. Thomas Nickles (ed.), *Scientific Discovery, Logic, and Rationality*. 1980.
57. Joseph Margolis, *Persons and Minds. The Prospects of Nonreductive Materialism*. 1977.
58. G. Radnitzky and G. Andersson (eds.), *Progress and Rationality in Science*, 1978.
59. Gerard Radnitzky and Gunnar Andersson (eds.), *The Structure and Development of Science*. 1979.
60. Thomas Nickles (ed.), *Scientific Discovery: Case Studies*. 1980.
61. Maurice A. Finocchiaro, *Galileo and the Art of Reasoning*. 1980.
62. William A. Wallace, *Prelude to Galileo*. 1981.
63. Friedrich Rapp, *Analytical Philosophy of Technology*. 1981.
64. Robert S. Cohen and Marx W. Wartofsky (eds.), *Hegel and the Sciences*. 1984.
65. Joseph Agassi, *Science and Society*. 1981.
66. Ladislav Tondl, *Problems of Semantics*. 1981.
67. Joseph Agassi and Robert S. Cohen (eds.), *Scientific Philosophy Today*. 1982.
68. Wuadysuaw Krajewski (ed.), *Polish Essays in the Philosophy of the Natural Sciences*. 1982.
69. James H. Fetzer, *Scientific Knowledge*. 1981.
70. Stephen Grossberg, *Studies of Mind and Brain*. 1982.
71. Robert S. Cohen and Marx W. Wartofsky (eds.), *Epistemology, Methodology, and the Social Sciences*. 1983.
72. Karel Berka, *Measurement*. 1983.
73. G. L. Pandit, *The Structure and Growth of Scientific Knowledge*. 1983.
74. A. A. Zinov'ev, *Logical Physics*. 1983.
75. Gilles-Gaston Granger, *Formal Thought and the Sciences of Man*. 1983.
76. R. S. Cohen and L. Laudan (eds.), *Physics, Philosophy and Psychoanalysis*. 1983.
77. G. Böhme et al., *Finalization in Science*, ed. by W. Schäfer. 1983.
78. D. Shapere, *Reason and the Search for Knowledge*. 1983.
79. G. Andersson, *Rationality in Science and Politics*. 1984.
80. P. T. Durbin and F. Rapp, *Philosophy and Technology*. 1984.
81. M. Marković, *Dialectical Theory of Meaning*. 1984.
82. R. S. Cohen and M. W. Wartofsky, *Physical Sciences and History of Physics*. 1984.
83. E. Meyerson, *The Relativistic Deduction*. 1985.
84. R. S. Cohen and M. W. Wartofsky, *Methodology, Metaphysics and the History of Sciences*. 1984.
85. György Tamás, *The Logic of Categories*. 1985.
86. Sergio L. de C. Fernandes, *Foundations of Objective Knowledge*. 1985.
87. Robert S. Cohen and Thomas Schnelle (eds.), *Cognition and Fact*. 1985.
88. Gideon Freudenthal, *Atom and Individual in the Age of Newton*. 1985.
89. A. Donagan, A. N. Perovich, Jr., and M. V. Wedin (eds.), *Human Nature and Natural Knowledge*. 1985.
90. C. Mitcham and A. Huning (eds.), *Philosophy and Technology II*. 1986.
91. M. Grene and D. Nails (eds.), *Spinoza and the Sciences*. 1986.
92. S. P. Turner, *The Search for a Methodology of Social Science*. 1986.
93. I. C. Jarvie, *Thinking About Society: Theory and Practice*. 1986.
94. Edna Ullmann-Margalit (ed.), *The Kaleidoscope of Science*. 1986.

95. Edna Ullmann-Margalit (ed.), *The Prism of Science.* 1986.
96. G. Markus, *Language and Production.* 1986.
97. F. Amrine, F. J. Zucker, and H. Wheeler (eds.), *Goethe and the Sciences: A Reappraisal.* 1987.
98. Joseph C. Pitt and Marcella Pera (eds.), *Rational Changes in Science.*
99. O. Costa de Beauregard, *Time, the Physical Magnitude.* 1987.
100. Abner Shimony and Debra Nails (eds.), *Naturalistic Epistemology: A Symposium of Two Decades.* 1987.
101. Nathan Rotenstreich, *Time and Meaning in History.* 1987.
102. David B. Zilberman (ed.), *The Birth of Meaning in Hindu Thought.* 1987.
103. Thomas F. Glick (ed.), *The Comparative Reception of Relativity.* 1987.
104. Zellig Harris *et al., The Form of Information in Science.* 1987
105. Frederick Burwick, *Approaches to Organic Form: Permutations in Science and Culture.* 1987.